Lecture Notes in Computer Science 8821

Commenced Publication in 1973
Founding and Former Series Editors:
Gerhard Goos, Juris Hartmanis, and Jan van Leeuwen

More information about this series at http://www.springer.com/series/7409

Agma Juci Machado Traina · Caetano Traina Jr.
Robson Leonardo Ferreira Cordeiro (Eds.)

Similarity Search and Applications

7th International Conference, SISAP 2014
Los Cabos, Mexico, October 29–31, 2014
Proceedings

 Springer

Editors
Agma Juci Machado Traina
Computer Science Department - ICMC
University of São Paulo at São Carlos
São Carlos
Brazil

Robson Leonardo Ferreira Cordeiro
Computer Science Department - ICMC
University of São Paulo at São Carlos
São Carlos
Brazil

Caetano Traina Jr.
Computer Science Department - ICMC
University of São Paulo at São Carlos
São Carlos
Brazil

ISSN 0302-9743
ISBN 978-3-319-11987-8
DOI 10.1007/978-3-319-11988-5

ISSN 1611-3349   (electronic)
ISBN 978-3-319-11988-5   (eBook)

Library of Congress Control Number: 2014950507

LNCS Sublibrary: SL3 – Information Systems and Applications, incl. Internet/Web and HCI

Springer Cham Heidelberg New York Dordrecht London

Printed on acid-free paper

Springer is part of Springer Science+Business Media (www.springer.com)

Preface

This volume contains the papers presented at the seventh International Conference on Similarity Search and Applications (SISAP 2014), held at Los Cabos, Mexico, during October 29–31, 2014.

The International Conference on Similarity Search and Applications (SISAP) is an annual forum for researchers and application developers in the area of similarity data management. It focuses on technological problems shared by many application domains, such as data mining, information retrieval, computer vision, pattern recognition, computational biology, geography, biometrics, machine learning, and many others that need similarity searching as a necessary supporting service.

Traditionally, SISAP conferences have put emphasis on distance-based searching, but in general the conference concerns both the effectiveness and efficiency aspects of any similarity search approach, welcoming contributions that range from theoretical aspects to innovative developments for which similarity search plays the central role.

The call for papers welcomed research papers (full or short papers) presenting previously unpublished research contributions, as well as case studies and application papers (short papers) describing existing applications of similarity search in real scenarios.

We received 45 complete submissions. The Program Committee (PC) comprised 53 researchers from 18 different countries. Each submission was assigned to at least three PC members. Reviews were discussed by the chairs and PC members when the reviews diverged and no consensus had been reached. The final selection of papers was made by the PC chairs based on the reviews received for each submission. Finally, the conference program includes 21 full papers and 6 short papers, which results in a 46.66% acceptance ratio.

The conference program and the proceedings are organized into five parts. The first part comprises papers proposing improvements to different methods and techniques for similarity search. A second part is devoted to papers dealing with efficient indexing solutions for similarity search and their application in real settings. The third part focuses on particular metrics and their effectiveness. The fourth part of the conference program includes papers dealing with new scenarios or presenting new approaches to similarity search. Finally, the last part comprises those papers devoted to solutions for similarity search in specific application domains, such as in streaming time series, image and audio retrieval and analysis, systems with CPU- and GPU-based processing, astroinformatics, computational neuroscience, and in particular types of recommender systems and search engines.

The conference program also includes two invited talks from outstanding scholars in the field. The first one, "Scalable Retrieval and Analysis of Simulation and Observation Data Sets" by Prof. K. Selçuk Candan, introduces and presents solutions to computational challenges that arise from the need to process, index, search, and analyze, in a scalable manner, large volumes of temporal data resulting from data-intensive simulations. The second one, "Visual Analytics for Interactive Subspace Similarity Search" by Prof. Daniel Keim, presents novel techniques that combine automated and visual methods to improve subspace search in high-dimensional data.

As in previous editions, the proceedings are published by Springer-Verlag in the Lecture Notes in Computer Science series. A selection of the best papers presented at the conference were recommended for publication in the journal Information Systems. The selection of best papers was made by the PC, based on the reviews received by each paper, and on the discussion during the conference.

SISAP conferences are organized by the SISAP initiative (www. sisap.org), which aims to become a forum to exchange real-world, challenging, and innovative examples of applications, new indexing techniques, common test-beds and benchmarks, source code, and up-to-date literature through its web page, serving the similarity search community.

We would like to thank all the authors who submitted papers to SISAP 2014. We would also like to thank all members of the PC and the external reviewers, for the enormous amount of work they have done. We would like to acknowledge the generous collaboration and financial support from Centro de Investigación Científica y de Educación Superior de Ensenada, B.C. (CICESE); the host institution, and from the Consejo Nacional de Ciencia y Tecnología (CONACyT); the Mexican public research agency. We want to express our gratitude to the PC members for their effort and contribution to the conference. All the submission, reviewing, and proceedings generation processes were carried out through the EasyChair platform.

October 2014
                                    Agma Juci Machado Traina
                                        Caetano Traina Jr.
                            Robson Leonardo Ferreira Cordeiro

Organization

Program Committee Chairs

Agma Juci Machado Traina   University of São Paulo, Brazil
Caetano Traina Jr.   University of São Paulo, Brazil

Program Committee Members

Agma Juci Machado Traina	University of São Paulo, Brazil
Ahmet Sacan	Drexel University, USA
Alberto Del Bimbo	Università degli Studi di Firenze, Italy
Alberto Laender	Federal University of Minas Gerais, Brazil
Altigran S. da Silva	Federal University of Amazonas, Brazil
Ambuj Singh	University of California at Santa Barbara, USA
Andre Balan	Federal University of ABC, Brazil
Andreas Zuefle	Ludwig-Maximilians-Universität München, Germany
Apostolos N. Papadopoulos	Aristotle University of Thessaloniki, Greece
Benjamin Bustos	University of Chile, Chile
Bjorn Thor Jonsson	Reykjavik University, Iceland
Caetano Traina Jr.	University of São Paulo, Brazil
Claudio Gennaro	ISTI-CNR, Italy
Christian Böhm	Ludwig-Maximilians-Universität München, Germany
Daniel Keim	University of Konstanz, Germany
Dimitrios Gunopulos	University of Athens, Greece
Divesh Srivastava	AT&T Labs-Research, USA
Dong Deng	Tsinghua University, China
Eamonn Keogh	University of California at Riverside, USA
Edgar Chavez	Universidad Nacional Autónoma de México, Mexico
Eduardo Valle	University of Campinas, Brazil
Elaine Parros Machado de Sousa	University of São Paulo, Brazil
Fabrizio Falchi	ISTI-CNR, Italy
Giuseppe Amato	ISTI-CNR, Italy
Gonzalo Navarro	University of Chile, Chile
Hanghang Tong	City College, CUNY, USA
Henning Müller	HES-SO, Switzerland
Jimeng Sun	Georgia Institute of Technology, USA

Joao Eduardo Ferreira	University of São Paulo, Brazil
Joe Tekli	Lebanese American University, Lebanon
Jose Oncina	University of Alicante, Spain
Luisa Mico	University of Alicante, Spain
Marcela Ribeiro	Federal University of São Carlos – UFSCar, Brazil
Marco Patella	University of Bologna, Italy
Nieves R. Brisaboa	University of A Coruña, Spain
Panagiotis Bouros	Humboldt-Universität zu Berlin, Germany
Paolo Ciaccia	University of Bologna, Italy
Pavel Zezula	Masaryk University, Czech Republic
Oscar Pedreira	University of A Coruña, Spain
Renata Galante	Federal University of Rio Grande do Sul, Brazil
Renato Fileto	Federal University of Santa Catarina, Brazil
Richard Connor	University of Strathclyde, UK
Richard Chbeir	IUT de Bayonne et du Pays Basque, France
Robson Leonardo Ferreira Cordeiro	University of São Paulo, Brazil
Rui Zhang	University of Melbourne, Australia
Simone Santini	Universidad Autómoma de Madrid, Spain
Thomas Seidl	RWTH Aachen University, Germany
Tomas Skopal	Charles University in Prague, Czech Republic
Vassilis Tsotras	University of California at Riverside, USA
Vincent Oria	NJIT, USA
Vladimir Pestov	University of Ottawa, Canada
Yasin Silva	Arizona State University, USA
Yoshiharu Ishikawa	Nagoya University, Japan

Additional Reviewers

Amelkin, Victor	Ma, Xiguo
Araujo, Samur	Marvulle, Valdecir
Bartoli, Federico	Prati, Ronaldo
Bartolini, Ilaria	Qi, Jianzhong
Calvo-Zaragoza, Jorge	Sun, Jichao
Ercoli, Simone	Taddesse, Fekade Getahun
Hoang, Minh	Tellez, Eric Sadit
Huang, Jin	Turchini, Francesco

Invited Talks (Abstracts)

Scalable Retrieval and Analysis of Simulation and Observation Data Sets⋆

K. Selçuk Candan

Professor of Computer Science and Engineering
Arizona State University

Abstract. Data- and model-driven computer simulations for under- stand-ing spatio-temporal dynamics of emerging phenomena are increasingly critical in various application domains, from predicting geo-temporal evolution of epidemics to helping reduce energy footprints of buildings leading to more sustainable building systems and architectural designs. These simulations track 10s or 100s of inter-dependent parameters, span-ning multiple information layers and spatio-temporal frames, affected by complex dynamic processes operating at different resolutions. Con-sequently, the key characteristics of data sets and models relevant to these data-intensive simulations often include the following: (a) volu-minous, (b) multi-variate, (c) multi-resolution, (d) spatio-temporal, and (e) inter-dependent. While very powerful and highly modular and flexi-ble simulation software exists, because of the volume and complexity of the simulation data, the varying spatial and temporal scales at which the key transmission processes operate and relevant observations are made, today experts lack the means to adequately and systematically interpret observations, understand the underlying processes, and re-use of existing simulation results in new settings. In this talk, I will introduce computa-tional challenges that arise from the need to process, index, search, and analyze, in a scalable manner, large volumes of temporal data resulting from data-intensive simulations and present some solutions.

Keywords: Time series, simulations, feature extration, analysis, indexing

---

⋆ This work is partially funded by NSF grants #1339835 ("E-SDMS: Energy Sim-ulation Data Management System Software"), #1318788 ("Data Management for Real-Time Data Driven Epidemic Spread Simulations"), #116394 ("RanKloud: Data Partitioning and Resource Allocation Strategies for Scalable Multimedia and Social Media Analysis"), #1016921 ("One Size Does Not Fit All: Empowering the User with User-Driven Integration"), and #1430144 ("Fraud Detection via Visual Analytics: An Infrastructure to Support Complex Financial Patterns (CFP)-based Real-Time Ser-vices Delivery"). This work is also supported in part by the NSF I/UCRC Center for Embedded Systems established through the NSF grant #0856090 in partnership with Johnson Controls Inc.

References

1. Candan, K.S., Rossini, R., Sapino, M.L., Wang, X.: SDTW: Computing DTW Distances using Locally Relevant Constraints based on Salient Feature Alignments. PVLDB 5(11), 1519–1530 (2012)
2. Candan, K.S., Rossini, R., Sapino, M.L., Wang, X.: STFMap: Query- and Feature-Driven Visualization of Large Time Series Data Sets. CIKM 2012, 2743–2745 (2012)
3. Chen, X., Candan, K.S.: LWI-SVD: Low-rank, Windowed, Incremental Singular Value Decompositions on Time-Evolving Data Sets. In: Accepted for Publication at the ACM SIGKDD Conference on Knowledge Discovery and Data Mining, KDD (2014)
4. Chen, X., Candan, K.S.: GI-NMF: Group Incremental Non-Negative Matrix Factorization on Data Streams. In: Accepted for Publication at the ACM International Conference on Conference on Information and Knowledge Management, CIKM (2014)
5. Huang, S., Li, X., Candan, K.S., Sapino, M.L.: Can you really trust that seed?": Reducing the Impact of Seed Noise in Personalized PageRank. In: Accepted for Publication at the International Conference on Advances in Social Network Analysis and Mining, ASONAM (2014)
6. Kim, M., Candan, K.S.: Efficient Static and Dynamic In-Database Tensor Decompositions on Chunk-Based Array Stores. In: Accepted for Publication at the ACM International Conference on Conference on Information and Knowledge Management, CIKM (2014)
7. Kim, M., Candan, K.S.: TensorDB: In-Database Tensor Manipulation with Tensor-Relational Query Plans. In: Accepted for Demonstration at the ACM International Conference on Conference on Information and Knowledge Management, CIKM (2014)
8. Kim, M., Selçuk Candan, K.: Pushing-down tensor decompositions over unions to promote reuse of materialized decompositions. In: Calders, T., Esposito, F., Hüllermeier, E., Meo, R. (eds.) ECML PKDD 2014, Part I. LNCS, vol. 8724, pp. 688–704. Springer, Heidelberg (2014)
9. Li, X., Huang, S., Candan, K.S., Sapino, M.L.: Focusing Decomposition Accuracy by Personalizing Tensor Decomposition (PTD). In: Accepted for Publication at the ACM International Conference on Conference on Information and Knowledge Management, CIKM (2014)
10. Nagendra, M., Candan, K.S.: SkySuite: A Framework of Skyline-Join Operators for Static and Stream Environments. In: Proceedings of the VLDB Endowment (PVLDB), vol. 6(12) (2013)
11. Nagendra, M., Candan, K.S.: Layered processing of skyline-window- join (SWJ) queries using iteration-fabric. In: IEEE International Conference on Data Engineering (ICDE), pp. 985–996 (2013)
12. Nagarkar, P., Candan, K.S.: HCS: Hierarchical Cut Selection for Efficiently Processing Queries on Data Columns using Hierarchical Bitmap Indices. In: International Conference on Extending Database Technology (EDBT), pp. 271–282 (2014)
13. Schifanella, C., Sapino, M.L., Candan, K.S.: On Context-Aware Co-Clustering with Metadata Support. J. Intell. Inf. Syst. 38(1), 209–239 (2012)
14. Wang, X., Candan, K.S., Sapino, M.L.: Leveraging Metadata for Identifying Local, Robust Multi-variate Temporal (RMT) Features. In: IEEE International Conference on Data Engineering (ICDE), pp. 388–399.

Visual Analytics for Interactive Subspace Similarity Search

Daniel Keim

Head of the Information Visualization and Data Analysis Research Group,
University of Konstanz, Germany

Abstract. In most similarity search applications, the data under consideration resides in high-dimensional data spaces, which often consist of combined features measuring different properties. In order to determine useful similarity measures, appropriate feature combinations (subspaces) of the data have to be taken into consideration, since they may show complementary, conjoint, or contradicting relations between the data items [3]. Which subspace is best in a given application context is difficult to determine by fully automatic methods, and therefore it is important to include the human in the process and combine the creativity and general knowledge of the human with the fast searching and analysis capabilities of the computer. Visual Analytics – the combination of automated and visual methods – can help to interactively determine the most relevant subspaces and define appropriate subspace similarity measures [4]. Subspace search algorithms guided by interestingness measures can be used to compute candidate sets of subspaces, which are then visualized to enable the user to compare and relate subspaces with respect to the involved dimensions and clusters of objects [1]. The approach helps the understanding of high-dimensional data from different perspectives and allows a flexible definition of subspace similarity measures [2].

Keywords: Visual Analytics, Interactive Similarity Search, Subspace Similarity, Interestingness Measures

References

1. Bertini, E., Tatu, A., Keim, D.: Quality metrics in high-dimensional data visualization:An overview and systematization. IEEE Transactions on Visualization and Computer Graphics 17(12), 2203–2212 (2011)
2. Tatu, A., Albuquerque, G., Eisemann, M., Bak, P., Theisel, H., Magnor, M., Keim, D.: Automated analytical methods to support visual exploration of high-dimensional data. IEEE Transactions on Visualization and Computer Graphics 17(5), 584–597 (2011)
3. Tatu, A., Maaß, F., Färber, I., Bertini, E., Schreck, T., Seidl, T., Keim, D.: Subspace search and visualization to make sense of alternative clusterings in high-dimensional data. In: 2012 IEEE Conference on Visual Analytics Science and Technology (VAST), pp. 63–72 (October 2012)

4. Tatu, A., Zhang, L., Bertini, E., Schreck, T., Keim, D., Bremm, S., von Landes-berger, T.: ClustNails: Visual analysis of subspace clusters. Tsinghua Science and Technology 17(4), 419–428 (2012)

Contents

Improving Similarity Search Methods and Techniques

Indexing and Applications

Metrics and Evaluation

New Scenarios and Approaches

Applications and Specific Domains

Efficient Algorithms for Similarity Search in Axis-Aligned Subspaces

Michael E. Houle[1], Xiguo Ma[2], Vincent Oria[2], and Jichao Sun[2]

[1] National Institute of Informatics, Tokyo 101-8430, Japan
meh@nii.ac.jp
[2] New Jersey Institute of Technology, Newark NJ 07102, USA
{xm23,oria,js87}@njit.edu

Abstract. Many applications — such as content-based image retrieval, subspace clustering, and feature selection — may benefit from efficient subspace similarity search. Given a query object, the goal of subspace similarity search is to retrieve the most similar objects from the database, where the similarity distance is defined over an arbitrary subset of dimensions (or features) — that is, an arbitrary axis-aligned projective subspace. Though much effort has been spent on similarity search in fixed subspaces, relatively little attention has been given to the problem of similarity search when the dimensions are specified at query time. In this paper, we propose several new methods for the subspace similarity search problem. Extensive experiments are provided showing very competitive performance relative to state-of-the-art solutions.

1 Introduction

Similarity search is of great importance to applications in many different areas, such as data mining, multimedia databases, information retrieval, statistics and pattern recognition. Specifically, a similarity query retrieves from the database those objects that most closely resemble a supplied query object, based on some measure of pairwise similarity (typically in the form of a distance function). Due to its importance, much effort has been spent on the efficient support of similarity search. However, most existing approaches consider search only with respect to a fixed feature space. In this paper, we focus on the subspace similarity search problem, in which the calculation of similarity values is restricted to a subset of dimensions specified along with the query object.

As with similarity search on fixed spaces, subspace similarity search may also have an impact in application areas where the feature set under consideration changes from operation to operation. Such changes could be due to a modification of query preferences (as in content-based image retrieval), or to the determination of the local structure at different locations within data (as in subspace clustering), or to a systematic exploration of feature subspaces (as in feature selection). In content-based image retrieval, images are often represented by feature vectors extracted based on color, shape, and texture descriptors. In an exploration of the data set, a query involving one combination of features (such as color) may be followed by a query on a different combination (such as shape). In subspace clustering [1], the formation of an individual cluster is generally assessed with respect to a subset of features that most closely describe the concept

A.J. Machado Traina et al. (Eds.): SISAP 2014, LNCS 8821, pp. 1–12, 2014.
DOI: 10.1007/978-3-319-11988-5_1 © Springer International Publishing Switzerland 2014

associated with the cluster. Since verification of a cluster requires the identification of a feature subset together with an object subset, the effectiveness of the overall clustering process may depend on the efficient processing of subspace similarity queries. Wrapper methods for feature selection [2] require an evaluation process, such as k-nearest neighbor (k-NN) classification, for the identification of effective combinations of features. Exploration of feature subspaces can be extremely time-consuming when the neighborhoods are determined using exhaustive search, due to the exponential number of potential combinations involved. To accelerate the process, the efficient support of subspace similarity search is needed.

Almost all existing similarity search indices require that the similarity measure and associated vector space both be specified before any preprocessing occurs. Traditional methods for fixed spaces (as surveyed in [3]) cannot be effectively applied for the subspace search problem: the subspaces to be searched are typically not known until query time, but even if they were known in advance, constructing an index for every possible query subspace would be prohibitively expensive. Of all the methods for similarity search appearing in the research literature, only very few have been specifically formulated for the subspace search problem; a survey of these methods will be presented in Sect. 2.1. In general, existing solutions for subspace similarity search suffer greatly in terms of the computational cost.

Of the two main types of similarity queries (k-NN queries and range queries), k-NN queries are often more important, due to the difficulty faced by the user in deciding range thresholds. This is especially the case for the search in subspaces, since the range values of interest will typically depend on the number of features associated with the subspace. In this paper, we focus only on k-NN queries.

We now formally define the subspace search problem for k-NN queries. Given an object domain $\mathcal{U}$, let $S \subseteq \mathcal{U}$ denote a set of database objects represented as feature vectors in $\mathbb{R}^D$. The set of features will be denoted simply as $F = \{1, 2, \cdots, D\}$, with feature $i \in F$ corresponding to the i-th coordinate in the vector representation. Let $d : \mathbb{R}^D \times \mathbb{R}^D \to \mathbb{R}$ be a distance function defined for the vector space. Given an object vector $u = (u_1, \ldots, u_{|F|}) \in S$, its projection with respect to a feature subset $F' \subseteq F$ is the vector $u' = (u'_1, \ldots, u'_{|F|})$ such that for all $i \in F$, $u'_i = u_i$ whenever $i \in F'$, and $u'_i = 0$ otherwise. The feature set F' thus indicates a unique axis-aligned projective subspace to which distance calculations can be restricted.

Definition 1 (Subspace k-NN Query). *Given a query object $q \in \mathcal{U}$, a query subspace $F' \subseteq F$, and a query neighborhood size k, a subspace k-NN query $\langle q, F', k \rangle$ returns the k objects of S most similar to q, for the distance function $d_{F'}(q, u) \triangleq d(q', u')$, where q' and u' are the projections of q and u with respect to F'.*

As an example of a subspace distance function, for any given $p \in [1, \infty)$, the L_p distance between two objects $q, u \in \mathcal{U}$ restricted to the axis-aligned projective subspace F' is defined as

$$d_{F'}(q, u) = \left(\sum_{i \in F'} |q_i - u_i|^p \right)^{\frac{1}{p}}.$$

In this paper, we present algorithms for subspace similarity search following the multi-step search strategy [4,5], utilizing 1-dimensional distances as lower bounds to efficiently prune the search space. The main contributions of this paper are:

- algorithms specifically tailored for subspace similarity search, both exact and approximate;
- a guide to the practical choice of an important algorithm parameter, based on a theoretical analysis of sample properties;
- an experimental evaluation across data sets of a variety of types and sizes, showing the efficiency and competitiveness of our algorithms.

The remainder of this paper is organized as follows. Sect. 2 discusses related work on subspace search and multi-step search algorithms. Our proposed algorithms are presented in Sect. 3. In Sect. 4, through experiments on several real-world datasets, we contrast the performance of our methods with those of existing methods. The discussion is concluded in Sect. 5.

2 Related Work

2.1 Subspace Similarity Search

Relatively few similarity search methods exist that are specifically designed for subspace search. In [6], the Partial VA-file (PVA) was proposed, which adapts the vector approximation file (VA-file) [7] to support subspace queries. The VA-file, designed for fixed-space similarity search, stores a compressed approximation of the data as a single file; at query time, the compressed approximation is scanned in its entirety, and uses the information for pruning the search within the original dataset. PVA, on the other hand, stores an approximation of data on each dimension separately, and processes the search using only those 1-dimensional VA-files that correspond to dimensions involved in the query. In [8], the Dimension-Merge Index (DMI) was developed, which combines multiple 1-dimensional index structures to answer subspace queries. DMI builds an index for each dimension separately (of any desired type), and utilizes those indexes with respect to the query dimensions to perform the search. The final query result is obtained by aggregation across neighborhoods associated with each of the query dimensions. In [9], the Projected R-Tree (PT) was proposed as a redefinition of the classical search structure R-tree [10] for subspace similarity search. Instead of integrating results of queries on 1-dimensional indices, PT utilizes a single index built on the full feature space (an R-tree) to answer queries with respect to subspaces. A best-first search heuristic is employed, subject to the restriction that only the query dimensions are considered for distance computations. PVA, DMI and PT all produce exact query results; however, as we shall see in Sect. 4, all tend to suffer greatly in terms of their computational cost.

Another approach to the subspace search problem was proposed in [11], for range queries. Here, the search space is reduced through the application of the triangle inequality on several pivot points. Since k-NN queries are not directly supported by this algorithm, for the experimental comparison in Sect. 4, we restrict our attention to PVA, DMI and PT.

2.2 Multi-step Search Algorithms

Our proposed solutions for the subspace search problem make use of multi-step search algorithms. Multi-step search was originally proposed for the adaptive similarity search problem, which aims to find the most similar objects to a query object from the database with respect to an adaptive similarity measure — one that can be determined by the user at query time. Multi-step search computes a query result using a fixed 'lower-bounding' distance function that is adapted to answer the same query with respect to a user-supplied 'target' distance function. The function d_l is a lower-bounding distance for the target distance d_t if $d_l(u,v) \leq d_t(u,v)$ for any two objects u,v drawn from a domain for which both d_l and d_t are defined.

The first multi-step k-NN search algorithm was proposed by Korn et al. [12]. Later, Seidl and Kriegel [4] proposed a more efficient multi-step algorithm. The algorithm scans the neighborhood list of the query object with respect to d_l to retrieve candidates for the query result, and stops when the candidate k-NN distance (target distance) is no larger than the lower-bounding distance currently maintained by the scan. The algorithm is optimal in that it produces the minimum number of candidates needed in order to guarantee a correct query result, given only a list of candidates ordered according to d_l. However, despite this performance guarantee, the algorithm may still be expensive in practice. Using the Seidl-Kriegel algorithm as a starting point, Houle et al. [5] designed an approximate multi-step algorithm, MAET+, with an early termination condition. MAET+ utilizes tests of a measure of the intrinsic dimensionality of the data, the *generalized expansion dimension* (GED) [13,5], to guide early termination decisions. In the remainder of this paper, we will refer to the Seidl-Kriegel algorithm as SK.

3   Algorithm

We now present our solutions to the subspace similarity search problem. Let us first introduce some additional notation. For any object $q \in \mathcal{U}$ and any subspace $F' \subseteq F$, let $N_{F'}(q,k)$ denote the set of k-nearest neighbors of q within database S with respect to subspace distance $d_{F'}$. Ties are broken arbitrarily but consistently. Let $\delta_{F'}(q,k)$ denote the k-th smallest subspace distance (with respect to F') from q to the objects in S.

The strategy underlying our methods involves the application of multi-step search, using a lower-bounding distance function to filter a candidate set from the database, and using the target distance function to refine the candidate set to obtain the final query result. The main concern here is the determination at query time of a lower-bounding distance function suitable for the indicated subspace. Due to the exponential number of possible subspaces, it is impossible to explicitly preprocess the data for every subspace. Instead, as potential lower-bounding distance functions, we consider only the 1-dimensional distance $d_{\{i\}}$ associated with each feature $i \in F$. Assuming that the lower-bounding property holds between $d_{\{i\}}$ and subspace distance $d_{F'}$ for all $i \in F'$ (which is the case for many practical distance measures, including the Euclidean distance), there are $|F'|$ lower-bounding distance functions that can be used in the search. However, practical performance may vary considerably according to the choice of $d_{\{i\}}$. In order to minimize the risk of choosing a poorly-performing lower-bounding distance, we select the distance function corresponding to the most discriminative query dimension. This is done by ranking the dimensions based on data variance, a simple yet effective ranking technique. Two ranking strategies are proposed in this paper: Single Ranking (SR) and Multiple Ranking (MR).

Algorithm **SK_SR** (*query q, subspace F', target neighborhood size k*)

    // Preprocessing step: obtain a single ranking of all dimensions.

1: **for** each dimension $i \in F$ **do**

2:    $\mu_i \leftarrow \frac{1}{|S|} \sum_{u \in S} u_i$.

3:    $\mathrm{Var}_i \leftarrow \frac{1}{|S|} \sum_{u \in S} (u_i - \mu_i)^2$.

4: **end for**

5: Rank all dimensions $i \in F$ in decreasing order of Var_i. Let $\Re(F)$ denote this ranking.

    // Query processing step: perform a multi-step search.

6: Among all the dimensions in subspace F', select the dimension i^* with the highest ranking according to $\Re(F)$.

7: Call SK(q, k) to produce the query result, with $d_{\{i^*\}}$ as the lower-bounding distance function, and $d_{F'}$ as the target distance function.

Fig. 1. The description of algorithm SK_SR

3.1 Single Ranking Strategy

The first of our proposed algorithms — SK_SR, described in Fig. 1 — employs a single overall ranking of dimensions based on variance. There are two main phases: a preprocessing phase and a query processing phase. In the preprocessing phase, the algorithm generates a single ranking of the dimensions, in terms of the variances of the data values computed separately for each of the dimensional coordinates — the larger the data variance for a given dimension, the higher the ranking of that dimension. In the query processing phase, as the lower-bounding distance function used in multi-step search, the algorithm chooses the dimension of highest rank from among the query dimensions. When Algorithm SK is used for performing the multi-step search (in Line 7), the query result is guaranteed to be correct. As an alternative, we may also utilize the approximate multi-step algorithm MAET+; this variant of subspace similarity search will be referred to as MAET+_SR. Specifically, we make a call to MAET+(q, k, t), where $t > 0$ is a parameter governing an early termination criterion. Larger choices of t can be expected to yield query results with higher accuracies at the possible expense of computational cost. In [5], a sampling method was designed for choosing t so that a desired proportion of potential queries can be correctly answered with high probability. For more details, we refer the reader to [5].

Note that like DMI, our search strategy requires the construction of a separate index for each of the dimensions. However, unlike DMI, our algorithms access only a single index per query, namely the most discriminative query dimension in terms of variance.

3.2 Multiple Ranking Strategy

The single ranking strategy has the advantage of being straightforward to apply. However, its effectiveness may be limited whenever the variance of a particular dimension differs greatly when restricted to the vicinity of differing query objects. For this reason, we have also designed a multiple ranking strategy that takes the query object into account when generating a ranking of dimensions.

Our multiple ranking strategy for subspace similarity search, SK_MR, is described in Fig. 2. In the preprocessing step, the algorithm first samples m reference points from

---

Algorithm **SK_MR** *(query q, subspace F', target neighborhood size k, sample size m, variance neighborhood size K)*

    // Preprocessing step: create multiple rankings of dimensions.

1: Create a reference set $R \subseteq S$ by sampling m points from the database, uniformly at random and without replacement.

2: **for** each reference point $v \in R$ **do**

3:    **for** each dimension $i \in F$ **do**

4:      $\mu_{v,i} \leftarrow \frac{1}{|K|} \sum_{u \in N_{\{i\}}(v,K)} u_i.$

5:      $\mathrm{Var}_{v,i} \leftarrow \frac{1}{|K|} \sum_{u \in N_{\{i\}}(v,K)} (u_i - \mu_{v,i})^2.$

6:    **end for**

7:    Rank all dimensions $i \in F$ in decreasing order of $\mathrm{Var}_{v,i}$. Let $\Re_v(F)$ denote this ranking.

8: **end for**

    // Query processing step: perform a multi-step search.

9: Linearly scan R to find v^*, the nearest reference point to q with respect to $d_{F'}$.

10: Select the query dimension $i^* \in F'$ with the highest ranking according to $\Re_{v^*}(F)$.

11: Call $\mathrm{SK}(q,k)$ to produce the query result, with $d_{\{i^*\}}$ being the lower-bounding distance function and $d_{F'}$ being the target distance function.

Fig. 2. The description of algorithm SK_MR

the database. Then, with respect to each reference point v, the algorithm determines a ranking (from highest to lowest) of all dimensions based on the variance of the coordinate values for the dimension in question, this time computed over a neighbor set of v (instead of over the entire dataset S). In the query processing step, the algorithm first finds the nearest reference point v^* of q in the query subspace (using sequential search within the reference set), and then uses the ranking of dimensions precomputed for v^* in the processing of query q. Again, we may replace SK with MAET+ to derive an approximation variant, MAET+_MR.

Two parameter choices must be considered when applying the multiple ranking strategy: the number of reference points m, and the size K of the neighborhoods within which data variance is computed. As will be shown in Sect. 4, the choice of K does not greatly affect the performance, provided that it is small relative to the dataset size $|S|$. On the other hand, the number of reference points m must be chosen with more care. If m is too large, the identification of the most discriminative query dimension may become unaffordable. If m is too small, the dimension i^* selected for multi-step search may not be very discriminative for the query. We next discuss how to choose a reasonable value for m.

Determining the Reference Set Size. For the multiple ranking strategy to be effective, for any given query point q, its nearest reference point v^* should be among the nearest neighbors of q within S (all with respect to the query subspace). Otherwise, the ranking of dimensions based at v^* may fail to approximate the ranking based at q. Fortunately, the following technical lemma shows that with even a relatively small number of reference points, v^* can lie in the local neighborhood of q with high probability.

Lemma 1 (Houle et al. [5]). *Let A be a set of positive integers, and let $A' \subseteq A$ be a subset sampled uniformly at random without replacement. Given a threshold τ, let a*

and a' refer to the number of elements in A and A', respectively, that are no greater than τ. Take η and η' to refer to the proportion of those elements within A and A', respectively. For any real number $\phi \geq 0$, we have $\Pr[|\eta - \eta'| \geq \phi] \leq 2e^{-2\phi^2|A'|}$.

Proof. Since A' is generated by uniform selection from A, random variable a' follows the hypergeometric distribution with expectation $\mathsf{E}[a'] = a|A'|/|A|$. In [14], Chvátal showed that random variable a' satisfies both $\Pr[\mathsf{E}[a'] \geq a' + \phi|A'|] \leq e^{-2\phi^2|A'|}$ and $\Pr[\mathsf{E}[a'] \leq a' - \phi|A'|] \leq e^{-2\phi^2|A'|}$. Both inequalities can be combined to yield the following error bound:

$$\Pr[|\eta - \eta'| \geq \phi] = \Pr\left[\left|\frac{a}{|A|} - \frac{a'}{|A'|}\right| \geq \phi\right] = \Pr\left[\left|\frac{\mathsf{E}[a']}{|A'|} - \frac{a'}{|A'|}\right| \geq \phi\right]$$
$$= \Pr[|\mathsf{E}[a'] - a'| \geq \phi|A'|] \leq 2e^{-2\phi^2|A'|}. \qquad \square$$

To apply this lemma to the analysis of the choice of reference set size, let $A = \{1, 2, 3, \ldots, |S|\}$ represent the ranks of all the objects in S with respect to a query object q, and let $A' \subseteq A$ store the ranks of all the reference points ($|A'| = m$). Also, let τ be the rank of the reference point v^*, which implies that $\eta' = 1/|A'|$. A small value of η would therefore indicate that v^* is in the local neighborhood of q, as desired. From Lemma 1, we know that the probability of η deviating from $\eta' = 1/|A'|$ by more than $\phi \geq 0$ is at most $2e^{-2\phi^2|A'|}$. That is, the probability of η being significantly larger than $1/|A'|$ vanishes quickly as the sample size $|A'|$ grows. In practice, even small sample sizes allow us to obtain reasonably small values of η with high probability. For example, if $|A'| = 5,000$ and $\phi = 0.02$, the lemma indicates that the probability of $\eta \geq 0.0202$ is at most 0.037, or equivalently, the probability of $\eta < 0.0202$ is at least 0.963.

4   Experimental Results

In this section, we present the results of our experimentation. We compared our algorithms with the state-of-the-art approaches PVA, PT and DMI.

4.1   Experimental Framework

Data Sets. Five publicly-available data sets were considered for the experimentation, so as to compare across a variety of set sizes, dimensions and data types.

- The Amsterdam Library of Object Images (ALOI) [15] consists of $110,250$ images of 1000 small objects taken from different viewpoints and illumination directions. The images are represented by 641-dimensional feature vectors based on color and texture histograms (for a detailed description of the image features, see [16]).
- The MNIST data set [17] consists of $70,000$ images of handwritten digits from 500 different writers, with each image represented by 784 gray-scale texture values.
- The Cortina data set [18] consists of $1,088,864$ images gathered from the World Wide Web. Each image is represented by a 74-dimensional feature vector based on homogeneous texture, dominant color and edge histograms.

- The Forest Cover Type set (FCT) [19] consists of $581,012$ data points, with each representing a 30×30 square meter area of forest. Each point is represented by 54 attributes, associated with elevation, aspect, slope and other geographical characteristics.
- The ANN_SIFT data set [20] consists of 10^7 SIFT descriptors [21] of 128 dimensions. The SIFT descriptors were extracted from approximately 10^6 general images.

Methodology. For each test, 1000 queries were generated at random, each consisting of an object q selected from the database, and a query subspace F'. Unless stated otherwise, the number of query dimensions was $|F'| = 8$, and the target neighborhood size was $k = 10$. Two quantities were measured for the evaluation: query result accuracy and execution time. The results were reported as averages over the 1000 queries performed. The execution time is shown as a proportion of the time needed for a sequential search of the entire dataset. For each query, the accuracy of its k-NN result is defined as the proportion of the result falling within the true k-NN (subspace) distance to q:

$$\frac{\mid \{v \in Y \mid d_{F'}(q, v) \le \delta_{F'}(q, k)\} \mid}{k},$$

where Y denotes the k-NN query result of q in subspace F' ($|Y| = k$). The Euclidean distance was used for all experiments.

4.2 Effects of Varying m and K on the Multiple Ranking Strategy

For the first set of experiments, for all of the datasets under consideration, we tested the effects on the multiple ranking strategy due to variation of the sample size m and variance neighborhood size K. When varying the sample size m, the variance neighborhood size K was chosen to be approximately 1% of the dataset size: specifically, the choices were $K = 10^3$ for ALOI and MNIST, $K = 10^4$ for Cortina and FCT, and $K = 10^5$ for ANN_SIFT. When varying K, the sample size m was fixed at 500 for all datasets tested. Since we observed similar trends in the results for all datasets, due to space limitations, in this version of the paper, we show the results of varying m and K only for the ALOI dataset.

The results for varying m are shown in Fig. 3(a). Here, we see that $m = 500$ is a sufficiently-large sample size for multiple ranking strategy to be effective, which is better than indicated by the theoretical analysis. From Lemma 1, we know that if $m = 500$, then for any dataset with any number of data points, the probability of $\eta < 0.062$ is at least 0.945 ($\phi = 0.06$). Recall that the effectiveness of the multiple ranking strategy is expected to increase as η diminishes. Our experimental findings show that the value of η in practice is typically much smaller than what the analysis indicates. In order to reduce the computational cost of the experimentation, we therefore set $m = 500$ for all remaining experiments.

Fig. 3(b) shows the results of varying K. As expected, the variance neighborhood size K does not greatly affect the performance, provided that it is set to reasonably small values relative to the dataset size. For all remaining experiments, we set $K = 10^3$ for ALOI and MNIST, $K = 10^4$ for Cortina and FCT, and $K = 10^5$ for ANN_SIFT.

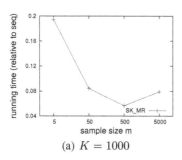
(a) $K = 1000$

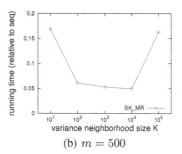
(b) $m = 500$

Fig. 3. The effects of varying m and K for the multiple ranking strategy, with dataset ALOI

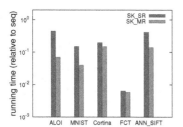

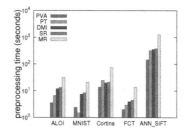

Fig. 4. The comparison of SR and MR on all datasets tested

Fig. 5. Preprocessing costs for all datasets tested

4.3 Comparison of Single Ranking and Multiple Ranking

We next compared the performance of the single ranking and multiple ranking strategies; the results are shown in Fig. 4. Unsurprisingly, multiple ranking outperformed single ranking for all datasets tested. Due to space limitations, in all experiments involving competing methods, we show a comparison of results only for multiple ranking.

4.4 Comparison with Other Methods

We conducted two sets of experiments for the comparison of our algorithms with competing methods, varying each of two parameters in turn: the number of subspace dimensions $|F'|$, and the target neighborhood size k. Specifically, we varied $|F'|$ from 2 to 32 while fixing $k = 10$, and varied k from 5 to 40 while fixing $|F'| = 8$.

The results of varying $|F'|$ are shown in Fig. 6. For all datasets and all choices of $|F'|$, our proposed methods generally outperform their competitors. Among all the methods tested, PVA is the most expensive, perhaps due to its use of sequential scan.

PT utilizes an R-tree built on the full-dimensional space to answer queries in subspaces; consequently, one would expect it to be less effective for subspaces in which $d_{F'}$ differs greatly from d_F. This can explain the improvement in the performance of PT as the number of subspace dimensions increases. Nevertheless, due to the limits on the performance of R-trees for spaces of even moderate dimensionality, PT will still become prohibitively expensive as the number of subspace dimensions grows.

DMI processes queries by aggregating partial results across neighborhoods with respect to every query dimension. The aggregation may become prohibitively expensive

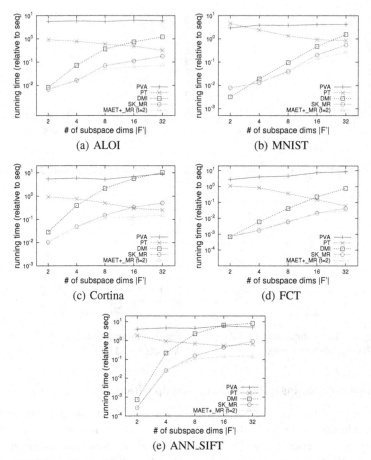

Fig. 6. The results of varying $|F'|$ on all tested datasets, with $k = 10$. The results are exact, except for those of MAET+_MR. The average accuracies of MAET+_MR with $t = 2$ are approximately 92%, 90%, 88%, 97% and 90% for ALOI, MNIST, Cortina, FCT and ANN_SIFT, respectively.

as the number of subspace dimensions increases. In contrast, our algorithms avoid expensive aggregation by restricting the processing to a single query dimension.

Relative to SK_MR, we observe that for high subspace dimensionality, MAET+_MR can achieve a significant improvement in running time while still achieving a high level of accuracy. We note that as the value of $|F'|$ increases, the computational cost of all tested methods must eventually tend to that of sequential search, as one would expect due to the curse of dimensionality.

Fig. 7 shows the results of varying k. Again, our proposed methods generally outperform their competitors, with MAET+_MR achieving a slight improvement in running time over SK_MR, at the cost of a slight loss of accuracy. We also observe that the behaviors of all tested methods are quite stable with respect to k.

Finally, Fig. 5 shows the preprocessing costs of all methods considered in our experimentation. While the preprocessing costs of our methods is substantial, the costs are justifiable in light of their improved performance at query time.

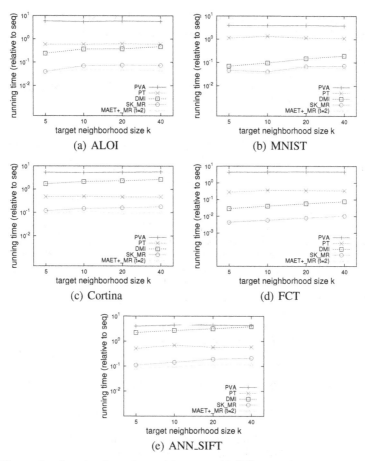

Fig. 7. The results of varying k on all tested datasets, with $|F'| = 8$. The results are exact, except for those of MAET+_MR. The average accuracies of MAET+_MR with $t = 2$ are approximately 88%, 96%, 89%, 98% and 92% for ALOI, MNIST, Cortina, FCT and ANN_SIFT, respectively.

5   Conclusion

We have presented new solutions for the subspace similarity search problem based on multi-step search, utilizing 1-dimensional lower-bounding distances for the efficient pruning of the search space. Our extensive experimental study showed that our algorithms are able to outperform their state-of-the-art competitors (PVA, PT and DMI) for a relatively wide range of subspace dimensions. We have also shown how practical choices of algorithm parameters can be guided by an analysis of sampling properties.

One possible direction for future research may include the investigation of multi-dimensional lower-bounding distances for pruning in multi-step subspace search. Although multi-dimensional distances could provide a tighter lower bound on the target distance, they cover fewer combinations of query dimensions, and thus may be only of limited practicality.

Acknowledgments. This work was partially supported by NSF under Grant 1241976 and by JSPS under Kakenhi Kiban (C) Research Grant 24500135.

References

1. Kriegel, H.P., Kröger, P., Zimek, A.: Subspace Clustering. WIREs Data Mining and Knowl. Discov. 2(4), 351–364 (2012)
2. Kohavi, R., John, G.H.: Wrappers for Feature Subset Selection. Artif. Intell. 97(1-2), 273–324 (1997)
3. Samet, H.: Foundations of Multidimensional and Metric Data Structures. Morgan Kaufmann (2006)
4. Seidl, T., Kriegel, H.P.: Optimal Multi-Step k-Nearest Neighbor Search. In: SIGMOD, pp. 154–165 (1998)
5. Houle, M., Ma, X., Nett, M., Oria, V.: Dimensional Testing for Multi-step Similarity Search. In: ICDM, pp. 299–308 (2012)
6. Kriegel, H.P., Kröger, P., Schubert, M., Zhu, Z.: Efficient Query Processing in Arbitrary Subspaces Using Vector Approximations. In: SSDBM, pp. 184–190 (2006)
7. Weber, R., Schek, H.J., Blott, S.: A Quantitative Analysis and Performance Study for Similarity-Search Methods in High-Dimensional Spaces. In: VLDB, pp. 194–205 (1998)
8. Bernecker, T., Emrich, T., Graf, F., Kriegel, H.P., Kröger, P., Renz, M., Schubert, E., Zimek, A.: Subspace Similarity Search Using the Ideas of Ranking and Top-k Retrieval. In: Proc. ICDE Workshop DBRank, pp. 4–9 (2010)
9. Bernecker, T., Emrich, T., Graf, F., Kriegel, H.-P., Kröger, P., Renz, M., Schubert, E., Zimek, A.: Subspace Similarity Search: Efficient k-NN Queries in Arbitrary Subspaces. In: Gertz, M., Ludäscher, B. (eds.) SSDBM 2010. LNCS, vol. 6187, pp. 555–564. Springer, Heidelberg (2010)
10. Guttman, A.: R-trees: a Dynamic Index Structure for Spatial Searching. In: SIGMOD, pp. 47–57 (1984)
11. Lian, X., Chen, L.: Similarity Search in Arbitrary Subspaces Under L_p-Norm. In: ICDE, pp. 317–326 (2008)
12. Korn, F., Sidiropoulos, N., Faloutsos, C., Siegel, E., Protopapas, Z.: Fast Nearest Neighbor Search in Medical Image Databases. In: VLDB, pp. 215–226 (1996)
13. Houle, M., Kashima, H., Nett, M.: Generalized Expansion Dimension. In: Proc. ICDM Workshop PTDM, pp. 587–594 (2012)
14. Chvátal, V.: The Tail of the Hypergeometric Distribution. Discrete Mathematics 25, 285–287 (1979)
15. Geusebroek, J.M., Burghouts, G.J., Smeulders, A.W.M.: The Amsterdam Library of Object Images. International Journal of Computer Vision 61(1), 103–112 (2005)
16. Boujemaa, N., Fauqueur, J., Ferecatu, M., Fleuret, F., Gouet, V., Saux, B.L., Sahbi, H.: IKONA: Interactive Generic and Specific Image Retrieval. In: Proc. Intern. Workshop on Multimedia Content-Based Indexing and Retrieval (2001)
17. LeCun, Y., Bottou, L., Bengio, Y., Haffner, P.: Gradient-Based Learning Applied to Document Recognition. Proc. IEEE 86(11), 2278–2324 (1998)
18. Rose, K., Manjunath, B.S.: The Cortina Data Set,
http://www.scl.ece.ucsb.edu/datasets/index.htm
19. Asuncion, A., Newman, D.J.: UCI Machine Learning Repository,
http://www.ics.uci.edu/~mlearn/MLRepository.html
20. Jégou, H., Tavenard, R., Douze, M., Amsaleg, L.: Searching in One Billion Vectors: Re-rank with Source Coding. In: ICASSP, pp. 861–864 (2011)
21. Lowe, D.G.: Distinctive Image Features from Scale-Invariant Keypoints. International Journal of Computer Vision 60(2), 91–110 (2004)

Partial Refinement for Similarity Search with Multiple Features

Marcel Zierenberg

Brandenburg University of Technology Cottbus – Senftenberg
Institute of Computer Science, Information and Media Technology
Chair of Database and Information Systems
P.O. Box 10 13 44, 03013 Cottbus, Germany
zieremar@tu-cottbus.de

Abstract. Filter refinement is an efficient and flexible indexing approach to similarity search with multiple features. However, the conventional refinement phase has one major drawback: when an object is refined, the partial distances to the query object are computed for *all features*. This frequently leads to more distance computations being executed than necessary to exclude an object. To address this problem, we introduce *partial refinement*, a simple, yet efficient improvement of the filter refinement approach. It incrementally replaces partial distance bounds with exact partial distances and updates the aggregated bounds accordingly each time. This enables us to exclude many objects before all of their partial distances have been computed exactly. Our experimental evaluation illustrates that partial refinement significantly reduces the number of required distance computations and the overall search time in comparison to conventional refinement and other state-of-the-art techniques.

1 Introduction

Similarity search with multiple features is an effective way of finding objects similar to a query object. Instead of using only a single *feature* for the comparison of objects (e.g., a single color histogram for the comparison of images), multiple features (e.g., color, edge and texture features) are utilized. A *distance function* assigned to each feature is employed to compute the respective *partial distances* (dissimilarities) between each of the compared objects' features. These partial distances are combined into an *aggregated distance* by means of an *aggregation function*. Finally, the most similar objects are determined according to the lowest aggregated distances to the query object.

Indexing approaches to similarity search [1, 2] aim to exclude as many objects as possible from the search to decrease CPU and I/O costs for the computation of distances.

Filter refinement is a well-known technique and utilized by several indexing approaches to multi-feature similarity search (e.g., [3, 4, 5, 6]). In general, the *filtering phase* aims to discard objects based on inexpensively computed approximations of the distance between the query and each database object (*bounds*). The *refinement phase* then computes the exact distances for the remaining candidates to determine the most similar objects. For search with multiple features, *partial bounds* for each feature are combined into an *aggregated bound*. The exclusion of objects in the filtering and refinement phase is based on those aggregated bounds.

A.J. Machado Traina et al. (Eds.): SISAP 2014, LNCS 8821, pp. 13–24, 2014.
DOI: 10.1007/978-3-319-11988-5_2 © Springer International Publishing Switzerland 2014

Unfortunately, the performance of filter refinement deteriorates with an increasing number of features. As the *(intrinsic) dimensionality* [7] of the aggregation function rises, the *approximation error* of the aggregated bounds increases as well. A higher approximation error results in less efficient search because fewer objects can be excluded.

1.1 Contribution

The main contribution of this paper is the improvement of the refinement step for filter refinement with multiple features. *Conventional refinement* (see Section 3) manages objects with the help of a candidate list sorted in ascending order according to the aggregated lower bounds. When an object on top of the candidate list is refined, all of the object's partial distances are computed exactly and combined into an aggregated distance. Unfortunately, this frequently leads to more partial distances being computed than necessary to exclude objects.

In contrast, our *partial refinement* approach (see Section 5) incrementally replaces partial distance bounds of objects with their exact partial distances, updates the aggregated bounds and reinserts the objects into the candidate list. This allows us to gradually tighten the aggregated bounds and to exclude many objects before all of their partial distances have been computed exactly.

Example 1. Consider for example a similarity search with two features as depicted in Figure 1a. The filtering phase produces a candidate list ordered according to the aggregated lower bounds. Conventional refinement requires three iterations and each iteration executes two distance computations (*fully refined*).

In contrast, partial refinement computes only one partial distance per iteration (*partially refined*), updates the aggregated bounds of the object and reinserts it into the candidate list. In this case, it permits a *direct* (case C1) and a *delayed exclusion* (case C2) of two objects, without computing all of their exact partial distances. While the conventional refinement approach requires six distance computations in this example, four distance computations are sufficient to determine the most similar object with partial refinement.

Example 2. Another example for the benefits of partial refinement is the *partial exclusion* of objects (case C3) for specific aggregation functions like the minimum or maximum function. If it becomes obvious that a specific partial distance does not influence the aggregated distance, it can be safely excluded from computation. For the example of the maximum function with three features in Figure 1b, two partial distance computations can be excluded because their upper bound (4 and 3) is lower than the exact partial distance of the first feature (5).

To demonstrate the efficiency of our approach, we experimentally compare partial refinement to the linear scan, conventional filter refinement [6], the *Onion-tree* [8] and the *Threshold Combiner Algorithm* [9] (see Section 6). The evaluation illustrates that partial refinement is able to significantly reduce the number of required distance computations and the overall search time.

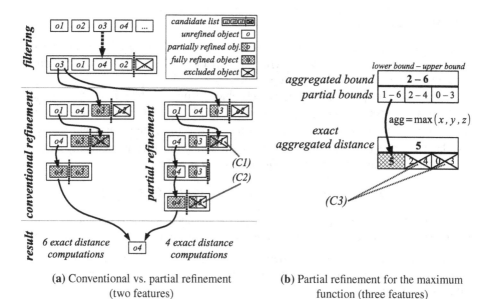

(a) Conventional vs. partial refinement
(two features)

(b) Partial refinement for the maximum
function (three features)

Fig. 1. Examples for direct (C1), delayed (C2) and partial exclusion (C3)

2   Preliminaries

This section defines the notations and terms used throughout this paper.

2.1   Nearest Neighbor Search

Similarity search can be performed by means of a *k-Nearest Neighbor query*. A kNN(q)-query in the universe of objects $\mathbb{U}$ returns k objects out of a database $DB = \{o^1, \ldots, o^n\} \subseteq \mathbb{U}$ that are closest (most similar) to the query object $q \in \mathbb{U}$. The distance between objects is computed by a distance function $\delta : \mathbb{U} \times \mathbb{U} \mapsto \mathbb{R}_{\geq 0}$ that operates on the features $\hat{q}$ and $\hat{o}^i$ extracted from the objects. The result is a (non-deterministic) set K with $|K| = k$ and $\forall o^i \in K, o^j \in DB \setminus K : \delta(q, o^i) \leq \delta(q, o^j)$.

A *multi-feature kNN-query* substitutes the single features $\hat{q}$ and $\hat{o}^i$ with m features $\hat{q} = (\hat{q}_1, \ldots, \hat{q}_m)$ and $\hat{o}^i = (\hat{o}^i_1, \ldots, \hat{o}^i_m)$. A distance function δ_j is assigned to each single feature to compute the *partial distances* $d^i_j = \delta_j(q, o^i)$. An *aggregation function* agg : $\mathbb{R}^m_{\geq 0} \mapsto \mathbb{R}_{\geq 0}$ combines all partial distances to an *aggregated distance* d^i_{agg} and the k nearest neighbors are then determined according to the aggregated distance.

An optional *weighting scheme* with weights $W = (w_1, \ldots, w_m)$ and $\forall w_j \in W : w_j \geq 0$ can be applied to the features of the aggregation function. These weights are typically unknown at the time of index construction. Instead, they are dynamically determined at query time in order to optimally adapt the aggregation function to the query object and the demands of the user [10].

2.2 Metric Indexing

A *metric* is a distance function with the properties *positivity* ($\forall x \neq y \in \mathbb{U} : \delta(x, y) > 0$), *symmetry* ($\forall x, y \in \mathbb{U} : \delta(x, y) = \delta(y, x)$), *reflexivity* ($\forall x \in \mathbb{U} : \delta(x, x) = 0$) and *triangle inequality* ($\forall x, y, z \in \mathbb{U} : \delta(x, z) \leq \delta(x, y) + \delta(y, z)$).

Metric indexing approaches [1, 2] exclude objects from search by computing *bounds* of the distance from the query object to database objects. The lower bound lb_j^i and upper bound ub_j^i of the exact partial distance $d_j^i = \delta_j(q, o)$ between query object q and database object o^i can be determined by exploiting the triangle inequality and the precomputed distance to a *reference object (pivot)* p as follows:

$$lb_j^i = |\delta_j(q, p) - \delta_j(p, o^i)| \leq \delta_j(q, o^i) \leq \delta_j(q, p) + \delta_j(p, o^i) = ub_j^i. \tag{1}$$

The *approximation error* $\epsilon_1^i, \ldots, \epsilon_m^i$ of the partial distance bounds is calculated by the weighted difference between the respective upper and lower bounds $\epsilon_j^i = w_j * (ub_j^i - lb_j^i)$.

The *intrinsic dimensionality* ρ is defined as $\rho = \frac{\mu^2}{2\sigma^2}$ where μ is the mean and σ^2 the variance of a distance distribution. It is frequently used as an estimator for the indexability of metric spaces [7].

2.3 Monotonicity and Aggregated Bounds

An aggregation function agg is *monotone increasing in the j-th argument* with $1 \leq j \leq m$, $d = (d_1, \ldots, d_j, \ldots, d_m)$ and $d' = (d_1, \ldots, d'_j, \ldots, d_m)$ iff:

$$\forall d, d' \in \mathbb{R}_{\geq 0}^m : d_j < d'_j \implies \text{agg}(d) \leq \text{agg}(d'). \tag{2}$$

This means, if all arguments except d_j are constant and d_j is increased to d'_j, the result of the aggregation function will either be constant or also increase.

An aggregation function agg is *globally monotone increasing* iff it contains only monotone increasing arguments. An example for a globally monotone increasing function is $\text{agg}(d_1, d_2) = d_1 + d_2$.

For the sake of simplicity, we consider only globally monotone increasing aggregation functions for distances (dissimilarity values) in the following. However, note that the stated results are easily adaptable to other types, like locally or flexible monotone aggregation functions and aggregation functions for similarity values [6].

Even though the aggregation function can be a metric if all partial distance functions are also metrics (e.g., arithmetic mean or maximum of L_1 distances), this is not necessarily the case. The minimum and the median function ($m > 2$) are examples for non-metric aggregation functions that are globally monotone increasing.

The aggregated lower (upper) bound lb_{agg}^i (ub_{agg}^i) on the exact aggregated distance $d_{agg}^i = \text{agg}\left(d_1^i, \ldots, d_m^i\right)$ of a globally monotone increasing aggregation function can be computed by inserting partial lower (upper) bounds for all features into the aggregation function:

$$lb_{agg}^i = \text{agg}\left(lb_1^i, \ldots, lb_m^i\right) \leq \text{agg}\left(d_1^i, \ldots, d_m^i\right) \leq \text{agg}\left(ub_1^i, \ldots, ub_m^i\right) = ub_{agg}^i. \tag{3}$$

The approximation error of the aggregated bounds is defined as $\epsilon_{agg}^i = ub_{agg}^i - lb_{agg}^i$.

Algorithm 1. Multi-feature kNN-query – filtering

Input: k, q, DB, agg, W

1 $t_{max} = \infty$;

2 **foreach** $o^i \in DB$ **do**

3 | Compute partial bounds lb^i_j and ub^i_j for each feature;   // Equation (1)

4 | Compute aggregated bounds lb^i_{agg} and ub^i_{agg};   // Equation (3)

5 | **if** $lb^i_{agg} > t_{max}$ **then continue**;   // exclude object?

6 | **else**

7 | | $candidates.\text{insert}(o^i)$;

8 | | $t_{max} = k$-th lowest ub^i_{agg};   // update threshold

9 | | $candidates.\text{cut}(t_{max})$;   // exclude objects with $lb^i_{agg} > t_{max}$

10 **return** $candidates$;

---

3 Filter Refinement

The following section briefly summarizes the conventional filter refinement approach to similarity search with multiple features.

To build the index, metric filter refinement approaches [4, 6] compute one matrix of distances between pivots and database objects per feature. Algorithm 1 depicts the filtering phase for a kNN-query with multiple features. At first, bounds for each partial distance are computed based on the precomputed distance matrices and Equation (1) (line 3). Subsequently, these partial bounds are combined into aggregated bounds by Equation (3) (line 4). Objects having a higher aggregated lower bound lb^i_{agg} than the k-th lowest aggregated upper bound ub^i_{agg} seen so far (t_{max}) are excluded from the search (lines 5 and 9). The remaining objects are managed by a candidate list sorted in ascending order according to lb^i_{agg} (priority queue).

In the conventional refinement phase (Algorithm 2) the previously determined candidate objects have to be refined. Starting with the candidate with the lowest aggregated lower bound lb^i_{agg}, we check if the object appeared at the top of the candidate list before (line 3). If not, the object was not refined yet and the exact aggregated distance d^i_{agg} has to be computed (line 5). Afterwards, the object is either excluded because its exact aggregated distance is larger than the current threshold value t_{max} (line 6) or it is reinserted into the candidate list.

If the object at the top of the candidate list was already refined before (line 3), the object is one of the k nearest neighbors because the object's exact aggregated distance d^i_{agg} is lower than the remaining objects' aggregated lower bounds lb^i_{agg}. Refinement is stopped as soon as k nearest neighbors were found.

4 Related Work

This section gives an insight into the state-of-the-art of indexing for similarity search with multiple features and filter refinement for multiple features in particular.

Algorithm 2. Multi-feature kNN-query – conventional refinement

---

Input: k, q, *candidates*, t_{max}, *agg*, W

```
1 repeat
2     oⁱ = candidates.pop();                    // get candidate with lowest lb_agg^j
3     if oⁱ is refined then results.insert(oⁱ);          // already refined?
4     else
5         Compute exact aggregated distance d_agg^i;          // refinement
6         if d_agg^i > t_max then continue;              // exclude object?
7         else …;                        // (lines 7 - 9 of Algorithm 1)
8 until results.size() = k;
9 return results;
```

If the aggregation function is a metric, an arbitrary (single-feature) metric index (e.g., *Onion-tree* [8]) can be build directly on top of the aggregated distances (*naïve approach*). Unfortunately, this solution prevents partial refinement and is inflexible because it requires the index to be rebuilt when the used aggregation function, features or weights are changed [6]. *Multi-metric indexing* [11] solves this problem partially. It defines a framework to transform arbitrary metric indexing approaches for single features into metric indices for multiple features with dynamic weighting. However, the restriction to metric aggregation functions remains.

The M^2-tree [12] is a multi-dimensional extension of the well-known *M-tree*. It supports dynamic weighting as well as metric and non-metric aggregation functions. However, it is not suitable for partial refinement and has the disadvantage that its clustering may be inefficient if only a subset of all indexed features is used for a query.

An index comprised of one matrix of distances to pivot objects per feature is described in [4]. This allows efficient queries with subsets of the indexed features and dynamic weighting since each matrix can be accessed individually. Filter refinement is used to exclude objects. However, the approach does not utilize a candidate list to determine the order of objects and objects that were not excluded are always fully refined.

Our previous research introduced *FlexiDex* [6], a flexible metric index for (logic-based) multi-feature similarity search. The index has to be created only once but can be efficiently used for different types of aggregation functions, numbers of features and weighting schemes. Originally, FlexiDex fully refines each object. However, in the course of our research we adapted it to incorporate all concepts of partial refinement.

Combiner algorithms (e.g., *Threshold Algorithm* (TA) [9]) merge the result lists of subqueries for each single feature into an aggregated result list. Once an object is seen in one of the lists, missing partial distances are computed by random access. This behavior resembles conventional refinement. However, filter refinement uses a single candidate list and sorts it based on the aggregated bounds. This allows it to adapt better to the aggregation function than combiner algorithms.

Our research focuses on metric indexing since it is more flexible and suffers less from the *curse of dimensionality* [2] than *spatial indexing* [1]. Nonetheless, partial refinement can be easily adapted to improve spatial indices that rely on filter refinement for multi-feature search (e.g., *GeVAS* [3] or *ASAP* [5]).

5 Partial Refinement

This section presents our main contribution, the *partial refinement approach*, which deals with the major drawback of conventional refinement to compute all partial distances of an object at once. We describe our concept in detail and give a pseudo-code implementation of the approach.

5.1 Exclusion of Objects

The main idea behind the concept of partial refinement is to exclude objects before all of their partial distances have been computed by gradually improving the quality of their aggregated bounds. This is accomplished by updating the aggregated bounds each time a partial distance of an object is computed exactly.

The order of partial distance computations for each individual object is determined in the filtering phase. The partial distances with the highest approximation error ϵ^i_j are computed first in order to quickly reduce the aggregated approximation error ϵ^i_{agg}. The following cases C1 – C3 are considered after every update of the aggregated bounds.

Direct exclusion (C1). If the updated aggregated lower bound lb^i_{agg} has increased above the current search threshold t_{max}, the object can be directly excluded from search without exactly computing the remaining partial distances.

Delayed exclusion (C2). If the updated aggregated lower bound lb^i_{agg} has not increased above the search threshold t_{max}, the object can currently not be excluded. The object is then reinserted into the candidate list and its position in the list is redetermined according to the updated aggregated lower bound. Now, if the search threshold t_{max} decreases below the updated aggregated lower bound lb^i_{agg} before the object reappears at the top of the candidate list, it can be excluded without exactly computing its remaining partial distances.

Partial exclusion (C3). For specific aggregation functions (e.g., minimum or maximum function) partial distance computations of an object can be excluded as soon as it becomes obvious that they do not influence the exact aggregated distance (*dominated distances*). This is achieved by comparing all partial distance bounds lb^i_j and ub^i_j of an object among each other.

5.2 Updating Aggregated Bounds

Partial refinement relies on the assumption that the computation of aggregated bounds and reinsertion into the candidate list is inexpensive in comparison to the computation of a partial distance.

In contrast to conventional refinement, which only needs to store the aggregated bounds for each object at query time, partial refinement additionally requires $2m$ partial bound values (m lower and m upper bounds) per object. Furthermore, a bit array b^i consisting of m bits per object is needed. Initially set to `false`, a bit b^i_j is set to `true` after the partial distance d^i_j for object o^i has been computed exactly.

It can easily be shown that replacing partial bounds with exact partial distances in Equation (3) can only result in tighter aggregated bounds. With each newly computed partial distance, the approximation error of the aggregated distance bounds ϵ_{agg}^i can be reduced. The *updated aggregated lower bounds* $\widehat{lb}_{agg}^i$ replace the old bounds after their computation and are defined as follows:

$$\widehat{lb}_j^i = \begin{cases} d_j^i, & \text{if } b_j^i = \texttt{true} \\ lb_j^i, & \text{otherwise} \end{cases}, \tag{4}$$

$$\widehat{lb}_{agg}^i = \text{agg}\left(\widehat{lb}_1^i, \ldots, \widehat{lb}_m^i\right) \geq lb_{agg}^i. \tag{5}$$

The *updated aggregated upper bounds* $\widehat{ub}_{agg}^i$ are defined analogously.

Obviously, after all partial distances of object o^i have been computed, the updated aggregated distance bounds are equal to the exact aggregated distance d_{agg}^i:

$$(b_1^i \wedge \ldots \wedge b_m^i) = \texttt{true} \implies \widehat{lb}_{agg}^i = d_{agg}^i = \widehat{ub}_{agg}^i. \tag{6}$$

5.3 Dominated Distances

Depending on the aggregation function (e.g., minimum or maximum function), it is not always necessary to compute all partial distances to determine the exact aggregated distance (case C3). In the following we will refer to those partial distances that are not needed as *dominated distances*.

For the example of the maximum function $\text{agg}_{\max}$, a partial distance d_j^i is dominated if a partial lower bound lb_x^i exists that is greater or equal to the partial upper bound ub_j^i:

$$\exists x \in \{1, \ldots, m\} : x \neq j \wedge lb_x^i \geq ub_j^i \implies$$

$$\text{agg}_{\max}\left(d_1^i, \ldots, d_j^i, \ldots, d_m^i\right) = \text{agg}_{\max}\left(d_1^i, \ldots, d_{j-1}^i, d_{j+1}^i, \ldots, d_m^i\right). \tag{7}$$

This means the partial distance d_j^i does not influence the aggregation result (maximum) as it cannot be the largest distance. We can therefore safely exclude the partial distance from computation. In this case, bit b_j^i is set to $\texttt{true}$ and d_j^i is set to the partial upper bound ub_j^i.

5.4 Partial Refinement Algorithm

Finally, we present the pseudo-code of partial refinement (see Algorithm 3). The conventional refinement of Algorithm 2 is adapted to incorporate the concepts presented in sections 5.1 – 5.3: the computation of a single partial distance (line 5), the detection of dominated distances (line 6), the update of the aggregated bounds (line 7) and the check for the object's exclusion or reinsertion into the candidate list, based on the updated aggregated bounds (lines 8 and 9).

Depending on the memory constraints of the system, disk-based or in-memory indexing can be utilized. A disk-based implementation of the filter refinement approach to multi-feature similarity search is described in [6] and also applicable to partial refinement. There, each distance matrix is compressed and stored in the form of a compact *signature file* that can be sequentially read from disk. For in-memory indexing all needed distance matrices are simply preloaded into main memory.

Algorithm 3. Multi-feature kNN-query – partial refinement

Input: k, q, *candidates*, t_{max}, *agg*, W

1 **repeat**

2      $o^i = $ *candidates*.pop();   `// get candidate with lowest` lb^i_{agg}

3   **if** $(b^i_1 \wedge \ldots \wedge b^i_m) = $**true then** *results*.insert(o^i);   `// Equation (6)`

4   **else**

5   Compute next exact partial distance d^i_j and set $b^i_j = $true;   `// partial ref.`

6   Check for dominated distances and update b^i accordingly;   `// Equation (7)`

7   Compute updated aggregated bounds $\widehat{lb}^i_{agg}$ and $\widehat{ub}^i_{agg}$;   `// Equation (5)`

8   **if** $\widehat{lb}^i_{agg} > t_{max}$ **then continue**;   `// exclude object?`

9   **else** $\ldots$;   `// (lines 7 - 9 of Algorithm 1)`

10 **until** *results*.size() $ = k$;

11 **return** *results*;

---

6   Evaluation

This section presents the experimental evaluation. We demonstrate that partial refinement can vastly reduce the number of required distance computations and the overall search time in comparison to conventional refinement and other state-of-the-art approaches.

6.1   Experimental Setup

Partial refinement was compared to the linear scan, conventional refinement [6], an Onion-tree [8] build on top of aggregated distances and the Threshold Combiner Algorithm [9] based on m (single-feature) Onion-trees in connection with the *HS-Algorithm* [13]. As recommended by the authors, all Onion-trees were built with the keep-small strategy [8].

All experiments were run on a 2×2.26 GHz Quad-Core Intel Xeon with 8 GB RAM and an HDD with 7,200 rpm. However, we restricted our experiments to a single CPU core since the provided implementation of the Onion-tree does not support parallelization.

We utilized the image collections *Caltech-256 Object Category Dataset* [14] (30,607 images) and *ImageCLEF WEBUPV Image Annotation Dataset* [15] (250,000 images) for our experiments. Efficiency was assessed by measuring the average number of distance computations and the average search time (wall-clock time) of kNN-queries for 100 randomly chosen query objects.

Features of varying intrinsic dimensionality and distance computation cost were chosen to examine the performance in distinct scenarios. Table 1 summarizes the used features and distance functions δ (Minkowski (L_p), Earth Mover's (EMD) and Quadratic Form (QFD)) and depicts the according intrinsic dimensionality ρ.

We used 64 pivot objects (randomly selected) per feature for filter refinement and kept all index data in main memory. Per feature, each object occupied 512 bytes of

Table 1. Used features, distance functions δ and intrinsic dimensionality ρ

Feature	δ	ρ per collection	
		Caltech256	WEBUPV
CEDD	L_2	12.05	11.70
FCTH	L_1	5.77	6.35
EdgeHistogram	weighted L_1	8.55	9.97
DominantColor	EMD + L_2	2.16	1.91
ColorHistogram	dynamic QFD	11.48	10.47

memory for the distances to the pivot objects (64×8 bytes; double precision), 16 bytes for the lower and upper partial distance bounds and 1 bit for the boolean flag b_j^i. Additional 16 bytes per object were required for the aggregated lower and upper bounds.

6.2 Aggregation Functions and Number of Features

The performance of partial refinement was investigated for various aggregation functions and numbers of features m. The features were added in the same order as given in Table 1 (from top to bottom).

Figures 2a and 2b show the number of required distance computations and search time for 10−NN-queries with the arithmetic mean. Obviously, the results of conventional and partial refinement were the same for a single feature. However, with an increasing number of features, partial refinement considerably outperformed all other approaches. It required up to 70 % less distance computations and up to 63 % less search time than conventional refinement. This means that the overhead of partial refinement (recomputing aggregated bounds and reinserting objects into the candidate list) is rather low in comparison to the time saved trough the reduced number of distance computations.

The number of required distance computations for 10-NN-queries with the median function is depicted in Figure 3a. Again, partial refinement was the optimal approach and computed up to 55 % less distances than conventional refinement. Note that the median function does not fulfill the triangle inequality for $m > 2$. Therefore, the Onion-tree frequently excluded objects that belonged to the correct query result.

In case of 10-NN-queries with the maximum function (Figure 3b), partial refinement slightly improved the already very good results of conventional refinement. The increase in the number of required distance computations per added feature was surprisingly low for partial refinement (≈ 200).

We conducted further experiments for other aggregation functions, like the minimum function, the geometric or the harmonic mean. However, these results are not shown as their behavior was mostly similar to the previous experiments.

6.3 Number of Result Objects and Collection Size

Figure 4a depicts the number of required distance computations of kNN-queries with the arithmetic mean for different numbers of result objects k. The Onion-tree and both filter refinement approaches were especially efficient for $k = 1$ because the query objects were elements of the collection. This allowed a very early termination of the search,

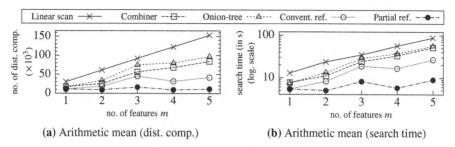

(a) Arithmetic mean (dist. comp.)   **(b)** Arithmetic mean (search time)

Fig. 2. Search performance for 10-NN-queries with arithmetic mean (Caltech256)

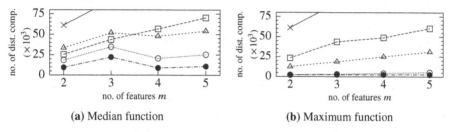

(a) Median function   **(b)** Maximum function

Fig. 3. Number of distance computations for 10-NN-queries (Caltech256)

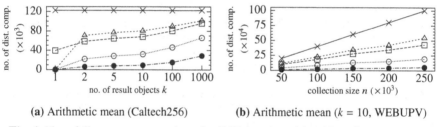

(a) Arithmetic mean (Caltech256)   **(b)** Arithmetic mean ($k = 10$, WEBUPV)

Fig. 4. No. of distance computations for kNN-queries with arithmetic mean ($m = 4$)

as soon as the respective query object was seen the first time. However, partial refinement was the optimal approach for greater numbers of result objects k and constantly required approximately 65 % less distance computations than conventional refinement.

The impact of the collection size n on the number of needed distance computations is presented in Figure 4b. Subsets of the WEBUPV image collection were obtained by dividing it into chunks of 50,000 images each. While the number of needed distance computations increased linearly with the collection size for all approaches, partial refinement exhibited the overall lowest increase.

7   Conclusions and Outlook

This paper introduced partial refinement, a simple, yet efficient improvement of the filter refinement approach to similarity search with multiple features. Partial refinement progressively replaces partial distance bounds with exact partial distances, updates the aggregated bounds accordingly and checks if objects can be excluded.

Our experimental evaluation has shown that partial refinement is able to significantly reduce the number of required distance computations and search time in comparison to conventional refinement and other state-of-the-art techniques.

Future research will focus on the introduction of new strategies to determine the optimal order of partial distance computations. Adapting the computation order to the used distance and aggregation functions can further improve the search performance.

References

[1] Samet, H.: Foundations of Multidimensional and Metric Data Structures. The Morgan Kaufmann Series in Computer Graphics and Geometric Modeling. Morgan Kaufmann Publishers Inc., San Francisco (2005)

[2] Zezula, P., Amato, G., Dohnal, V., Batko, M.: Similarity Search: The Metric Space Approach. Advances in Database Systems, vol. 32, pp. 1–191. Springer-Verlag New York Inc., Secaucus (2006)

[3] Böhm, K., Mlivoncic, M., Schek, H.-J., Weber, R.: Fast Evaluation Techniques for Complex Similarity Queries. In: Proc. of the 27th International Conference on Very Large Data Bases, VLDB 2001, pp. 211–220. Morgan Kaufmann Publishers Inc., San Francisco (2001)

[4] Bustos, B., Keim, D., Schreck, T.: A Pivot-Based Index Structure for Combination of Feature Vectors. In: Proc. of the 2005 ACM Symposium on Applied Computing, SAC 2005, pp. 1180–1184. ACM, New York (2005)

[5] Jagadish, H.V., Ooi, B.C., Shen, H.T., Tan, K.-L.: Toward Efficient Multifeature Query Processing. IEEE Trans. on Knowl. and Data Eng. 18, 350–362 (2006)

[6] Zierenberg, M., Bertram, M.: FlexiDex: Flexible Indexing for Similarity Search with Logic-Based Query Models. In: Catania, B., Guerrini, G., Pokorný, J. (eds.) ADBIS 2013. LNCS, vol. 8133, pp. 274–287. Springer, Heidelberg (2013)

[7] Chávez, E., Navarro, G., Baeza-Yates, R., Marroquín, J.L.: Searching in Metric Spaces. ACM Comput. Surv. 33, 273–321 (2001)

[8] Carélo, C.C.M., Pola, I.R.V., Ciferri, R.R., Traina, A.J.M., Traina Jr., C., de Aguiar Ciferri, C.D.: Slicing the Metric Space to Provide Quick Indexing of Complex Data in the Main Memory. Inf. Syst. 36(1), 79–98 (2011)

[9] Fagin, R., Lotem, A., Naor, M.: Optimal Aggregation Algorithms for Middleware. In: Proc. of the 20th ACM SIGMOD-SIGACT-SIGART Symposium on Principles of Database Systems, PODS 2001, pp. 102–113. ACM, New York (2001)

[10] Zellhöfer, D., Schmitt, I.: A Preference-Based Approach for Interactive Weight Learning: Learning Weights Within a Logic-Based Query Language. Distributed and Parallel Databases 27, 31–51 (2010)

[11] Bustos, B., Kreft, S., Skopal, T.: Adapting Metric Indexes for Searching in Multi-Metric Spaces. Multimedia Tools Appl. 58(3), 467–496 (2012)

[12] Ciaccia, P., Patella, M.: The M2-tree: Processing Complex Multi-Feature Queries with Just One Index. In: DELOS Workshop: Information Seeking, Searching and Querying in Digital Libraries (2000)

[13] Hjaltason, G.R., Samet, H.: Ranking in Spatial Databases. In: Egenhofer, M., Herring, J.R. (eds.) SSD 1995. LNCS, vol. 951, pp. 83–95. Springer, Heidelberg (1995)

[14] Griffin, G., Holub, A., Perona, P.: Caltech-256 Object Category Dataset. Tech. rep. 7694. California Institute of Technology (2007)

[15] Villegas, M., Paredes, R., Thomee, B.: Overview of the ImageCLEF 2013 Scalable Concept Image Annotation Subtask. In: CLEF 2013 Evaluation Labs and Workshop, Online Working Notes, Valencia, Spain (2013)

Video Retrieval with Feature Signature Sketches

Adam Blažek, Jakub Lokoč, and Tomáš Skopal

SIRET Research Group, Department of Software Engineering,
Faculty of Mathematics and Physics, Charles University in Prague
blazekada@gmail.com, {lokoc,skopal}@ksi.mff.cuni.cz

Abstract. In this paper, we present an effective yet efficient approach for known-item search in video data. The approach employs feature signatures based on color distribution to represent video key-frames. At the same time, the feature signatures enable users to intuitively draw simple colored sketches of the desired scene. We describe in detail the video retrieval model and also discuss and carefully optimize its parameters. Furthermore, several indexing techniques suitable for the model are presented and their performance is empirically evaluated in the experiments. Apart from that, we also investigate a bounding-sphere pruning technique suitable for similarity search in vector spaces.

1 Introduction

The volume of video data has been increasing rapidly over the last years which challenges the state-of-the-art video management and retrieval systems. Independently on the volume and the nature of the data, users still expect fast and accurate responses as well as simple user interfaces to specify a query and to intuitively browse the results. These demands are making the design of a system for video indexing and retrieval a true challenge.

A large amount of attention has been paid to the systems based on semantic annotation [3, 7], allowing users to specify text queries. To deal with the lack of annotation, complex concept and event detectors are being employed, but despite the progress made in the last years, the *semantic gap* still persists. If we consider just the reliably detectable common concepts (e.g., human faces or cars) we may end up with zero annotation, thus we cannot rely on them exclusively.

For these reasons the general-purpose content-based methods are getting more popular. Many visual descriptors [19, 22] were introduced to enable fast extraction, indexing and searching in large scale video archives. The systems [16] based on these descriptors usually demand an example as a query; however, such an example may not be always available. In such case, the user has to put an effort into obtaining the example, say using the Google Images, which can be time consuming or even impossible in some cases. Let us give an example: We are searching for a shot containing a particular TV studio interior, filmed from an unusual angle while we do not have an example. If we try to find the example in an independent image database with the phrase "TV studio" we will probably retrieve plenty of results and it might be hard to find such image that is visually close to the searched scene.

A.J. Machado Traina et al. (Eds.): SISAP 2014, LNCS 8821, pp. 25–36, 2014.
DOI: 10.1007/978-3-319-11988-5_3 © Springer International Publishing Switzerland 2014

This scenario matches the problem of the so-called *known-item search* (KIS), where the user "knows" what objects in images she is searching for (by imagination and/or textual description), however, she has no example to run a traditional query (an example video shot/key-frame in our case). By allowing users to specify the desired shot directly, for example using a sketch [6], the need of example can be eliminated. It is crucial, however, to keep the user interface as simple as possible. Only such descriptor shall be utilized that is descriptive enough, is understandable to users, and can be easily specified (sketched/drawn).

Respecting the mentioned demands, we utilize the feature signatures [18] where a video key-frame is represented by a set of color regions. Such simple representation enables users to specify these regions directly in simple sketches. Unlike the fixed grid or dominant color features, the feature signatures are able to capture even fewer significant color regions and adapt to the complexity of a key-frame. Moreover, the resulting feature space has only 5 dimensions and is suitable for usage of the Euclidean distance which makes the retrieval process efficient.

We already introduced a simple tool [9] based on the feature signatures at the Video Browser Showdown (VBS) 2014 workshop [1, 20] and, by winning 3 out of 4 categories, it was demonstrated that the feature signatures alone can form a model which is able to compete with and even outperform the current state-of-the-art methods. In this paper, we describe in detail an improved model (and tool) that is able to deal with large amount of data. In order to preserve properties of the tool such as instant responses and high effectiveness, a proper indexing technique has to be introduced. We evaluate the performance of a simple grid index as well as the state-of-the-art of indexing metric spaces - M-Index [15]. We employ a Bounding-sphere based pruning technique in addition to other techniques [11, 24].

First, we discuss other approaches in the field of the KIS in video (Section 2). Then, we describe in detail the feature signatures video retrieval model and propose several index variants suitable for our feature space (Section 3). In the experiments (Section 4) we evaluate the performance of the proposed index variants as well as optimize the parameters of the retrieval model. Finally, we conclude the paper and propose a possible future work (Section 5).

2   Related Work

Tools for the KIS in video are being evaluated at various multimedia retrieval events, like the VBS workshop at the Multimedia Modeling (MMM) conference series. In this particular case, the usage of a textual query is prohibited, participants are forced to introduce innovative and interactive interfaces to their tools which makes the event even more interesting. Visual as well as textual KIS tasks are evaluated in a single video document and in a large video archive. We shortly describe the tools of all the participants of VBS 2014. The results will be briefly reviewed later.

David Scott et al. [21] participated with a tool based on automated semantic annotation of both audio and video data. In particular, occurrences of 60 visual concepts were identified and indexed using the current state-of-the-art methods such as SIFT, SVM and BoVW. In addition, the tool supported "face browsing" where all the faces found in the video were presented and users could list the shots in which selected faces appeared.

Similar approach was followed by Anastasia Moumtzidou et al. [13]. More than 300 concepts were detected and visual similarity search among the results was supported by MPEG-7 and SURF features. Agglomerative hierarchical clustering of the detected shots were employed in order to provide a hierarchical view of the results.

A very innovative tool was introduced by Claudiu Cob Arza et al. [5] which exploited advantages of collaborative search. In contrast to the previous approaches, only simple descriptors such as MPEG-7 color layout and motion histogram [19] were extracted. Users could specify the desired scene (via dominant color, background and foreground movement, scene duration, etc.) simultaneously on several devices such as tablets or smart-phones. Promising results could be marked for further examination by any of the collaborators.

A tool benefiting from both concept detectors and simple color descriptors was presented by the team from NII and UIT [14]. Training data for concept classifiers were obtained from Google Images. In addition, a simple 4x3 grid of the dominant colors for each video segment was calculated. Users could specify a sequence of patterns comprising a concept occurrence and grid-like color sketch to filter out the irrelevant segments of a video.

Werner Bailer et al. [2] introduced a video browsing tool originally created for media production where a high redundancy is expected. Low-level features such as a global color distribution, camera motion and object trajectories are extracted and aggregated into MPEG-7 descriptors in addition to SURF descriptors. Both descriptor types are used to cluster the video segments and to enable a visual similarity search.

Finally, we participated with a tool Signature-Based Video Browser [9] which extended version is introduced in this paper.

3   Retrieval Model

In this section, we describe in detail the employed retrieval model for searching key-frames of user interest along with suitable indexing technique.

3.1   Video Representation

In our approach, we assume users can easily memorize simple color stimuli from the observed video clip and thus we focus on position-color feature signatures [18] that can flexibly aggregate and simply represent the color distribution of the contents of the key-frames. In order to extract a feature signature from a given key-frame, the extraction algorithm maps all pixels of the key-frame into 5-dimensional feature space[1] $\mathbb{F} \subset \mathbb{R}^5$ and then performs an adaptive variant of the k-means algorithm [8]. The k-means algorithm results in the set of *centroids* of the detected clusters (ideally centers of distinct color regions in the extracted key-frame) forming feature signature $\mathbb{FS} \subset \mathbb{F}$, where the initial set of centers for the k-means is distributed uniformly. Due to the adaptive nature of the utilized k-means algorithm, the feature signatures vary in the number of *centroids* respecting the complexity of the key-frames. Beside the color and the position of *centroids*, the weight (i.e., the number of pixels contributing to the cluster) could be extracted; however, in this work we do not utilize this information in the retrieval model. An example of a feature signature is shown in Fig. 1, where we may observe the utilized feature signatures can be simply and intuitively interpreted as a rough approximation of the original image. Such simple colored circles can be directly sketched by users trying to define their query intent which can substitute the uncomfortable need of an example query image.

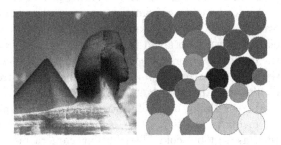

Fig. 1. A key-frame and the visualization of the feature signature. For each *centroid* a circle with the corresponding color and position is drawn. The weight of the centroid is depicted as the diameter of the circle.

The video retrieval techniques also try to reduce the number of key-frames by selecting only the representative ones. However, employing a scene detection and representing the detected scenes with only one key-frame would bring a possibility of not capturing some of the less significant color regions. For this reason, we decided to select simply every k-th frame for the feature signatures extraction. Although this method may introduce a noticeable overhead, it gives the desired robustness to the retrieval model. Furthermore, the extraction is still reasonably fast allowing to extract the feature signatures from an hour of video within a few minutes (on a low-end desktop PC using single core). One feature signature comprises tens of *centroids* and it is sufficient to reserve only 3 bytes for the color and 4 bytes for the position. As a result, the memory demands are reasonably low[2].

---

[1] The feature space is formed by coordinates (x, y, L, a, b), where x, y denotes the position of the pixel and L, a, b represent its color in the CIE LAB color space [23].

[2] In practice, it is sufficient to process 2 frames per a second of the video. Assuming the proposed 7-byte representation and 25 *centroids* per frame, an hour of video produces less than 2Mb of descriptor data.

3.2   Retrieval Algorithm

Let us assume that we have already extracted the feature signatures $FS_i = \{r_{ij}\}$ for the selected key-frames F_i where r_{ij} denotes the j-th *centroid* of the i-th feature signature. Users are enabled to specify a simple sketch of the viewed scene with colored circles (i.e., user-defined *centroids*) which allows to represent a query in the same way as the extracted feature signatures. Since users may memorize only the most distinct color regions from the desired scene, we expect only a few query *centroids* to be specified; hence, the model uses local instead of global matching. Let us explain the ranking scheme on a user defined query sketch FS_u comprising *centroids* r_{uv}. The *centroid* distance $dist_{uvi}$ between query *centroid* r_{uv} and a key-frame F_i is estimated as the distance to the nearest *centroid* from FS_i (1), L_2 denoting the Euclidean distance.

$$dist_{uvi} = \min_{\forall r_{ij} \in FS_i} L_2(r_{ij}, r_{uv}) \tag{1}$$

To obtain the sketch ranking $rank_{ui}$ of the key-frame F_i the *centroid* distances are scaled to the $[0;1]$ interval and averaged according to (2) where $D_{uv} = \{dist_{uvi} \mid \forall i\}$ (i.e., all the distances from *centroid* r_{uv} to the key-frames).

$$rank_{ui} = \underset{\forall r_{uv} \in FS_u}{\text{avg}} \frac{(dist_{uvi} - \min D_{uv})}{(\max D_{uv} - \min D_{uv})} \tag{2}$$

As the searched shot may consist of more than one visually discriminative scene, users are enabled to specify two (and possibly more) time-ordered query sketches. For a two-sketch query FS_u followed by FS_w the sketch rankings $rank_{ui}$ and $rank_{wi}$ are obtained and the overall ranking $rank_i$ of the key-frame F_i is calculated according to (3) where ε denotes a user-defined time range. Note the second query sketch is treated differently than the first – the most similar key-frame within a consecutive time neighborhood[3] is used.

$$rank_i = rank_{ui} + \min_{t=i+1}^{i+\varepsilon} rank_{wt} \tag{3}$$

Once we obtain the overall rankings, it is desirable to merge near-duplicate results generated by the dense key-frame representation. We accomplish that with the Alg. 1 where the function Neighbor returns all the key-frames from the predefined neighborhood, say 10 seconds around the popped key-frame. Note that the algorithm produces properly sorted results. The final results are presented as a simple list of the matched key-frames, each surrounded with adjacent key-frames in order to make the identification of the searched scene easier.

The model presented so far demands the database to be fully scanned. Although the computation of the distances is not expensive, the processing of a query is not feasible once the dataset grows significantly. It is thus desirable to omit as much database *centroids* from the ranking process as possible and consequently to avoid computing the distances. As long as we do not omit the

---

[3] Note t is logical time, i.e., the order of key-frames.

Algorithm 1. Results merging

QUEUE ← *overall rankings*   ▷ A priority queue with respect to the ranking
MERGED ← *empty*   ▷ A list of the results after merging
while QUEUE.NonEmpty() **do**
    RESULT ← QUEUE.Pop()
    MERGED.Add(RESULT)
    QUEUE.Remove(Neighbor(RESULT))
end while
return MERGED

---

centroids from the searched key-frames[4], their ranking will not be affected nega-
tively and the effectiveness will be preserved. We will derive later (Sec. 4.3) that
this paradigm leads to standard *range queries.*

Now, a key-frame F_i having all the *centroids* omitted has not properly defined
the *centroid* distance $dist_{uvi}$. In such case, we set it to max D_{uv}, i.e., the worst
match.

3.3 Indexing the Feature Space

In order to efficiently process *range queries* in the utilized 5-dimensional vector
space (x, y, L, a, b) using the Euclidean distance, we investigate both spatial and
metric indexing approaches, each represented by a suitable method. Since the
utilized feature extraction does not favor any key-frame region, the distribution
of the position coordinates shows high degree of uniformity and thus we have
selected a grid index as the representative of spatial indexing methods. As a
metric space method, we have selected the current state-of-the-art technique –
the M-Index [15]. As both techniques are well described in the literature, we just
simply remember their most important properties.

Grid Index. The grid index (Grid) can be used to divide the feature space into
uniform cuboid-like bins, where the number of bins grows exponentially with the
space dimension which limits this approach only for low-dimensional spaces. The
advantage of the uniform grid index is that the bin where an indexed *centroid*
belongs can be directly computed. When the index is queried with a *range query*
(q, r) every bin having non-empty intersection with the sphere defined by q and
r has to be examined.

M-Index. The M-Index is a member of permutation-based index family where
the pivots (selected objects from the metric space) help to dynamically cluster
the feature space with respect to the data distribution. More specifically, the
M-index uses the repetitive Voronoi-based partitioning to define the dynamic

---

[4] In fact, even some *centroids* from the searched key-frames can be omitted since (1)
implies that only the closest *centroid* from the key-frame contributes to the ranking.

cluster tree structure used for efficient *range query* processing where all possible metric filtering principles are combined (for more details, see [15]). Besides the original M-Index, we employ the *Cut-region extension* (M-Index CR)[11] enabling more efficient filtering already in the upper levels of the cluster tree. Since pivot selection technique is crucial for the M-Index performance [4] as well as the number of pivots and other parameters, we carefully determine the optimal setup for each M-Index variant separately.

Bounding Sphere Constraint. Since we utilize vector spaces, we can improve filtering power of the utilized indexes by tight bounding spheres[5] evaluated dynamically for each bin/cluster separately. The motivation is the additional bounding can describe the region more tightly than grid bin or voronoi-based cell cut-off by rings centered in global pivots. On the other side, the creation and maintenance of the bounding spheres (finding centers and radii) can be a costly indexing overhead and thus we employ efficient approximate algorithm [17] suitable also for dynamic indexing. We utilized the *Bounding Sphere Constraint* for both the Grid (Grid BS) and M-Index (M-Index BS and M-Index BS + CR).

4   Experiments

Firstly, the optimal parameters of the proposed retrieval model are established basing on the users behavior. In particular we determine the importance of the color and position and the minimal relevant neighborhood of a query *centroid* so that the model remains effective. Secondly, the performance of the proposed index variants is evaluated under user-defined queries and a real dataset. We also include a short overview of the VBS 2014 results.

4.1   Settings

The experiments were realized on an EBU MIM-SCAIE video dataset [12] from which we have selected 27 hours of diverse video content. The resulting database contained total of 4.8 millions *centroids*.

More than 40 users were told to find a randomly selected short clip within five minutes. Almost 100 successfully found clips along with user-defined sketches were gathered and were used to optimize the parameters.

In order to evaluate the performance of the indexing techniques, 300 user-defined query *centroids* were collected and used for querying the index. Since we have utilized just cheap Euclidean distance, we have focused mainly on the overall time needed to process a query. The measurements were performed on an Intel Xeon CPU @2.80 GHz in a single thread. The data structures occupied only a few Mb of memory (excluding key-frame images) and were kept in the RAM memory.

---

[5] Referred also as ball-regions or enclosing balls.

The best performing parameters for each index variant were explored and used for the comparison of the indexing techniques.

4.2   The Importance of the Position and Color

In this section, we focus on the relation between the color and position coordinates, i.e., what is more important for finding the relevant key-frames. This relation is typically modeled by additional weighs for the color and position coordinates, for example, using weighted Euclidean metric. However, for fixed weights, the same effect can be achieved by scaling the feature space prior to the indexing and searching. In the following text, we fix the color coordinates and scale only the position coordinates with a position-color ratio (PCR)[6].

To determine the optimal value of the PCR, we have performed following steps. For every searched clip c, we have queried the system with the user-defined query sketch under different values of the PCR and tracked the position of the clip c in the results. For PCR $= x$, the impairment ranking was obtained according to (4) where P_{cx} denotes the position of the searched clip in the results (less is better). In this way, the imp_{cx} is the relative impairment of the position with respect to the best and worst possibility within the evaluated interval of the PCR.

$$imp_{cx} = \frac{(P_{cx} - \min P_c)}{(\max P_c - \min P_c)} \quad \text{where} \quad P_c = \{P_{cx} \mid \forall x\} \tag{4}$$

The average position impairment from all the user searches is depicted in Fig. 2a. The optimal PCR found in this process depends of course on the initial setup. In our case, the color coordinates were scaled to the interval 0-255 and the position coordinates followed the proportions of the key-frames (350 x 200). Finding the optimum at 1.35, we scaled the position coordinates by this number.

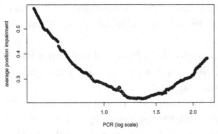 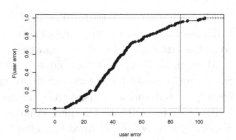

(a) The dependency of the position impairment on the PCR.

(b) The empirical distribution function of the *user errors*.

Fig. 2. The position-PCR dependency and the distribution of the *user errors*

---

[6] Note that the sketch rankings computed in (2) are scaled which allows us to optimize only the PCR.

4.3   Relevant Neighborhood of a Query *Centroid*

It is desirable to delimit the minimal neighborhood of a query *centroid* in the database that has to be examined in order to retrieve the wanted clip. Users are of course inaccurate in specifying the query *centroids* (e.g., the color of the intended region may be in fact darker) and the model has to tolerate these inaccuracies. We define the *user error* for a query *centroid* as the distance to the nearest database *centroid* from the searched clip. More formally, for a query *centroid* r specified by a user searching for a clip c the *user error* e_r is calculated according to (5) where FS_c stands for all the feature signatures extracted from the searched clip c.

$$e_r = \min_{\forall r_{ij} \in FS_c} L_2\left(r, r_{ij}\right) \tag{5}$$

The empirical distribution function of the *user errors* is depicted in Fig. 2b, where the vertical red line marks the 95% quantile which is ca 87 in our case. Please note that this number may vary with different datasets and users. We can see that if we omit the database *centroids* beyond the range of 87 (i.e., a *range query*), the ranking of the searched clip remains unaffected in most of the cases.

4.4   Index Performance

The performance of the index variants were evaluated under the growing database of up to 4.8 millions of *centroids*. The results are depicted in Fig. 3a, where we can observe that the *BS Constraint* brings decent performance improvement to the Grid as well as to the M-Index and M-Index CR.

Since the Grid uses fixed data structures, visiting bins demands a constant time, independently on the database size. This is not the case of the M-Index where the number of buckets grows with the database size causing the cost of their traversing to reflect this trend. As a result, for smaller databases with less than 500 000 *centroids* (i.e., almost 4 hours of video) the M-Index BS + CR is the best option, despite the need of more distances to be computed (Fig. 3b), while for larger ones the Grid BS takes the lead.

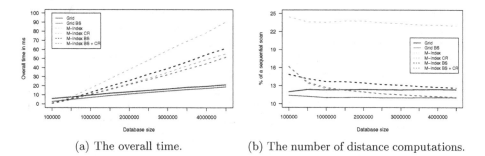

(a) The overall time.   (b) The number of distance computations.

Fig. 3. The performance of the index variants under growing database

4.5   VBS 2014 Results

The full analysis of the VBS 2014 results is out of the scope of this paper; however, we include a brief overview in order to support the statements about the effectiveness of our approach. As mentioned earlier, the tools competed in four categories: visual and textual KIS in single video and visual and textual KIS in video archive. The score

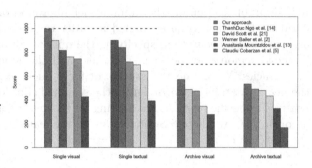

Fig. 4. The results of VBS 2014. The dashed lines mark the maximal possible score.

for each KIS task was calculated as $\frac{50+50\frac{limit-t}{limit}}{penalty}$ where t, $limit$ and $penalty$ denotes the time of the correct submission, the time limit for the KIS task and the number of incorrect submissions (or 1 if there is none). The results are depicted in Fig. 4.

5   Conclusion

In this paper we presented the feature signatures based video retrieval model along with suitable indexing techniques. It was shown at VBS 2014 that such a simple approach can compete with and even outperform current the state-of-the-art complex approaches in the known-item search in video. Although we optimized both the parameters of the model and the indexing technique, it is questionable whether the model is capable of preserving both efficiency and effectiveness in scope of hundreds or thousands of hours of video. We believe, however, that the model forms solid basis for content-based video retrieval systems and can be further improved.

It may seem that the M-Index is not suitable for our scenario since is outperformed by the Grid index; however, it should be stressed out that the M-Index is faster in very small databases. If we have many small video files indexed separately and users are enabled to filter the files (e.g., by the creation date or length) prior to the search itself, the M-Index would be better choice.

We also investigated the *BS Constraint* which can help to prune a vector feature space more efficiently when processing *range queries*. Approximate bounding spheres can be computed efficiently and dynamically and we demonstrated that even roughly computed bounding spheres can improve the filtering power.

5.1   Future Work

Video data carry way more information than the color distribution in static key-frames. For example, the motion of the background and foreground can be possibly detected and associated with particular *centroids*. It is also desirable to utilize the weights of *centroids*. In our opinion, this information may play an important role and we plan to incorporate it in the retrieval model.

Next, we believe that enhancing the results browsing can improve effectiveness greatly. The results can be organized in an advanced way [10] or the standard query-by-example paradigm might be employed once users retrieve some initial results.

Acknowdledgement. This research has been supported in part by Czech Science Foundation projects P202/11/0968 and P202/12/P297.

References

[1] Bailer, W., Schoeffmann, K.: Video search showcase (formely video browser showdown) (May 2014), http://www.videobrowsershowdown.org/

[2] Bailer, W., Weiss, W., Schober, C., Thallinger, G.: Browsing linked video collections for media production. In: Gurrin, C., Hopfgartner, F., Hurst, W., Johansen, H., Lee, H., O'Connor, N. (eds.) MMM 2014, Part II. LNCS, vol. 8326, pp. 407–410. Springer, Heidelberg (2014)

[3] Bartolini, I., Patella, M., Romani, C.: Shiatsu: tagging and retrieving videos without worries. Multimedia Tools and Applications 63(2), 357–385 (2013)

[4] Bustos, B., Navarro, G., Chavez, E.: Pivot selection techniques for proximity searching in metric spaces. In: Proceedings of the XXI Internatinal Conference of the Chilean Computer Science Society, SCCC 2001, pp. 33–40 (2001)

[5] Cobârzan, C., Hudelist, M.A., Del Fabro, M.: Content-based video browsing with collaborating mobile clients. In: Gurrin, C., Hopfgartner, F., Hurst, W., Johansen, H., Lee, H., O'Connor, N. (eds.) MMM 2014, Part II. LNCS, vol. 8326, pp. 402–406. Springer, Heidelberg (2014)

[6] Hu, R., Collomosse, J.: Motion-sketch based video retrieval using a trellis levenshtein distance. In: 2010 20th International Conference on Pattern Recognition (ICPR), pp. 121–124 (August 2010)

[7] Inoue, N., Kamishima, Y., Wada, T., Shinoda, K., Sato, S.: Tokyotech+ canon at trecvid 2011. In: Proceedings of NIST TRECVID Workshop (2011)

[8] Kruliš, M., Lokoč, J., Skopal, T.: Efficient extraction of feature signatures using multi-gpu architecture. In: Li, S., El Saddik, A., Wang, M., Mei, T., Sebe, N., Yan, S., Hong, R., Gurrin, C. (eds.) MMM 2013, Part II. LNCS, vol. 7733, pp. 446–456. Springer, Heidelberg (2013)

[9] Lokoč, J., Blažek, A., Skopal, T.: Signature-based video browser. In: Gurrin, C., Hopfgartner, F., Hurst, W., Johansen, H., Lee, H., O'Connor, N. (eds.) MMM 2014, Part II. LNCS, vol. 8326, pp. 415–418. Springer, Heidelberg (2014)

[10] Lokoč, J., Grošup, T., Skopal, T.: Sir: The smart image retrieval engine. In: Navarro, G., Pestov, V. (eds.) SISAP 2012. LNCS, vol. 7404, pp. 240–241. Springer, Heidelberg (2012)

[11] Lokoč, J., Moško, J., Čech, P., Skopal, T.: On indexing metric spaces using cutregions. Information Systems 43(0), 1–19 (2014)

[12] Matton, M., Messina, A., Bailer, W., Évain, J.-P.: The ebu mim-scaie content set for automatic information extraction on broadcast media. In: Proceedings of the 5th ACM Multimedia Systems Conference, MMSys 2014, pp. 13–17. ACM, New York (2014)

[13] Moumtzidou, A., et al.: Verge: An interactive search engine for browsing video collections. In: Gurrin, C., Hopfgartner, F., Hurst, W., Johansen, H., Lee, H., O'Connor, N. (eds.) MMM 2014, Part II. LNCS, vol. 8326, pp. 411–414. Springer, Heidelberg (2014)

[14] Ngo, T.D., Nguyen, V.H., Lam, V., Phan, S., Le, D.-D., Duong, D.A., Satoh, S.: Nii-uit: A tool for known item search by sequential pattern filtering. In: Gurrin, C., Hopfgartner, F., Hurst, W., Johansen, H., Lee, H., O'Connor, N. (eds.) MMM 2014, Part II. LNCS, vol. 8326, pp. 419–422. Springer, Heidelberg (2014)

[15] Novak, D., Batko, M.: Metric index: An efficient and scalable solution for similarity search. In: Second International Workshop on Similarity Search and Applications, SISAP 2009, pp. 65–73 (August 2009)

[16] Padmakala, S., AnandhaMala, G.S., Shalini, M.: An effective content based video retrieval utilizing texture, color and optimal key frame features. In: 2011 International Conference on Image Information Processing (ICIIP), pp. 1–6 (November 2011)

[17] Ritter, J.: An efficient bounding sphere. In: Graphics Gems, pp. 301–303. Morgan Kaufmann, San Diego (1990)

[18] Rubner, Y., Tomasi, C.: Perceptual metrics for image database navigation, vol. 1. Springer (2000)

[19] Schoeffmann, K., Lux, M., Taschwer, M., Boeszoermenyi, L.: Visualization of video motion in context of video browsing. In: IEEE International Conference on Multimedia and Expo, ICME 2009, pp. 658–661 (June 2009)

[20] Schoeffmann, K., Ahlström, D., Bailer, W., Cobârzan, C., Hopfgartner, F., McGuinness, K., et al.: The video browser showdown: a live evaluation of interactive video search tools. International Journal of Multimedia Information Retrieval 3(2), 113–127 (2014)

[21] Scott, D., et al.: Audio-visual classification video browser. In: Gurrin, C., Hopfgartner, F., Hurst, W., Johansen, H., Lee, H., O'Connor, N. (eds.) MMM 2014, Part II. LNCS, vol. 8326, pp. 398–401. Springer, Heidelberg (2014)

[22] Sikora, T.: The mpeg-7 visual standard for content description-an overview. IEEE Transactions on Circuits and Systems for Video Technology 11(6), 696–702 (2001)

[23] Tkalcic, M., Tasic, J.F.: Colour spaces: perceptual, historical and applicational background. In: Eurocon (2003)

[24] Zezula, P., Amato, G., Dohnal, V., Batko, M.: Similarity search: the metric space approach, vol. 32. Springer (2006)

Some Theoretical and Experimental Observations on Permutation Spaces and Similarity Search

Giuseppe Amato, Fabrizio Falchi, Fausto Rabitti, and Lucia Vadicamo

Istituto di Scienza e Tecnologie dell'Informazione "A. Faedo",
via G. Moruzzi 1, Pisa 56124, Italy
{firstname.lastname}@isti.cnr.it

Abstract. Permutation based approaches represent data objects as ordered lists of predefined reference objects. Similarity queries are executed by searching for data objects whose permutation representation is similar to the query one. Various permutation-based indexes have been recently proposed. They typically allow high efficiency with acceptable effectiveness. Moreover, various parameters can be set in order to find an optimal trade-off between quality of results and costs.

In this paper we studied the permutation space without referring to any particular index structure focusing on both theoretical and experimental aspects. We used both synthetic and real-word datasets for our experiments. The results of this work are relevant in both developing and setting parameters of permutation-based similarity searching approaches.

Keywords: permutation-based indexing, similarity search, content based image retrieval.

1 Introduction

Representing dataset objects as lists of preselected pivots ordered by their closeness to each object is a recent approach that have been proved to be very useful in many recent approximate similarity search techniques [3,8,14,20]. These approaches share the intuition that similarity between objects can be approximated by comparing their representation in terms of permutations. The quality of the obtained results have proved that whenever the permutations of two objects are similar then the two objects are likely to be similar also with respect to the original distance function.

In this paper, we studied the permutation space withouth relying on any specific indexing structure with the goal of making theoretical and experimental observations that can be of help in both setting parameters of existing permutation based approaches and developing new one.

2 Related Work

Predicting the closeness between objects on the basis of ranked lists of a set of pivots was originally and independently proposed in [8] and [4]. In [8] data

A.J. Machado Traina et al. (Eds.): SISAP 2014, LNCS 8821, pp. 37–49, 2014.
DOI: 10.1007/978-3-319-11988-5_4 © Springer International Publishing Switzerland 2014

objects and queries are represented as appropriate permutations of a set of reference objects, called *permutants*, and their similarity is approximated by comparing their representations in term of permutations. As distance between permutations, Spearman rho, Kendall Tau and Spearman Footrule were tested. Spearman rho revealed better performance.

The MI-File approach [4,3] uses an inverted file to store relationships between permutations. Spearman Footrule Distance is used to estimate the similarity between the query and the database objects. To reduce the storage, each object is encoded using the only nearest reference points and further approximations and optimizations are adopted to improve both efficiency and effectiveness.

The Permutation Prefix Index (PP-Index), was proposed in [13,14]. PP-Index associates each indexed object with a short prefix of predefined length of the full permutation. The prefixes are indexed by a *prefix tree* kept in main memory and all the relevant information relative to the indexed objects are serialized sequentially in a *data storage* kept on disk. PP-index uses the permutations prefixes in order to quickly retrieve a candidate set of objects that are likely to be at close distance to the query. The result set is then obtained using the original distance function by a sequential scan of the candidate set.

In [20], the concept of Locality-sensitive Hashing (LSH) was extend to a general metric space by using a permutation approach. In [19], a quantized representation of the permutation lists with its related data structure was proposed and a specific data structed, namely the Metric Permutation Table, was also defined. In [22] authors presented the *neighboord approximation* (NAPP) techinique whose main idea is to represent each object by the set of its nearest pivots and approximate the similarity between objects on the basis of the number of shared pivots. Three strategies for parallelization of permutation-based indexes using inverted files were presented in [18]. Posting lists decomposition, reference points decomposition, and multiple independent inverted files were studied and compared.

In [2], various pivot selection techniques were tested on three permutation-based indexing approaches (i.e., [8,3,14]). The results revealed that each indexing approach has its own best selection strategies but also that the random selection of pivots, even if never the best, results in good performance.

In [17,1] a Surrogate Text Representation (STR) derivated from the MI-File has been proposed. The conversion of the permutations in a textual form allows using off-the-shelf text search engines for similarity search.

3   Permutation-Based Representation

Given a a domain $\mathcal{D}$, a *distance function* $d : \mathcal{D} \times \mathcal{D} \to \mathbb{R}$ and a fixed set of objects $P = \{p_1 \ldots p_n\} \subset \mathcal{D}$ that we call pivots, we define a permutation-based representation Π_o (briefly permutation) of an object $o \in \mathcal{D}$ as the list of pivots identifiers ordered by their closeness to o, with the pivots being a fixed set of objects.

Formally, the permutation-based representation $\Pi_o = (\Pi_o(1), \Pi_o(2), ..., \Pi_o(n))$ lists the pivot identifiers in an order such that $\forall j \in \{1, 2, ..., n-1\}$, $d(o, p_{\Pi_o(j)}) \leq d(o, p_{\Pi_o(j+1)})$, where $p_{\Pi_o(j)}$ indicates the pivot at position j in the permutation associated with object o.

Denoting the position of a pivot p_i, in the permutation of an object $o \in \mathcal{D}$, as $\Pi_o^{-1}(i)$ so that $\Pi_o(\Pi_o^{-1}(i)) = i$, we obtain an equivalent representation Π_o^{-1}:

$$\Pi_o^{-1} = (\Pi_o^{-1}(1), \Pi_o^{-1}(2), ..., \Pi_o^{-1}(n))$$

This representation is very useful for essentially two reasons: first, $\Pi_o^{-1} \in \mathbb{R}^n$ allowing representing permutation in the Cartesian coordinate system; second, the Euclidean distance between two objects x, y represented as Π_x^{-1} and Π_y^{-1} is equivalent to the Spearman rho distance between Π_x and Π_y (see Section 3.1).

3.1   Comparing Permutations

The idea of approximating the distance $d(x, y)$ between any two objects $x, y \in \mathcal{D}$ by comparing their permutation-based representation $\Pi_x, \Pi y$ was originally proposed in [8]. As distance between permutations, Spearman rho, Kendall Tau and Spearman Footrule were tested. Spearman rho revealed better performance. Given two permutations Π_x and Π_y, Spearman rho is defined as:

$$S_\rho(\Pi_x, \Pi_y) = \sqrt{\sum_{1 \leq i \leq n} (\Pi_x^{-1}(i) - \Pi_y^{-1}(i))^2}$$

Following the intuition that the most relevant information of the permutation Π_o is in the very first, i.e. nearest, pivots, Spearman rho distance with location parameter $S_{\rho,l}$ defined in [15], intended for the comparison of top-l lists, has been also proposed.

$S_{\rho,l}$ differs from S_ρ for the use of an inverted truncated permutation $\tilde{\Pi}_o^{-1}$ that assumes that pivots further than $p_{\Pi_o(l)}$ from o being at position $l + 1$. Formally, $\tilde{\Pi}_o^{-1}(i) = \Pi_o^{-1}(i)$ if $\Pi_o^{-1}(i) \leq l$ and $\tilde{\Pi}_o^{-1}(i) = l + 1$ otherwise.

It is worth to note that only the first l elements of the permutation Π_o are needed, in order to compare any two objects with the $S_{\rho,l}$.

4   Theoretical Observations

As mentioned in Section 3, the permutation-space representation Π_o^{-1} belongs to $\mathbb{R}^n$. Moreover, the Spearman rho distance between two permutations Π_x and Π_y results in a Euclidean distance between Π_x^{-1} and Π_y^{-1}. In the following we consider the Π_o^{-1} representation in a Cartesian coordinate system.

If we consider the case $n = 3$, the set of all possible permutation-based representation (i.e., the set of all permutations on 3 elements) is formed by $\{(1, 2, 3), (1, 3, 2), (2, 1, 3), (2, 3, 1), (3, 1, 2), (3, 2, 1)\}$. It is easy to see that all this points lie on the plane $x + y + z = 6$ and represent the vertices of a regular hexagon as depicted in Figure 1.

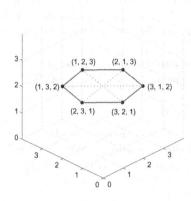

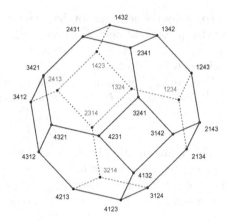

Fig. 1. The six points in $\mathbb{R}^3$ obtained by permuting the coordinate of the vector $(1, 2, 3)$

Fig. 2. Permutahedron with $4! = 24$ vertices

Consider now the $n = 4$ case: the vectors of all possible Π_o^{-1} lie in a three-dimensional subspace of $\mathbb{R}^4$ and are the vertices of a truncated octahedron (see Figure 2).

In general, the $n!$ points x obtained by permuting the coordinates of the vector $(1, 2, \ldots, n)$, form the vertices of a $(n-1)$-dimensional polytope embedded in a n-dimensional space, referred to as *permutahedron* (also spelled *permutohedron*) [23,16]. In fact, given that both the sum of vector values x_i (i.e., $\Pi_o^{-1}(i)$) and their squared values are fixed, all the vertices lie on both the hyperplane

$$x_1 + x_2 + \cdots + x_n = \frac{n(n+1)}{2}$$

the n−sphere

$$x_1^2 + x_2^2 + \cdots + x_n^2 = \frac{n(n+1)(2n+1)}{6}.$$

That is they lie on the intersection between an hyperplane and a sphere both in $\mathbb{R}^n$, i.e., on a $n-1$ sphere residing in n-dimensional space.

The permutahedron is a very interesting convex polytope. It is centrally symmetric and its vertices can be identified with the permutation of n objects in such a way that two vertices are connected by an edge if and only if the corresponding permutations differ by an adjacent transposition. It is rather easy to see that the squared Euclidean distance between any two vertices is an even integer, moreover, for $n > 4$, the squared distances constitute every even integer up through the maximum possible value, that is $\frac{1}{3}(n^3 - n)$ [21,23].

As observed in [21], standing on any vertex of a permutahedron and looking around at neighbouring vertices, the view of the surrounding space is the same: there would be $n-1$ adjacent vertices evenly distributed around the observation vertex, which Euclidean distance is $\sqrt{2}$. Furthermore, the number of vertices and their relative positions within a generic ϵ-ball neighbourhood is independent of the observation vertex.

The *permutahedron* precisely illustrate how the permutation-based representation are positioned in the space were the Euclidean distance is equivalent to the Spearman rho. It is worth to mention that the Spearman Footrule, sometimes used in permutation based-indexing, results in a L1 (also Manattan) distance in the same space. However, it does not help very much in understanding the distance distribution.

In order to understand the Spearman rho distance distribution it is useful to use its not-squared root variant (S_ρ^2) because of its interesting distribution properties. In [11] it was shown that S_ρ^2 distance has:

- mean: $\frac{1}{6}(n^3 - n)$
- variance: $\frac{1}{36}n^2(n-1)(n+1)^2$
- maximum value: $\frac{1}{3}(n^3 - n)$

Unfortunately, S_ρ^2 is not a metric. However, due to the monotony of the square root function, there are not changes in the order of the results of a k-NN search with respect to the ones that can be obtained with S_ρ. Moreover, normalized by its means and variance, S_ρ^2 has a limiting normal distribution [12]. Chávez's intrinsic dimensionality [10] of the permutation space with squared Spearman rho distance is $\frac{1}{2}(n-1)$.

5   Performance Evaluation of the Permutation Space

For our experiments we did not use any specific index approach. In fact, we performed sequential scan of permutation-based representation archives in order to retrieve most similar objects with respect to the query by using the Spearman rho distance function.

5.1   Datasets and Groundtruth

Random Float Vectors. As synthetic dataset we considered random generated vectors of floats of various dimensionalities d between 2 and 10. For each dimension we randomly generated float between 0 and 1. As distance measure for comparing any two vectors we used the Euclidean distance.

CoPhIR. As real-word dataset we used CoPhIR dataset [7], which is the largest multimedia metadata collection available for research purposes. It consists of 106 millions images crawled from Flickr. We run experiments by using as distance function d a linear combination of the five distance functions for the five MPEG-7 descriptors that have been extracted from each image. We adopted the weights proposed in [5]. As the ground truth, we have randomly selected 100 objects from the dataset as test queries and we have sequentially scanned the CoPhIR to compute the exact results. The queries were removed from the dataset itself.

5.2   Pivots Selection

For the CoPhIR dataset we randomly selected 10,000 pivots from the whole 106M objects collection. We then created subsets of this first selection. In the following we report experiments obtained on a subset of the entire CoPhIR collection. Thus it happens that some pivots are also in the dataset while some are not.

Pivots for the random float vectors were randomly generated without selecting between the objects in the dataset.

Variuos pivots selection strategies have been proposed for permutation-based indexing [2]. Experimental results have shown that while each specific index strategies have its own best selection approach, the random selection is always a good choice.

5.3   Parameters

In this section we summarize the parameters that have to be set for each specific experiment.

d - Float Vectors Dimensionality. This parameter is only necessary to indicate which random float vector dataset was used for the specific experiment. Experiments are reported for $d = 2, 4, 6, 8$.

m - Dataset Size. For both the synthetic and the CoPhIR dataset we recursively selected a subset of the collection. We performed experiments up to 1M and 10M objects for the random float vectors and CoPhIR datasets respectively.

n - Number of Pivots. The max number of pivots we used was 10,000. The smallest set of pivots have been obtained recursively selecting a subset of the larger collection.

l - Permutation Length. Various values of l for the Spearman rho with location parameter (see Section 3.1) where tested. Please note that $l = n$ results in the standard Spearman rho distance.

a - Amplification Factor. When a k-NN search is performed, a candidate set of results of size $k' = a * k$ is retrieved considering the similarity of the permutations based on S_ρ. This set is then reordered considering the original distance $d : \mathcal{D} \times \mathcal{D} \to \mathbb{R}$.

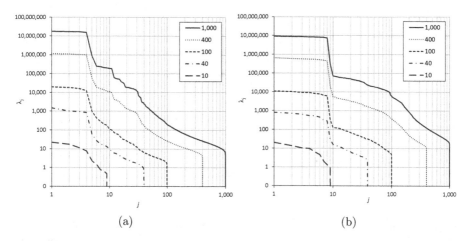

Fig. 3. Variances (eigenvalues) $\lambda_1 \geq \lambda_2 \geq \cdots \geq \lambda_n$, for various number of pivots n, corresponding to each principal component of the permutation obtained from the random float vectors of dimensionality 4 (a), 8 (b)

5.4   Evaluation Measure

Permutation-based indexing approaches, typically re-rank a set of approximate results using the original distance. In this work we did the same. Thus, if the k-NN results list $\tilde{\mathcal{R}}_k$ returned by a search technique has an intersection with the ground truth $\mathcal{R}_k$, the objects in the intersection are ranked consistently in both lists. The most appropriate measure to use is then the *recall*: $|\tilde{\mathcal{R}}_k \cap \mathcal{R}_k|/k$. In the experiments we fixed the number of results k to 10.

5.5   Principal Component Analysis

While PCA can not be performed on a generic domain $\mathcal{D}$ that can have a non metric distance and/or being a non vector space, once the permutation-based representation has been obtained it is always possible to run PCA on the Π_o^{-1}. We did this for both the random float vectors and CoPhIR dataset.

In Figure 3, we show the eigenvalues of each principal component of the permutations obtained for various number of pivots n. The dimensionality of the float vectors was 4 for (a) and 8 for (b). Please note, that both axes have log-arithmic scale. With 1,000 pivots it is clear in both cases what the original dimensionality of the vector space was. In fact, there is a large drop in the eigen-values passing from the 4th and 5th eigenvectors in (a), and from 8th and 9th in (b). The results also show that with more pivots we obtain a permutation-based representation that better fix the original data complexity.

We did the same for the CoPhIR dataset reporting the results in Figure 4. It is interesting to see that, in the logarithmic scale, the eigenvalues linearly decrease. However, CoPhIR did not reveal any specific dimensionality.

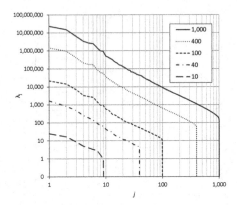

Fig. 4. Variances (eigenvalues) $\lambda_1 \geq \lambda_2 \geq \cdots \geq \lambda_n$, for various number of pivots n, corresponding to each principal component of the data obtained from the mapping in the permutation space of the CoPhIR 30,000 objects

In [6], it was shown that the combined distance function that we are also using in our experiments, results on the CoPhIR dataset in a near normal distribution with an intrinsic dimensionality, measured following the approach presented in [9], of about 13. Unfortunately, the same information can't be induced from Figure 4. Some non-linearity can be seen around 6 and 9, but performing PCA on the CoPhIR doesn't allow to understand the intrinsic dimensionality of the dataset as well as it allowed to understand the real dimensionality of the random generated float vectors.

5.6   Recall

In this section we relate the various parameters presented in 5.3 to the *recall* obtained on k-NN searching for $k = 10$. As mentioned before, results were obtained sequentially scanning archives of permutations by using the Spearman rho with and without location parameter l. Please note that $l = n$, i.e. for location parameter l equal to the number of pivots, the Spearman rho with an without location parameter are equivalent.

In Figure 5, we report the *recall* obtained on the random float vector datasets of 2 (a), 4 (b), 6 (c), 8 (d) dimensionalities, varying the location parameter l and for various number n of pivots. In these experiments we fixed the amplification factor $a = 1$. The most interesting result is that for small dimensionalities (2 and 4) there is a maximum *recall* that can be obtained varying l. In other words, $l = n$ it is not always the best solution, but there is an optimal l that appears not to vary for $n > l$. It also interesting to see that this optimal l varies significantly with the dimensionality of the original vector space. For 8 dimension vectors we are not even able to see this effect in the results. Probably, in this case the optimal l is well above 10,000 which is the max number of pivots we tried. Another important

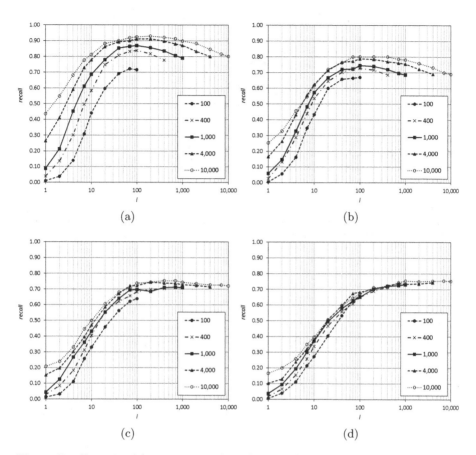

Fig. 5. *Recall* varying l for various number of pivots obtained on 100,000 random float vectors of dimensionality 2 (a), 4 (b), 6 (c), 8 (d)

observation is that the differences between the recall obtained by the various set of pivots tend to be smaller for higher d.

The same type of experiments were conducted on the CoPhIR dataset. In Figure 6, we report the recall obtained for $a = 1$ (a) and $a = 10$ (b). As for the random float vectors, it appears to be an optimal l that does not vary significantly with n. While the amplification factor does significantly impact the overall *recall*, the optimal l still remain almost the same. These results are consistent with the ones obtained on the random float vectors for dimensionality of about 4. In terms of indexability with respect to the permutation-based approach, CoPhIR appears to be as complex as random generated vectors of dimensionality between 4 and 6. In fact, we shown in Figure 5 that for random float vectors of 8 dimensions, the optimal l equals the number of pivots.

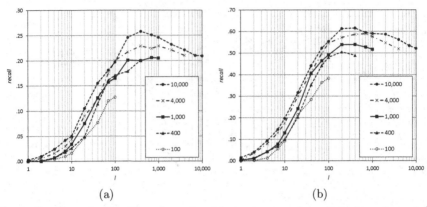

Fig. 6. *Recall* varying location parameter l for various number of pivots and $a = 1$ (a) and 10 (b) on CoPhIR 10M objects

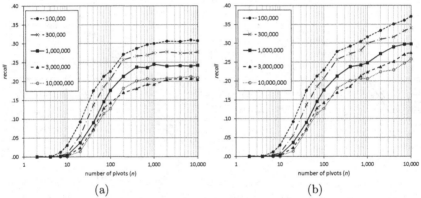

Fig. 7. *Recall* varying the number of pivots for various dataset sizes (all subsets of CoPhIR) obtained without location paramter l (a) and with the optimum l (b)

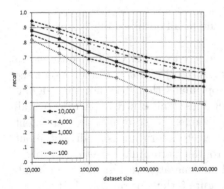
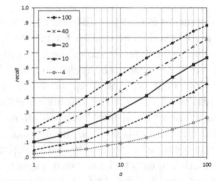

Fig. 8. *Recall* varying dataset size (all subsets of CoPhIR) for various l and $a = 10$

Fig. 9. *Recall* varying a for various l, number of pivots $n = 10,000$ and size of the CoPhIR subsets 10M

In Figure 7, the *recall* obtained varying the number of pivots for various size of CoPhIR subsets is reported. In this case we use $a = 1$. In Figure 7 (a), we show the results obtained for $l = n$ (i.e., the standard Spearman rho). In Figure 7 (b), we report the recall obtained for the optimal l which depends on both n and dataset size. Comparing these two figures it is evident that higher *recall* can be obtained increasing the number of pivots only if the optimal location parameter l is used. However, in our experiments, we had very near optimal results by using $l = 200$ (as can be seen for 10M objects in Figure 6). The intuition is that after a certain number of pivots, information regarding distant pivots is not only useless but distracting. Pleas note that the experiments performed on the random vectors indicate that the distant pivots are useful when the dimensionality of the dataset is above 8 (up to 10,000 pivots). Thus, while the observations made on the CoPhIR datasets are useful for understanding its characteristics and the fact that it exists an optimal l for a specific dataset, $l = 200$ is a near optimal solution only for the CoPhIR dataset and it probably reflects its complexity which appears to be lower than the intrinsic dimensionality evaluated in [6] following the [9] approach.

In Figure 8, we show the *recall* obtained varying the size of the CoPhIR subset for various number of pivots, optimal l (different for each combination of number of pivots n and dataset size) and $a - 10$. This graph is useful for understanding the loss in *recall* when the dataset increase. The results show that there is almost a linear dependency between the number of pivots needed to achieve a given quality of results and the dataset size.

In Figure 9, we fixed both the number of pivots (10, 000) and the dataset size (10M) reporting the *recall* varying a for various l. As obvious, the larger the amplifier factor a the better the quality of the results. Please note that l and a are the most relevant parameters in trading efficiency versus effectiveness in permutation based indexes. In fact, the shorter the permutation Π_o, the fewer the information to be stored for each object. Moreover, the less the amplification factor a, the smaller the number of objects to be retrieved for each search.

6   Conclusion

In this work we studied the permutation space focusing on both theoretical and experimental aspects not relying on any specific index structure. We used both synthetic and the CoPhIR dataset for the experiments varying various parameters that are typically used for trading-off between efficiency and effectiveness.

We first made some observations on the permutation space generating random permutations in order to understand its specific characteristic. We showed that the points are vertices of a permuthaedron, that using a squared Spearman rho results in Gaussian distance distribution.

The experiments conducted using random float vectors of various dimensionality shown that the complexity of the dataset affects the optimal value of l in terms of *recall* and that the dimensionality of the original vector space can be argued by performing PCA on the permutation space.

Also in the case of the CoPhIR dataset we found that it exists an optimal l for each specific number of pivots. Moreover, this optimal l is very stable and typically around 200. Thus, we believe that the optimal length of the permutations is in relation with the intrinsic complexity of the dataset even if this complexity can not be clearly seen performing PCA on the permutation space.

The experiments also revealed a linear dependency between the number of pivots and dataset size. Other results were shown considering l and amplifier factor a combination considering that they are the most useful parameters in trading-off efficiency and effectiveness in permutation indexes.

References

1. Amato, G., Bolettieri, P., Falchi, F., Gennaro, C., Rabitti, F.: Combining local and global visual feature similarity using a text search engine. In: 2011 9th International Workshop on Content-Based Multimedia Indexing (CBMI), pp. 49–54. IEEE Computer Society (2011)
2. Amato, G., Esuli, A., Falchi, F.: Pivot selection strategies for permutation-based similarity search. In: Brisaboa, N., Pedreira, O., Zezula, P. (eds.) SISAP 2013. LNCS, vol. 8199, pp. 91–102. Springer, Heidelberg (2013)
3. Amato, G., Gennaro, C., Savino, P.: Mi-file: using inverted files for scalable approximate similarity search. In: Multimedia Tools and Applications, pp. 1–30 (2012)
4. Amato, G., Savino, P.: Approximate similarity search in metric spaces using inverted files. In: Proceedings of the 3rd International Conference on Scalable Information Systems, InfoScale 2008, pp. 28:1–28:10. ICST (Institute for Computer Sciences, Social-Informatics and Telecommunications Engineering) (2008)
5. Batko, M., Falchi, F., Lucchese, C., Novak, D., Perego, R., Rabitti, F., Sedmidubsky, J., Zezula, P.: Building a web-scale image similarity search system. Multimedia Tools and Applications 47(3), 599–629 (2010)
6. Batko, M., Kohoutková, P., Novak, D.: CoPhIR image collection under the microscope. In: Skopal, T., Zezula, P. (eds.) Second International Workshop on Similarity Search and Applications, SISAP 2009, pp. 47–54. IEEE Computer Society (2009)
7. Bolettieri, P., Esuli, A., Falchi, F., Lucchese, C., Perego, R., Piccioli, T., Rabitti, F.: CoPhIR: a test collection for content-based image retrieval. CoRR abs/0905.4627 (2009)
8. Chávez, E., Figueroa, K., Navarro, G.: Effective proximity retrieval by ordering permutations. IEEE Transactions on Pattern Analysis and Machine Intelligence 30(9), 1647–1658 (2008)
9. Chávez, E., Navarro, G.: Measuring the dimensionality of general metric spaces. Department of Computer Science, University of Chile, Tech. Rep. TR/DCC-00-1 (2000)
10. Chávez, E., Navarro, G., Baeza-Yates, R., Marroquín, J.L.: Searching in metric spaces. ACM Computing Surveys 33(3), 273–321 (2001)
11. Diaconis, P.: Group representations in probability and statistics. Lecture Notes-Monograph Series, vol. 11. Institute of Mathematical Statistics (1988)
12. Diaconis, P., Graham, R.L.: Spearman's footrule as a measure of disarray. Journal of the Royal Statistical Society. Series B (Methodological) 39(2), 262–268 (1977)
13. Esuli, A.: MiPai: Using the PP-index to build an efficient and scalable similarity search system. In: Skopal, T., Zezula, P. (eds.) Second International Workshop on Similarity Search and Applications, SISAP 2009, pp. 146–148. IEEE Computer Society (2009)

14. Esuli, A.: Use of permutation prefixes for efficient and scalable approximate simi-larity search. Information Processing & Management 48(5), 889–902 (2012)
15. Fagin, R., Kumar, R., Sivakumar, D.: Comparing top k lists. In: Proceedings of the Fourteenth Annual ACM-SIAM Symposium on Discrete Algorithms, SODA 2003, pp. 28–36. Society for Industrial and Applied Mathematics (2003)
16. Gaiha, P., Gupta, S.K.: Adjacent vertices on a permutohedron. SIAM Journal on Applied Mathematics 32(2), 323–327 (1977)
17. Gennaro, C., Amato, G., Bolettieri, P., Savino, P.: An approach to content-based image retrieval based on the lucene search engine library. In: Lalmas, M., Jose, J., Rauber, A., Sebastiani, F., Frommholz, I. (eds.) ECDL 2010. LNCS, vol. 6273, pp. 55–66. Springer, Heidelberg (2010)
18. Mohamed, H., Marchand-Maillet, S.: Parallel approaches to permutation-based indexing using inverted files. In: Navarro, G., Pestov, V. (eds.) SISAP 2012. LNCS, vol. 7404, pp. 148–161. Springer, Heidelberg (2012)
19. Mohamed, H., Marchand-Maillet, S.: Quantized ranking for permutation-based indexing. In: Brisaboa, N., Pedreira, O., Zezula, P. (eds.) SISAP 2013. LNCS, vol. 8199, pp. 103–114. Springer, Heidelberg (2013)
20. Novak, D., Kyselak, M., Zezula, P.: On locality-sensitive indexing in generic metric spaces. In: Proceedings of the Third International Conference on Similarity Search and Applications, SISAP 2010, pp. 59–66. ACM (2010)
21. Santmyer, J.: For all possible distances look to the permutohedron. Mathematics Magazine 80(2), 120–125 (2007)
22. Tellez, E.S., Chavez, E., Navarro, G.: Succinct nearest neighbor search. Information Systems 38(7), 1019–1030 (2013)
23. Ziegler, G.M.: Lectures on Polytopes. Graduate Texts in Mathematics. Springer, New York (1995)

Metric Space Searching Based on Random Bisectors and Binary Fingerprints[*]

José María Andrade, César A. Astudillo, and Rodrigo Paredes

Departamento de Ciencias de la Computación, Universidad de Talca, Chile
jandrade@alumnos.utalca.cl, {castudillo,raparede}@utalca.cl

Abstract. We present a novel index for approximate searching in metric spaces based on random bisectors and binary fingerprints. The aim is to deal with scenarios where the main memory available is small. The method was tested on synthetic and real-world metric spaces. Our results show that our scheme outperforms the standard permutant-based index in scenarios where memory is scarce.

Keywords: Similarity search, Random bisectors, Binary signatures.

1 Introduction

Similarity search is an extension of exact searching, motivated by data types that cannot be queried by exact matching. This problem consists in finding elements within a given dataset that are similar to a given query according to a similarity criterion. There is a wide range of applications where the exact comparison is of little use. For instance, consider the case when a person is asked to scan its fingerprint so as to retrieve medical records. The system will obtain a different version of the fingerprint depending on the amount of pressure the person places on the sensor. In these situations, the only way of retrieving relevant objects — that is, objects that are similar to the query— is by tolerating small variations between objects. Other applications include multimedia databases containing images, audio, video, documents, and so on [4].

Proximity queries can be formalized using the metric space model [4,8,10,11]. Essentially, this model considers a pair $(\mathbb{X}, d)$, where $\mathbb{X}$ is a universe of objects and $d : \mathbb{X} \times \mathbb{X} \to R^+ \cup \{0\}$ is a nonnegative distance function defined among them. Objects in $\mathbb{X}$ do not necessarily have coordinates (think, for instance, in strings). On the other hand, the function d provides a dissimilarity criterion to compare objects from $\mathbb{X}$. In general, the smaller the distance between two objects, the more "similar" they are. The function d satisfies the metric properties, namely: *positiveness* $d(x, y) \geq 0$, *symmetry* $d(x, y) = d(y, x)$, *reflexivity* $d(x, x) = 0$, and *triangle inequality* $d(x, z) \leq d(x, y) + d(y, z)$, for every $x, y, z \in \mathbb{X}$.

The standard scenario of proximity searching considers a finite database of interest $\mathbb{U} \subset \mathbb{X}$, of size n. Later, when a new query object $q \in \mathbb{X} \setminus \mathbb{U}$ arrives, its proximity query consists in retrieving relevant objects from $\mathbb{U}$.

[*] This work is partially funded by Fondecyt grants 11121350 and 1131044, Chile.

A.J. Machado Traina et al. (Eds.): SISAP 2014, LNCS 8821, pp. 50–57, 2014.
DOI: 10.1007/978-3-319-11988-5_5 © Springer International Publishing Switzerland 2014

There are two basic queries, namely, *range* and *k-nearest neighbor* ones. The range query (q, r) retrieves all the elements in $\mathbb{U}$ within distance r to q. The *k*-nearest neighbor query $NN_k(q)$ retrieves the k elements in $\mathbb{U}$ that are closest to q. Both queries can be trivially answered by exhaustively scanning the database, requiring n distance evaluations. However, as the distance function is assumed to be expensive to compute (e.g., when comparing two fingerprints), frequently the complexity of the search is defined in terms of the total number of distance evaluations performed, instead of using other indicators such as CPU or I/O time. Thus, the ultimate goal is to build an offline index that, hopefully, will accelerate the process of solving online queries.

In this paper, we show a novel metric space index based on random bisectors and binary fingerprints to approximately solve the similarity search problem. An advantage of our index is that only requires a marginal amount of space. As we detail in the experimental section, when solving the $NN_1(q)$ in the hard metric space of uniformly distributed vectors in $\mathbb{R}^{128}$ under Euclidean distance, our method is able to retrieve 98% of the true answer by analyzing only 10% of the dataset, and this is achieved by using only 288 bits per element in the index. In the same experimental setup, our index overcomes the state-of-the-art Permutation Based Index (PBI) [3], as the later only retrieves 77% of the answer.

2   Related Work

In this section we briefly explain the *compact-partition based algorithms* and the PBI. Then, we describe two concepts that are central for the present study, namely, the Hamming distance and locality-sensitive hashing.

Compact-Partition Based Indices. These methods split the space into zones as compact as possible. For each partition, they store a representative object and extra information that permits the exclusion of that partition at query time.

This family of methods can be divided into the *Voronoi partition* and *covering radius* schemes. A *Voronoi Partition* method selects a subset of representative objects, called centers, denoted as $\{c_1, \ldots, c_m\}$, associating the remainder of the objects according to their proximity to its closest center. At query time, the distances $(d(q, c_1), \ldots, d(q, c_m))$ are evaluated and the closest center c is identified. Those regions satisfying the inequality $d(q, c_i) > d(q, c) + 2r$ can be safely discarded because they never intersects the query ball. On the other hand, the *covering radius*, $cr(c_i)$, corresponds to the distance between its center and the farthest element in its respective zone. So, at query time, when $d(q, c_i) - r > cr(c_i)$, then the zone i can be safely discarded.

The Permutation Based Index. Let $\mathbb{P} \subset \mathbb{U}$ be a subset of permutants. Each element $u \in \mathbb{U}$ computes the distance towards all the permutants $p_1, \ldots, p_{|\mathbb{P}|} \in \mathbb{P}$. The PBI *does not store distances*. Instead, for each $u \in \mathbb{U}$, it stores a sequence of permutant identifiers $\Pi_u = i_1, i_2, \ldots, i_{|\mathbb{P}|}$, called the permutation of u. Each permutation Π_u stores the identifiers in increasing order of distance, so

$d(u, \mathbb{P}_{i_j}) \leq d(u, \mathbb{P}_{i_{j+1}})$. Permutants at the same distance take an arbitrary but consistent order. Thus, a simple implementation needs $n|\mathbb{P}|$ space. Observe that it is possible to compact several permutant identifiers in a single machine word.

The crux of the PBI is that two equal objects are associated to the same permutation, while similar objects are, hopefully, related to similar permutations. In this sense, when Π_u is similar to Π_q one expects that u is close to q. The similarity between the permutations can be measured by Kendall Tau K_τ, Spearman Footrule S_F, or Spearman Rho S_ρ metric [5], among others. As these three distances have similar retrieval performance [3], for simplicity we use S_F, defined as $S_F(\Pi_u, \Pi_q) = \sum_{j=[1,|\mathbb{P}|]} |\Pi_u^{-1}(i_j) - \Pi_q^{-1}(i_j)|$, where $\Pi_u^{-1}(i_j)$ denotes the position of permutant p_{i_j} in the permutation Π_u. For example, if we have two permutations $\Pi_u = (42153)$ and $\Pi_v = (32154)$, then $S_F(\Pi_u, \Pi_v) = 8$.

Finally, at query time, we compute Π_q and compare it with all the permutations stored in the PBI. Next, $\mathbb{U}$ is traversed in increasing permutation dissimilarity. If we limit the number of distance computations, we obtain a probabilistic search algorithm that is able to find the right answer with some probability.

Hamming Distance. Given two binary sequences of equal length, the Hamming distance is the number of positions at which the corresponding symbols differ [7].

Locality-Sensitive Hashing. The Locality-Sensitive Hashing (LSH) is a family of techniques that map the input data into a set of buckets using several hash functions. The overall goal is that, with high probability, similar objects are mapped to the same bucket, and simultaneously, different objects are assigned to different buckets. This concept differs from the usual approach of hash functions, instead of avoiding collisions between similar objects, LSH encourages them.

The key point of LSH is to define the hash function family. The authors of [1] survey several alternatives for the vector space. One of those alternatives is related to our bisector approach. That idea is to pick random unit-length vectors $u \in \mathbb{R}^D$ and then define $h_u(v) = sign(u \cdot v)$. Using many random vectors, it is possible to build a binary sequence for each object. This hash function family was devised to approximate the cosine distance between two vectors in $\mathbb{R}^D$.

As far as we know, there is only one application of LSH to metric spaces. In [9], the authors use LSH to avoid the sequential scanning of the PBI.

3   Random Bisectors and Binary Fingerprints

In this section, we detail our proposed index for approximated similarity searching in metric spaces and the corresponding algorithms for solving similarity queries. The index uses the concepts of *virtual* random bisectors and binary fingerprints (RBBF) to build the data structure. We use *virtual* bisectors, since in general metric spaces, objects do not necessarily possess Cartesian coordinates.

Our index, called the Random Bisectors and Binary Fingerprints 1 (RBBF1), represents the objects using binary fingerprints. RBBF1 can be classified as a

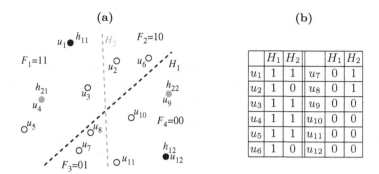

Fig. 1. Randomly generated data points used to show the behavior of our index. In (a), we see the objects partitioned using two virtual bisector hyperplanes H_1 and H_2. In (b), we see the fingerprint associated to each object.

new LSH approach for metric space searching, where each bisector hyperplane is a member in the hash function family.

For lack of space, in the following we only sketch the algorithms. A longer explanation, including pseudo-codes can be found in [2] (in Spanish).

Construction. RBBF1 only stores the binary fingerprint F_j for each object $u_j \in \mathbb{U}$. One of the main advantages of this philosophy is that it demands very little memory. To compute the fingerprint of an object, we need to simulate several bisectors. In this context, a bisector is understood as a *virtual* hyperplane which is orthogonal to the imaginary segment connecting the two endpoints and which intersects the midpoint of the segment.

If we were considering the vector space, each bisector hyperplane can be actually computed. However, in general metric spaces, objects do not necessarily have coordinates. Instead, what is available is the dissimilarity function d. Hence, when randomly picking two objects $h_{i1}, h_{i2} \in \mathbb{U}$, we can implicitly separate the space into two regions (by closeness to h_{i1} or h_{i2}) in an analogous manner as the above-mentioned bisector. Therefore, the method computes the distance between each object $u \in \mathbb{U}$ to the objects h_{i1} and h_{i2}. The index then determines which of two object is the closest, identifying the corresponding region with a bit.

Fig. 1a illustrates the RBBF1. It shows a scatter plot with 12 random objects and two random bisectors. The first bisector, H_1, is induced by the objects $h_{11} = u_1$ and $h_{12} = u_{12}$. The first component of the fingerprint refers to H_1, and is set to one for those elements closer to u_1 and set to zero when they are closer to u_{12}. Analogously, the second bisector, H_2, is generated from the objects $h_{21} = u_4$ and $h_{22} = u_9$. Fig. 1b shows resultant fingerprints for each object.

As we manage several hyperplanes, we store all the location information in a binary matrix F as follows. For all the objects $u_j \in \mathbb{U}$, and for all the λ hyperplanes, if $d(u_j, h_{i1}) \leq d(u_j, h_{i2})$ then $F_{ji} \leftarrow 1$, otherwise, $F_{ji} \leftarrow 0$.

The j-th row of the matrix F is called the *fingerprint* for instance j, and contains λ bits, one for each bisector, respectively. Naturally, the construction cost of RBBF1 is $O(n\lambda)$ both in evaluations of the distance function and CPU time.

Solving Similarity Queries with RBBF1. We use the RBBF1 index to speed up both k-nearest neighbor and range search queries, as explained below.

Our assumption is that two objects that are equivalent (i.e., with distance equal to zero) possess the same binary fingerprint, and that similar objects should be associated to fingerprints that differ in few bits. More in detail, if the fingerprint F_q is similar to the F_u, we expect that the object u is close to q. Note that two neighboring regions in the Voronoi diagram differ in just one bit in their respective fingerprints. We decided to follow the intuitive idea of traversing the dataset following the order induced by the increasing Hamming distance between the query fingerprint F_q and the fingerprint of every object in $\mathbb{U}$.

RBBF1 does not allow the exclusion of objects at query time, which in our opinion is not an inconvenient, because in high-dimensional metric spaces almost all the exact algorithms resort to sequential scanning. Fortunately, as the order induced by the Hamming distance is so promissory, we can stop the searching after reviewing a fraction αn of the objects in the dataset, as a workload, and obtain a really good answer. Naturally, the bigger the workload, the fewer relevant elements are lost by the technique. This desirable property is verified in Section 4 with strings and vectors.

The search mechanism starts by computing the fingerprint F_q of the query object q. Next, the method ranks all the elements within $\mathbb{U}$ by increasing Hamming distance with respect to F_q so as to compute the promissory review order. Subsequently, we use the workload to compare the best ranked objects with the query using the real distance of the metric space.

When solving range queries, the method reports any object in the workload within a distance r with respect to the query object q. On the other hand, in the case of k-nearest neighbors, the k closets objects in the workload are reported.

4   Experimental Evaluation

We tested our method on strings using the edit distance, and also uniformly distributed vectors in $\mathbb{R}^D$, for dimensions $D = 32$, 64, and 128 using Euclidean distance. The experiments were run on an Intel i5 of 2.6 GHz (two physical and four virtual cores), with 2GB of RAM, local drive and MS Windows 8.1 Professional of 64 bits, using JDK version 1.7.0_45. In the construction experiment, we only measure the CPU as RBBF1 needs $2n\lambda$ distance evaluations to build the index. The results shown correspond to averages after 10 constructions.

To test the search method, we measure the percentage of query retrieval varying λ and the workload, and also the percentage of retrieval for a fixed workload varying λ. We compare our approach with the standard PBI in the compact version, that is, we pack several permutant identifiers in the same machine word.

The permutations are compared using the Spearman Footrule (see Section 2). Additionally, we do not measure distance evaluations as they are limited by the workload. The results include the average of 100 NN_k queries. A longer experimental evaluation, including range query results is available in [2].

4.1 String under Edit Distance

We tested RBBF1 using a dictionary called `Dutch.dic`, obtained from the Metric Space Library [6], which contains an unsorted set of 229,328 words belonging to the Dutch language.

Construction. RBBF1 pre-calculates the distances between each object of the dataset to the set of λ pair of objects, thus demanding $O(n\lambda)$ time, with a correlation coefficient $R^2 \geq 0.966$.

Searching. Fig. 2a shows a summary of the best retrieval results for NN_1 and NN_{20} queries using the optimal value of λ for this dataset. RBBF1 presents a good retrieval performance. For instance, reviewing just a 5% of the dataset, RBBF1 retrieves 96% of the true answers in NN_1 queries, and 82% in NN_{20} queries. This is achieved by requiring *only* a single integer per object. For space constrains, we omit range query plots. However, we verify that the performance is similar. For instance, a query $(q, 1)$ retrieves 1.38 objects in average, and the optimal value of λ is also 24, retrieving a 91.3% of the true answers.

Because 8 identifiers can be packed using 32 bits, it is fair to compare RBBF1 with the standard PBI using 8 permutants. We observed that RBBF1 outperforms the PBI index with respect to retrieval ratio using the same space. We allowed more permutants in the PBI until its performance matches the one obtained by RBBF1 (these curves are omitted in the plot). This occurred when PBI employed 14 permutants, necessitating 63% more memory than RBBF1.

4.2 Uniformly Distributed Vectors under Euclidean Distance

We also performed tests by generating 30,000 random vectors from a uniform distribution in the range $[0, 1]^D$. We randomly selected 100 query items and computed the respective percentage of retrieval. Our aim is to observe the performance of RBBF1 in different dimensional spaces.

Searching. Figs. 2b, 2c and 2d summarize the best retrieval results for NN_1 and NN_{20} queries for D equals to 32, 64, and 128, respectively. RBBF1 and PBI are compared using the same memory requirements (i.e., 8 permutants). We also allow that the PBI uses more memory to match the performance of RBBF. This occurs when the PBI uses 63% to 100% more memory than the RBBF.

We observe that for the case of $\mathbb{R}^{32}$, the RBBF1 index with a workload of reviewing 10% of the dataset reaches to 82% and 74% of retrieval for NN_1 and NN_{20}, respectively. Since RBBF1 consistently improves with the available space, we run experiments using longer signatures.

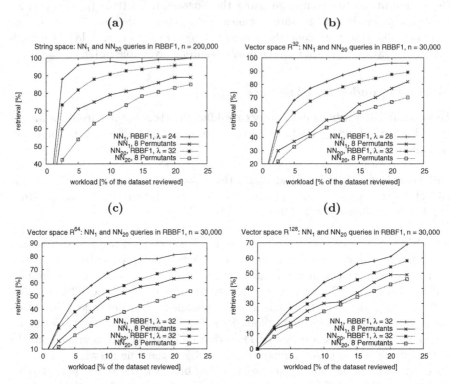

Fig. 2. Comparison of RBBF1 and PBI using NN_k queries on strings and vectors.

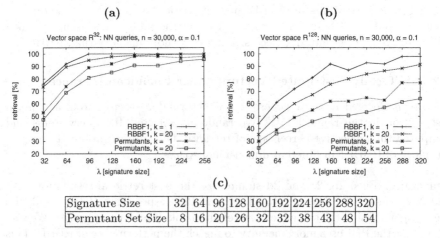

Fig. 3. NN_k queries of uniformly distributed vectors in $\mathbb{R}^D$, for $D = \{32, 128\}$, using an increasing number of fingerprints. In (c) the equivalence between the number of fingerprints and the cardinality of the permutant set.

Table 3c shows the size of a permutant set that uses the same memory than the allowed for the long signatures. Figure 3 shows NN_k retrieval for $k = 1$ and 20 reviewing 10% of the database (that is, with $\alpha = 0.1$). We note that RBBF1 effectively uses the extra space in order to improve the retrieval, and also use the space more efficiently than standard PBI.

5 Conclusions and Future Work

The paper presented a novel index for approximate searching that relies on random bisectors and binary fingerprints. The method was tested on strings and vectors, comparing the respective performances with the Permutant-Based Index (PBI). The experimental results show that our index outperformed the PBI scheme. We believe that this occurs because of the sorting mechanism, based on neighboring regions between fingerprints. Remarkably, a marginal amount of memory is required for storing the index. In fact, we use a single integer per object for the experiments with strings and nine for the case of vectors.

Avenues to be explored include the pattern recognition applications of the RBBF and the concept of maintaining a graph of neighborhood between signatures that, hopefully, would improve the results.

References

1. Andoni, A., Indyk, P.: Near-optimal hashing algorithms for approximate nearest neighbor in high dimensions. Commun. ACM 51(1), 117–122 (2008)
2. Andrade, J.M.: Diseño y desarrollo de un índice basado en hiperplanos para búsqueda en espacios métricos, memoria de Título del Departamento de Ciencias de la Computación, Universidad de Talca (Mayo 2014) (in Spanish)
3. Chávez, E., Figueroa, K., Navarro, G.: Effective proximity retrieval by ordering permutations. IEEE Trans. on Pattern Analysis and Machine Intelligence (TPAMI) 30(9), 1647–1658 (2009)
4. Chávez, E., Navarro, G., Baeza-Yates, R., Marroquín, J.: Searching in metric spaces. ACM Computing Surveys 33(3), 273–321 (2001)
5. Fagin, R., Kumar, R., Sivakumar, D.: Comparing top k lists. SIAM J. Discrete Math. 17(1), 134–160 (2003)
6. Figueroa, K., Navarro, G., Chávez, E.: Metric spaces library (2007), http://www.sisap.org/Metric_Space_Library.html
7. Hamming, R.W.: Error detecting and error correcting codes. The Bell System Technical Journal 29(2), 147–160 (1950)
8. Hjaltason, G., Samet, H.: Index-driven similarity search in metric spaces. ACM Transactions Database Systems 28(4), 517–580 (2003)
9. Sadit Tellez, E., Chavez, E.: On locality sensitive hashing in metric spaces. In: Proc. 3rd Intl. Conf. on Similarity Search and Applications (SISAP 2010), pp. 67–74. ACM Press (2010)
10. Samet, H.: Foundations of Multidimensional and Metric Data Structures. Morgan Kaufmann (2006)
11. Zezula, P., Amato, G., Dohnal, V., Batko, M.: Similarity Search – The Metric Space Approach. Advances in Database System, vol. 32. Springer (2006)

Faster Proximity Searching with the Distal SAT

Edgar Chávez[1], Verónica Ludueña[2],
Nora Reyes[2], and Patricia Roggero[2]

[1] CICESE, Ensenada, México
elchavez@cicese.mx
[2] Departamento de Informática Universidad Nacional de San Luis, Argentina
{vlud,nreyes,proggero}@unsl.edu.ar

Abstract. In this paper we present the *Distal Spatial Approximation Tree (DiSAT)*, an algorithmic improvement of *SAT*. Our improvement increases the discarding power of the *SAT* by selecting distal nodes instead of the proximal nodes proposed in the original paper. Our approach is parameter free and it was the most competitive in an extensive benchmarking, from two to forty times faster than the *SAT*, and faster than the List of Clusters (*LC*) which is considered the state of the art for main memory, linear sized indexes in the model of distance computations.

In summary, we obtained an index more resistant to the curse of dimensionality, establishing a new benchmark in performance, faster to build than the *LC* and with a small memory footprint. Our strategies can be used in any version of the *SAT*, either in main or secondary memory.

1 Introduction

Proximity searching consists in finding objects from a collection near a given query. The literature is vast and there are many specializations of the problem. We will fix our attention in *exact* queries under metric distances. A metric database is a finite subset $S \subseteq \mathbb{U}$. Distances are computed with a function $d : \mathbb{U} \times \mathbb{U} \to \mathbb{R}$, such that for any $x, y, z \in \mathbb{U}$, $d(x,y) > 0$, $d(x,y) = 0 \iff x = y$, $d(x,y) = d(y,x)$ (symmetry), and obeying the triangle inequality: $d(x,z) + d(z,y) \geq d(x,y)$. For a query $q \in \mathbb{U}$ and $r \in \mathbb{R}^+$, $(q,r)_d = \{x \in S \mid d(q,x) \leq r\}$ denote a *range query*. $kNN_d(q)$ denote the K-nearest neighbors of q, say $R \subseteq S$ such that $|R| = k$ and $\forall u \in R, v \in S - R$, $d(q,u) \leq d(q,v)$. If the database S is large and/or the distance function is expensive to compute, than a sequential scan to answer queries does not scale and an index should be used.

Complexity Model. The problem at hand has been elusive for the analysis. A cost model allowing worst case guarantees for known indexing techniques is still pending in the literature. The folklore among specialists sustains that metric axioms are too weak to produce even a usable notion of complexity for the problem. However, it is well documented the existence of instances of metric databases hard to index, all data algorithms will end up reviewing the entire database even for selective queries. This is known as the *curse of dimensionality* even if

A.J. Machado Traina et al. (Eds.): SISAP 2014, LNCS 8821, pp. 58–69, 2014.
DOI: 10.1007/978-3-319-11988-5_6 © Springer International Publishing Switzerland 2014

a proper notion of dimensionality is elusive [1]. Complementarily, more progress have been done in the approximate setup, where probabilistic guarantees have been provided for the accuracy, when the memory, the speed and a notion of dimensionality are bounded, as in [2] and references therein. In view of the above discouraging panorama, our algorithmic improvement proposal for indexing will be tested experimentally. In this regard, only a few tricks are known and used for indexing. In a way those tricks are derived from the triangle inequality. Surveying all of them is beyond the scope of this paper. Much more details are found in surveys and books on the topic, such as [3–5].

Pivot tables are well known, generic approaches to indexing. Another alternative is to partition the space in compact zones, usually in a recursive manner, storing a representative object (a "center") c_i for each zone plus a few extra data that permits quickly discarding the zone at query time. The general idea is to have coherent clusters of objects. During search, entire zones can be discarded depending on the distance from their cluster center c_i to the query q. Two criteria can be used to delimit a zone. Representative techniques are: *Geometric Near-neighbor Access Tree (GNAT)* [6], *List of Clusters (LC)* [7] , the *Spatial Approximation Tree (SAT and DSAT)* [8,9].

Some data structures combine both ideas by dividing the space into compact partitions, and at the same time storing distances to pivots. The *D–index* [10,11] divides the space into *separable* partitions of data blocks and combines this with pivot-based strategies to decrease I/O costs and distance evaluations performed during searches. It supports disk storage and it is dynamic. Adapting the *D–index* to particular applications requires a non-trivial parameterization process. Another example in this group is obtained by adding pivots to some clustering-based data structure, as the *PM–tree* [12] does on top of the *M–tree* [13].

2   The Spatial Approximation Tree

Since our approach is an improvement of all the versions of *SAT* we will include a detailed discussion of this data structure. The *Spatial Approximation Tree (SAT)* [8] is a data structure aiming at approaching the query spatially, that is, start at the root and get iteratively closer to the query navigating the tree. The *SAT* is build as follows. An element a is selected as the root, and it is connected to a set of *neighbors* $N(a)$, defined as a subset of elements $x \in \mathbb{U}$ such that x is closer to a than to any other element in $N(a)$. The other elements (not in $N(a) \cup \{a\}$) are assigned to their closest element in $N(a)$. Each element in $N(a)$ is recursively the root of a new subtree containing the elements assigned to it. For each node a the covering radius is stored, that is, the maximum distance $R(a)$ between a and any element in the subtree rooted at a. Fig. 1 shows an example *SAT* and the search path for a query.

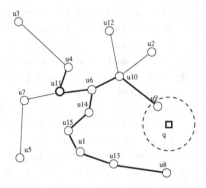

Fig. 1. Example of a SAT and the traversal towards a query q, starting at u_{11}. From [5].

BuildTree(Node a, Set of nodes S)
1. $N(a) \leftarrow \emptyset$ /* neighbors of a */
2. $R(a) \leftarrow 0$ /* covering radius */
3. **For** $v \in S$ in increasing distance to a **Do**
4. $R(a) \leftarrow \max(R(a), d(v, a))$
5.   **If** $\forall b \in N(a)$, $d(v, a) < d(v, b)$ **Then**
6. $N(a) \leftarrow N(a) \cup \{v\}$
7. **For** $b \in N(a)$ **Do** $S(b) \leftarrow \emptyset$
8. **For** $v \in S - N(a)$ **Do**
9. $c \leftarrow \text{argmin}_{b \in N(a)} d(v, b)$
10. $S(c) \leftarrow S(c) \cup \{v\}$
11. **For** $b \in N(a)$ **Do** **BuildTree**$(b, S(b))$

Algorithm 1. Algorithm to build a SAT for $S \cup \{a\}$ with root a

Algorithm 1 depicts the construction process.

It is first invoked as **BuildTree**$(a, S - \{a\})$ where a is a random element of S selected as its root.

Note the construction process do not enforce a balanced data structure. While it is a disadvantage in exact searching, it seems that unbalancing does speed up searching in metric data structures [7]. In fact, the most competitive indexing algorithm in high dimensions, is precisely the *List of Clusters* (LC) which can be seen as an extremely unbalanced tree. The LC is considered the state of the art for indexing. We will see in the experimental part that our data structure outperforms the LC both in construction and searching time in all but a few cases, establishing a new benchmark.

One key aspect of SAT is that a greedy search will find all the objects previously inserted. For a query $(q, r)_d$, in each node a it is determined the closest element c of q among $a \cup N(a)$, then we use the same greedy search

RangeSearch(Node a, Query q, Radius r,
                     Distance d_{min})
1. If $d(a,q) \le R(a) + r$ Then
2.   If $d(a,q) \le r$ Then Report a
3. $d_{min} \leftarrow$ min $\{d(c,q),\ c \in N(a)\} \cup \{d_{min}\}$
4.   For $b \in N(a)$ Do
5.   If $d(b,q) \le d_{min} + 2r$ Then
6.   **RangeSearch**(b,q,r,d_{min})

Algorithm 2. The algorithm to search for $(q,r)_d$ in a SAT with root a

entering all the nodes $b \in N(a)$ such that $d(q,b) \le d(q,c) + 2r$ because any element $x \in (q,r)_d$, can differ from q by at most r at any distance evaluation, so it could have been inserted inside any of those b nodes. In the process, we report all the nodes x founded close enough to q.

Algorithm 2 RangeSearch$(a,q,r,d(a,q))$ describes the process. Here a is the tree root, r the range of the search and q the query object.

2.1   Dynamic Spatial Approximation Trees

If the objects to be indexed are not known beforehand, the SAT cannot be built with Algorithm 1. Instead of examining all possible objects to decide which of them fulfill the near condition, the neighbors are selected in a first-come-first-serve basis. There are several strategies to maintain an arbitrary arity in the tree, and to support also deletions as described in [9]. The arity was thought to play the lead role in the efficiency of searching, in this paper we have found a different factor accounting for the efficiency.

It has been shown that $DSAT$ outperforms the static version for certain arity combinations. In [9] the authors proposed a couple of practical rules based on experiments: a) Low arities are good for low intrinsic dimensions or small search radii, and b) Large arities can be used for high intrinsic dimensions. From an algorithmic perspective this is an odd behavior, because a *static* data structure may have *all* the information of the data instance, while a dynamic data structure have *limited knowledge* about the data. In this paper we have found the underlying reason of this behavior. We describe our findings below.

3   The Distal Spatial Approximation Tree

From the definition of the SAT in algorithm 1, the starting set for neighbors of the root a, $N(a)$ is empty. This implies we can select *any* database element as the first neighbor. Once this element is fixed the database is split in two halves by the hyperplane defined by proximity to a and the recently selected neighbor. Any one of the elements in the a side can be selected as the second neighbor. While the zone of the root (those database elements closer to the root than the previous neighbors) is not empty, it is possible to continue with the subsequent neighbor selection.

BuildTree(Node a, Set of nodes S)
1. $N(a) \leftarrow \emptyset$ /* neighbors of a */
2. $R(a) \leftarrow 0$ /* covering radius */
3. Fix an order π in the set S
4. For $v \in S$ according to order π Do
5. $R(a) \leftarrow \max(R(a), d(v,a))$
6.   If $\forall b \in N(a),\ d(v,a) < d(v,b)$ Then
7. $N(a) \leftarrow N(a) \cup \{v\}$
8. For $b \in N(a)$ Do $S(b) \leftarrow \emptyset$
9. For $v \in S - N(a)$ Do
10. $c \leftarrow \text{argmin}_{b \in N(a)} d(v,b)$
11. $S(c) \leftarrow S(c) \cup \{v\}$
12. For $b \in N(a)$ Do **BuildTree**$(b, S(b))$

Algorithm 3. Algorithm to build a SAT^+ for $S \cup \{a\}$ with root a

Sorting the elements in increasing order of distance to the root is just one of the $n!$ possible permutations of the database elements. Each database permutation can be used as an order for the SAT construction. Each insertion order will produce a *correct* version of the SAT, and the same searching algorithm can be used. It is very likely that the performance at search time will be different for each permutation, one natural question is: *What is the best permutation for a given database?* Instead of blindly trying every permutation we try to optimize the discarding rules of the SAT. A subtree is avoided using two rules, hyperplanes and covering radius. The key aspect in the hyperplane discarding rule is the separation between the two defining points, because the query ball is more likely to fall completely in either side of the hyperplane and all the objects in the opposite side can be discarded. A good hyperplane separation in the upper levels of the tree also implies small covering radius in the lower levels of the tree. We exploit this two observations using several heuristics in our *DiSAT* data structure. Interestingly enough, the original policy for SAT works exactly in the opposite direction of this improvement strategy. Even a *random* selection of the insertion order outperform the original SAT; this explains the dynamic version being better than the static version.

3.1   The SAT$^+$ Strategy

Algorithm 3 gives a formal description of the construction of our data structure. The difference is in selecting the insertion order π in line 3. We tried farthest-to-nearest order from the root . Searching is done with the standard procedure described in Algorithm2.

A random permutation, or equivalently a random order, for the construction of the SAT is similar to inserting elements online in the $DSAT$. The difference will be to have a *natural* number of neighbors instead of an arbitrary arity to be tuned up. We call this the SAT^{Rand} in the experiments. We tested this construction mainly to explain the behavior of the $DSAT$.

When working with hyperplanes to perform data separation it is advisable to use object pairs far from each other as documented in [5] for the $GNAT$ and GHT data structures. Using the above observations, we can ensure a good separation of the implicit hyperplanes by selecting the first neighbor as the farthest element to the root. Clearly it is advisable to do this recursively, at every node of the tree. Please note that this heuristic is the exact opposite of the original ordering in the construction of the SAT.

3.2   The SAT^{Glob} Strategy

Sorting elements by distance at every level can be time consuming. We tried a fixed insertion order π by sorting elements for distance to the tree root, farthest first. This fixed order π is used in all the following levels. Therefore, SAT^{Glob} and SAT^+ are similar only at the first level of the tree, on the following levels the order π already determined is used without performing any new sort. This also serves to probe for the recursive need to select good hyperplanes at each tree level.

3.3   The SAT^{Out} Strategy

So far we have selected a random element as the tree root. Since we are aiming at maximizing the hyperplane separation, it makes a lot of sense to select the fathest pair as the root and the first neighbor respectively. This way there will be a lot of room for farthest pair selection in the lower levels of the tree.

The "farthest pair problem" is well known. We want objects $x, y \in S$, such that $d(x, y) \geq d(z, v), \forall z, v \in S$. This can be doing by comparing all against all the elements of the database, this is prohibitively expensive since it involves $O(n^2)$ operations. A randomized version is very effective and uses only $O(n)$ operations. The idea is to select a random starting point u_0, locate its farthest neighbor u_1 and repeat to find u_2, etc. A few iterations will get a good approximation of the farthest pair.

4   Experimental Results

For our first experiment we selected three widely used benchmark databases, all from the SISAP Metric Library www.sisap.org, NASA images, Strings and Color Histograms. We use euclidean distance for NASA images and Color Histograms, and edit distance for Strings. In all cases, we built the indexes with 90% of the points and used the other 10% (randomly chosen) as queries. All results are averaged over 10 index constructions using different permutations of the datasets. We have considered range queries retrieving on average 0.01%, 0.1% and 1% of the dataset. Given the existence of range-optimal algorithms for k-nearest neighbor searching it is enough to consider only range searches in the experimental part. The source code SAT and $DSAT$ is available in www.sisap.org, we submitted the code for *DiSAT*. The arity parameter of the *DSAT* was selected using the recommendation in [9][1].

Fig. 2 (Subfigs. 2(a)) contains the results of construction costs obtained in the experiments for the three metric spaces. We show the comparison of the construction costs for the original *SAT*, for the *DSAT*, and for the new SAT^{Rand}, SAT^+, SAT^{Glob}, and SAT^{Out} built using the new construction criterions. As it can be seen, the SAT^+ gets the worst construction costs. It can be explained because the arity in this case is the largest, and as it was shown in [9] construction cost grows with the tree arity. Moreover, despite of SAT^{Out} uses the same neighbor selection policy as SAT^+, SAT^{Out} achieves better construction costs

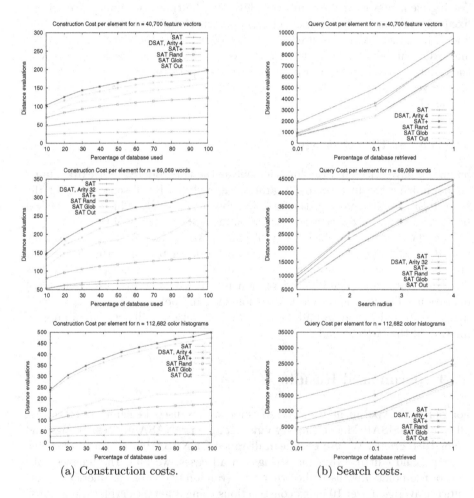

(a) Construction costs.   (b) Search costs.

Fig. 2. Comparison of construction and search costs

---

[1] The best arity for the NASA images and for Color histograms is of 4, and arity 32 for the Strings.

because the maximum arity tree is significantly lower than SAT^+ because we selected the root more properly. Fig. 2 (Subfigs. 2(b)) depicts that the new SAT^+, SAT^{Glob} and SAT^{Out} significantly improve searching costs with respect to other ones, and they are very similar between them. However, SAT^{Out} achieves the lowest search costs.

We postulate that in the new indexes the neighbors of the root represent a more accurate sample of the different zones in the metric space and produce better hyperplane separation in two senses, the inter-sibling separation and the root-node separation. These two conditions also imply small covering radii. This in turn produces a more compact partition of the space, improving the search cost.

5   Comparison with Other Indexes

Among all the exact indexes AESA [14] stands as the lower bound in distance computations; however, it uses a quadratic amount of space. In this version of the paper we will only compare with linear size indexes. We have performed an exhaustive comparison with other approaches and confirmed the *DiSAT* as a competitive, standing as a new efficiency benchmark. Due to space restrictions, in this version of the paper we only compare with the *List of Clusters* (em LC), as this data structure currently holds the benchmark for exact searching. In [8,9] *SAT* and *DSAT* were compared with several competitive indexes, so transitively we show that *DiSAT* is a very efficient index because is a better option than *SAT* and *DSAT*.

5.1   List of Clusters

The *List of Clusters* (*LC*) [7], with a proper parameter selection stands as the most competitive exact index when counting distance computations as the complexity measure. As we have improved the original *SAT*, with our SAT^+ and SAT^{Out}, we want to test how competitive is our approach against the state of the art. One drawback of the *LC* is the construction cost, another is the manually selected cluster size.

Fig. 3 compares construction and search costs, Subfigs. 3(a) and Subfigs. 3(b) respectively, of the SAT^+, SAT^{Out}, and *LC*. We test different values of m (*LC(m)*), some of them are presented in this comparison. We select values that allow us to show the behavior of *LC* at a similar construction cost and a similar or even better search cost with respect to our indexes.

For the NASA images database, SAT^+ and SAT^{Out} beat *LC* for all search radii, even with a value of $m = 25$ for *LC* that implies approximately 5 times our construction costs. Moreover, in this database we could not get any cluster size that would enable *LC* to be superior to our indexes, , even if we disregard construction costs. Nevertheless, *LC* outperforms our indexes with $m = 100$ in all radii considered for Strings database, but it needs 2.5 to 3 times our construction costs. Moreover, in this database *LC* with similar construction costs

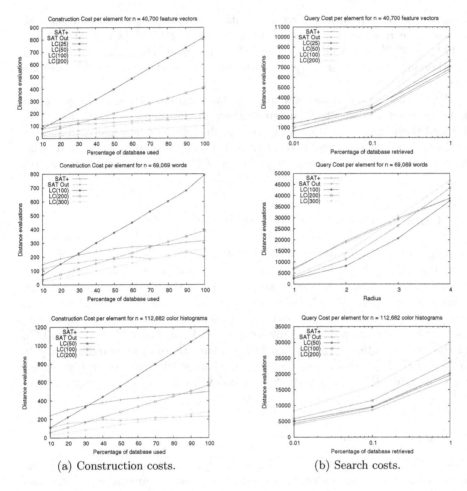

(a) Construction costs.   (b) Search costs.

Fig. 3. Comparison of construction and search costs against the LC, considering different cluster sizes

(with $m = 200, 300$) beats us, but for large radii. For Color histograms database, again it can be seen that our our SAT^+ and SAT^{Out} surpass LC for all radii considered. However, for $m = 50$ LC achieves slightly higher search costs than ours, but it needs to pay almost 6 times more than our cost of construction.

Please notice that as the size of the database grows, the increase in construction cost per element is not significant. It is also apparent that SAT^+ and SAT^{Out} have a good tolerance to large radii without needing parameter tuning. In the case of the LC a wrong parameter imply poor performance and/or large construction cost.

Scalibility. We also experimented with a larger database to test the scalability of our approach, and at the same time to compare with the LC. For this experiments, we use a 10 million images subset of the **COPHIR** database. For the *List of Clusters* we use a cluster size of 2048. We build the indexes on increasing sizes of the database in order to evaluate how is the behavior of all indexes as the database size grows. We started in 100,000 objects, doubling the size of the database up to 10 million objects. We reserved 200 objects, which would not be indexed, to be used as queries. In all sizes we use the same threshold r of 200 for the range queries, with $r = 200$ we retrieve in average more than 100 objects. Please notice that the items retrieved decrease with the database size, not retrieving any object in the sizes range of 100,000 to 400,000. Fig. 4 shows the construction costs obtained, and Fig. 5 depicts the search costs for the three indexes compared.

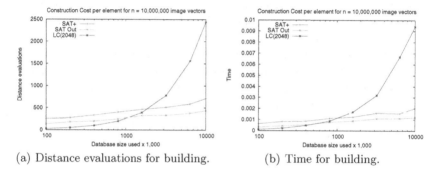

(a) Distance evaluations for building.   (b) Time for building.

Fig. 4. Comparison of construction costs for increasing subsets of COPHIR database

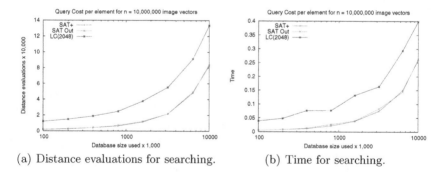

(a) Distance evaluations for searching.   (b) Time for searching.

Fig. 5. Comparison of search costs for increasing subsets of COPHIR database

As it can be noticed, our indexes outperform significantly the *List of Clusters*, in both construction and search time. Although construction costs are higher in lower sizes of the database on our indexes, our costs do not change too much as the database size grows, while with the *List of Clusters* it grows very quickly. During searches is even more remarkable that our indexes are better than the

List of Clusters obtaining better search costs for all sizes considered. Therefore, these experiments allow empirically to demonstrate that SAT^+ and SAT^{Out} are very scalable indexes.

6   Conclusions and Future Work

We have presented a new heuristic for constructing the SAT. The rule is counterintuitive and consist in selecting distal instead of proximal nodes. With this approach our proposed index $DiSAT$ stands as the new efficiency benchmark, supported by exhaustive experimentation. It improves the construction and searching times w.r.t. LC and other data structures.

Distal node selection can be used in static, dynamic and secondary memory versions of the SAT and produce more compact subtrees, inducing more locality to the implicit partitions of the subtrees. This factor will impact IO operations in secondary memory versions of $DiSAT$.

One possible consequence of a compact underlying partition, induced by a small covering radius is the possibility of producing coherent clusters suitable for statistics, mining, pattern recognition and machine learning purposes. One aspect of the putative clustering procedure is to produce a stable clustering (independent of the choice of the root, for example), or alternatively detecting natural, parameter free clusters.

References

1. Navarro, G.: Analyzing metric space indexes: What for? In: Second International Workshop on Similarity Search and Applications, SISAP 2009, pp. 3–10. IEEE (2009)
2. Houle, M.E., Nett, M.: Rank cover trees for nearest neighbor search. In: Brisaboa, N., Pedreira, O., Zezula, P. (eds.) SISAP 2013. LNCS, vol. 8199, pp. 16–29. Springer, Heidelberg (2013)
3. Zezula, P., Amato, G., Dohnal, V., Batko, M.: Similarity Search: The Metric Space Approach. Advances in Database Systems, vol. 32. Springer (2006)
4. Samet, H.: Foundations of Multidimensional and Metric Data Structures (The Morgan Kaufmann Series in Computer Graphics and Geometric Modeling). Morgan Kaufmann Publishers Inc., San Francisco (2005)
5. Chávez, E., Navarro, G., Baeza-Yates, R., Marroquín, J.: Searching in metric spaces. ACM Computing Surveys 33(3), 273–321 (2001)
6. Brin, S.: Near neighbor search in large metric spaces. In: Proc. 21st Conference on Very Large Databases (VLDB 1995), pp. 574–584 (1995)
7. Chávez, E., Navarro, G.: A compact space decomposition for effective metric indexing. Pattern Recognition Letters 26(9), 1363–1376 (2005)
8. Navarro, G.: Searching in metric spaces by spatial approximation. The Very Large Databases Journal (VLDBJ) 11(1), 28–46 (2002)
9. Navarro, G., Reyes, N.: Dynamic spatial approximation trees. Journal of Experimental Algorithmics 12, 1–68 (2008)
10. Dohnal, V., Gennaro, C., Savino, P., Zezula, P.: D-index: Distance searching index for metric data sets. Multimedia Tools and Applications 21(1), 9–33 (2003)

11. Dohnal, V.: An access structure for similarity search in metric spaces. In: Lindner, W., Fischer, F., Türker, C., Tzitzikas, Y., Vakali, A.I. (eds.) EDBT 2004. LNCS, vol. 3268, pp. 133–143. Springer, Heidelberg (2004)
12. Skopal, T., Pokorný, J., Snásel, V.: PM-tree: Pivoting metric tree for similarity search in multimedia databases. In: ADBIS (Local Proceedings) (2004)
13. Ciaccia, P., Patella, M., Zezula, P.: M-tree: an efficient access method for similarity search in metric spaces. In: Proc. of the 23rd Conference on Very Large Databases (VLDB 1997), pp. 426–435 (1997)
14. Vidal Ruiz, E.: An algorithm for finding nearest neighbours in (approximately) constant average time. Pattern Recognition Letters 4, 145–157 (1986)

A Dynamic Pivoting Algorithm Based on Spatial Approximation Indexes

Diego Arroyuelo

Department of Informatics, Universidad Técnica Federico Santa María
Yahoo! Labs Santiago, Chile
darroyue@inf.utfsm.cl

Abstract. Metric indexes aim at reducing the amount of distance evaluations carried out when searching a metric space. Spatial approximation trees (*sa-trees* for short), in particular, are efficient data structures, which have shown to be competitive in metric spaces of medium to high difficulty, or queries with low selectivity. *Sa-trees* can be also made dynamic, and can use the available space to improve the query performance adding pivot information. In this paper we extend previous work on dynamic *sa-trees* with pivots, and show how the pivot information can be used to a full extent to improve the search performance. The result is a technique that allows one to traverse a dynamic *sa-tree* without necessarily comparing all traversed nodes against the query object. As a result, the novel algorithm makes a much better use of the available space, yielding a saving of distance computations of about 10% to 70%, compared with previous *sa-tree* schemes that use pivot information.

1 Introduction

The classical way of searching a database has been that of finding those database records whose search attribute (or *key*) has a given value. However, this is not suitable when searching non-traditional databases, such as multimedia databases (e.g., image, video, or audio), multidimensional vector spaces (which has applications in GIS), and digital libraries, among others. In such cases, one would want to find the database objects that are "similar" to a given query object.

The similarity search problem is usually modeled as *proximity searches in metric spaces*. In the *metric space model* [2], there is a universe $\mathcal{U}$ of objects, and a positive real-valued distance function $d : \mathcal{U} \times \mathcal{U} \rightarrow \mathbb{R}^+$ defined among them. The distance between two objects models their similarity: the smaller the distance is, the more similar the objects are. We assume that the distance satisfy the three axioms that make the set a metric space:

Strict positiveness: $\forall x, y \in \mathcal{U}, \ d(x,y) = 0 \Leftrightarrow x = y$;
Symmetry: $\forall x, y \in \mathcal{U}, \ d(x,y) = d(y,x)$; and
Triangle inequality: $\forall x, y, z \in \mathcal{U}, \ d(x,z) \leqslant d(x,y) + d(y,z)$.

The triangle inequality property is used to save comparisons in a proximity query. The distance function is usually expensive to compute, hence we define the search complexity as the number of distance evaluations performed.

A.J. Machado Traina et al. (Eds.): SISAP 2014, LNCS 8821, pp. 70–81, 2014.
DOI: 10.1007/978-3-319-11988-5_7 © Springer International Publishing Switzerland 2014

We are given (in advance to queries) a database $S \subseteq \mathcal{U}$ of size $|S| = n$. Proximity-search algorithms are allowed to build an *index* of the database, avoiding exhaustive searches at query time [2]. Building an index is usually an expensive process. However, this cost is amortized after enough queries have been issued. At query time, given a query object $q \in \mathcal{U}$, we must retrieve all similar elements found in S. There are two typical queries of this kind:

Range Queries: retrieve all elements in S within distance r to q. That is, the set $\{x \in S, \ d(x, q) \leqslant r\}$.

Nearest-Neighbor Queries: retrieve the k closest elements to q in S. That is, a set $A \subseteq S$ such that $|A| = k$ and $\forall x \in A, y \in (S - A), \ d(x, q) \leqslant d(y, q)$.

In this paper we focus on range queries only.

Algorithms to search in general metric spaces can be divided into two large areas [2]: *pivoting* algorithms, and *compact partitions* algorithms. Pivoting algorithms are better suited for low dimensional (or easy) metric spaces, whereas compact partitions algorithms deal better with high dimensional (or hard) metric spaces. Although pivoting algorithms can use the available memory to improve the query performance, they need to use more memory to beat the latter as dimension grows. On the other hand, indexes based on compact partitions use a fixed amount of memory and cannot be improved by giving them more space.

Since a time ago, there are also data structures that combine both approaches, as for instance the memory-adaptive dynamic spatial approximation trees from [1]. These are basically dynamic spatial approximation trees (*dsa-trees*) [3], on which pivot information is added. Hence, they are able to trade memory space for a better query performance. However, pivots on *dsa-trees* are not used to prune the search nor to discard traversed elements. They are used just to save (in some cases) distance evaluations when the stopping criterion of *dsa-trees* determines that a given branch of the tree must be pruned. Every traversed node is inevitably compared against the query, even though it is not contained within the query radius. This obviously increases the query cost. Our research question is: What are the consequences for the spatial approximation approach, if we use pivot information to avoid comparing traversed elements?

In this paper we extend previous work [1] on *dsa-trees* with pivots, and show how the pivot information can be used to a full extent to improve the search performance. Basically, we adapt the search approach [1] such that pivots are used to avoid distance evaluation on traversed nodes. The resulting algorithms allows one to traverse a dynamic *sa-tree* without necessarily comparing all traversed nodes against the query. Avoiding such distance evaluations does not necessarily represents an improvement of query performance: we have less information for the spatial approximation, hence probably more tree branches would be visited. Ours is a compromise which probably traverses more tree branches, yet using pivots to avoid distance evaluations. We will show experimental results indicating that our approach uses the available memory more efficiently than previous work [1]: our search algorithm makes a better use of the available space, yielding a saving of distance computations of about 10% to 70%.

2 Preliminary Concepts on Metric Space Indexing

Indexing metric spaces is key for achieving efficient search performance in similarity search applications. We review in this section the most important indexing approaches needed to understand our work.

2.1 Pivoting Algorithms

Pivoting algorithms choose a set $\mathcal{P} = \{p_1, \ldots, p_k\}$ of *pivots* from the database S. They precompute and store all distances $d(a, p_1), \ldots, d(a, p_k)$ for all $a \in S$. Given a query $q \in \mathcal{U}$, pivoting algorithms compute the distances $d(q, p_1), \ldots, d(q, p_k)$ against the pivots. By using the information stored for every database object and the distances between the pivots and the query, we define:

Definition 1. *Given a query element $q \in \mathcal{U}$, the* pivot distance *between $a \in S$ and q gets defined as:*

$$\mathcal{D}(a, q) = \max_{p_i \in \mathcal{P}} |d(a, p_i) - d(q, p_i)|.$$

It can be proven that $\mathcal{D}(a, q) \leqslant d(a, q)$ for any $a \in S$, $q \in \mathcal{U}$. The pivot distance $\mathcal{D}$ is an estimation of the actual distance d, which is used to save distance evaluations: each a such that $\mathcal{D}(a, q) > r$ can be discarded because we deduce $d(a, q) > r$. All the elements that cannot be discarded in this way are directly compared against q.

Pivoting schemes perform better as more pivots are used, this way beating any other index. They are, however, better suited to "easy" metric spaces [2]. In hard spaces they need too many pivots to beat other algorithms.

2.2 Dynamic Spatial Approximation Trees

We briefly outline in this subsection how dynamic spatial approximation trees (*dsa-trees*) work, as we build on this data structure. See [3] for further details and proofs of correctness of the algorithms.

Insertion Algorithm. The *dsa-tree* is built incrementally, via insertions. The tree has a maximum arity A. Each tree node a stores a timestamp of its insertion time, $time(a)$, its covering radius, $R(a)$, and its set of children $N(a)$ (the so-called *neighbors* of a). To insert a new element x, its point of insertion is sought starting at the tree root and moving to the neighbor closest to x, updating $R(a)$ in the way. We finally insert x as a new (leaf) child of a if *(1)* x is closer to a than to any $b \in N(a)$, and *(2)* the arity of a, $|N(a)|$, is not already maximal. In other case, we insert x in the subtree of the closest element $b \in N(a)$. Neighbors are stored left to right in increasing timestamp order. Note that the parent is always older than its children.

Range Search Algorithm. The idea is to replicate the insertion process of the elements to be retrieved. Given a query q and a radius r, we act as if we wanted to insert q but keep in mind that relevant elements may be at distance up to r from q, so in each decision for simulating the insertion of q we permit a tolerance of $\pm r$. So it may be that relevant elements were inserted in different children of the current node, and backtracking is necessary.

Note that, at the time an element x was inserted, a node a may not have been chosen as its parent because its arity was already maximal. So, at query time, we must choose the minimum distance to x only among $N(a)$. Note also that, when x was inserted, elements with higher timestamp were not yet present in the tree, so x could choose its closest neighbor only among older elements. Hence, we consider the neighbors $\{b_1, \ldots, b_k\}$ of a from oldest to newest, disregarding a, and perform the minimization as we traverse the list. That is, we enter into subtree b_i if $d(q, b_i) \leqslant \min \{d(q, b_1), \ldots, d(q, b_{i-1})\} + 2r$.

We use timestamps to reduce the work inside older neighbors. Say that $d(q, b_i) > d(q, b_{i+j}) + 2r$. We have to enter subtree b_i anyway because b_i is older. However, only the elements with timestamp smaller than $time(b_{i+j})$ should be considered when searching inside b_i; younger elements have seen b_{i+j} and they cannot be interesting for the search if they are inside b_i. As parent nodes are older than their descendants, as soon as we find a node inside subtree b_i with timestamp larger than $time(b_{i+j})$ we can stop the search in that branch.

Algorithm 1 performs range searching on a *dsa-tree*. Note that, except in the first invocation, $d(a, q)$ (lines 1 and 2) is already known from the invoking process, so it must no be recomputed in a real implementation.

Algorithm 1. *dsat* Search(Node a, Query q, Radius r, Timestamp t).

1: **if** $time(a) < t \wedge d(a, q) \leqslant R(a) + r$ **then**
2:     **if** $d(a, q) \leqslant r$ **then**
3:         report a
4:     **end if**
5:     $d_{min} \leftarrow +\infty$
6:     **for** $b_i \in N(a)$ in increasing timestamp order **do**
7:         **if** $d(b_i, q) \leqslant d_{min} + 2r$ **then**
8:             $k \leftarrow \min \{j > i, \ d(b_i, q) > d(b_j, q) + 2r\}$
9:             *dsat* Search($b_i, q, r, time(b_k)$)
10:        **end if**
11:        $d_{min} \leftarrow \min \{d_{min}, \ d(b_i, q)\}$
12:    **end for**
13: **end if**

2.3   DSA-Trees with Pivots

Previous work [1] showed how to use the available memory to improve the search performance of *dsa-trees*. We associate a set of pivots to every tree node. At insertion time, in order to decide that a new element x must be added as a children (or neighbor) of an already existing node a, note that x has been already compared against the set $A(x)$ of ancestors of x, and also against the siblings of the ancestors. Some of these distances are used as pivot information, without introducing extra distance computations. See the original work [1], which shows how these pivots are computed at insertion time. From now on, we assume that each node x of a *dsa-tree* has a set $P(x)$ of pivots. The resulting data structure is called *hybrid dsa-tree* (H-DSAT for short).

Range Search Algorithm. *dsa-tree* Algorithm 1 is modified to use the set $P(x)$ stored at each tree node x. Recall that, given a set of pivots, $D(a, q)$ is a lower bound for $d(a, q)$. Consider again Algorithm 1. If at line 1 it holds that $D(a, q) > R(a) + r$, then surely $d(a, q) > R(a) + r$, and hence we can stop the search at node a without actually evaluating $d(a, q)$. This leads to the following definition.

Definition 2. *An element a in S is said to be* covering radius feasible *(cr-feasible for short) for query q if $D(a, q) \leqslant R(a) + r$. The set of* cr-feasible neighbors *of a node a is a subset of $N(a)$, and will be denoted by* cr-$F(a)$.

Also, we use D along with the *hyperplane criterion* to save distance computations at search time: for any cr-feasible element b_i such that $D(b_i, q) > d_{min} + 2r$, it holds that $d(b_i, q) > d_{min} + 2r$. Hence, we can stop the search in the cr-feasible node b_i without evaluating $d(b_i, q)$ (at line 5 of Algorithm 1).

Definition 3. *Let* cr-$F(a)$ *be the set $\{b_1, \ldots, b_k\}$, in increasing order of timestamp. An element $b_i \in$ cr-$F(a)$ is said to be* hyperplane feasible *(h-feasible for short) for query q if $D(b_i, q) \leqslant d_{min} + 2r$, where d_{min} is minimized using only the distances $d(b_1, q), \ldots, d(b_{i-1}, q)$ that have been computed in the current query.*

Definition 4. *The* feasible neighbors *of node a, denoted $F(a)$, are the cr-feasible plus the h-feasible neighbors $b \in N(a)$. The other neighbors of a are said to be* infeasibles.

Note that only feasible neighbors of a node a must be taken into account when processing a query. The remaining subtrees can be discarded completely using D rather than d. However, it does not immediately follow that we obtain for sure an improvement in search performance. The reason is that infeasible nodes still serve to reduce d_{min} in Algorithm 1, which in turn may save us entering into younger siblings. Hence, by saving computations against infeasible nodes, we may have to enter into new siblings later. This is an intrinsic price of our method. At search time, $D(a, q)$ can be computed without additional evaluations of d for any a in the data structure. A query stack is used to maintain the distances between the query object and the pivots as we backtrack the tree (see [1] for details). Algorithm 2 shows the first basic approach for range search on a H-DSAT.

However, in order to use timestamp information as much as possible in line 8, we run into the risk of comparing infeasible elements against q. this reduces the benefits of pivots in the data structure. Some improvements to this weakness were presented [1], being the best one as follows.

Algorithm 2. H-DSAT Search(Node a, Query q, Radius r, Timestamp t)

```
 1: if time(a) < t ∧ d(a, q) ≤ R(a) + r then
 2:    if d(a, q) ≤ r then
 3:       report a
 4:    end if
 5:    d_min ← +∞
 6:    cr-F(a) ← {b ∈ N(a), D(b, q) ≤ R(b) + r}
 7:    for b_i ∈ N(a) in increasing timestamp order do
 8:       if b_i ∈ cr-F(a) ∧ D(b_i, q) ≤ d_min + 2r then
 9:          if d(b_i, q) ≤ d_min + 2r then
10:             k ← min {j > i,  d(b_i, q) > d(b_j, q) + 2r}
11:             H-DSAT Search(b_i, q, r, time(b_k))
12:          end if
13:       end if
14:       if d(b_i, q) has already been computed then
15:          d_min ← min {d_min,  d(b_i, q)}
16:       end if
17:    end for
18: end if
```

Using Timestamps of Feasible Neighbors. The use of timestamps is not essential for the correctness of the algorithms. Any larger value would work, although the optimal choice is to use the smallest correct timestamp. Another alternative is to compute a safe approximation to the correct timestamp, but ensuring that no infeasible elements are ever compared against q. Note that every feasible neighbor of a node will be compared against q inevitably. If for $b_i \in F(a)$ it holds that $d(b_i, q) \leq d_{min} + 2r$, then we compute the oldest timestamp t among the reduced set $\{b_{i+j} \in F(a),\ d(b_i, q) > d(b_{i+j}, q) + 2r\}$, and stop the search inside b_i at nodes whose timestamp is newer than t. This ensures that only feasible elements are compared against q, and under that condition it uses as much timestamping information as possible. This alternative is called H-DSATF.

3   Reducing the Cost of Traversing an H-DSAT

H-DSATs [1] use the available memory space to improve the search performance of *dsa-trees*. However, their search algorithms use pivots only to check the spatial approximation stopping criteria (that is, the covering-radius and hyperplane feasibility, see line 8 of Algorithm 2). This means that all traversed nodes are inevitably compared against q, even though for some element x in the search path it holds that $D(x, q) > r$. The question is, therefore, whether we can improve the search cost if we avoid comparing elements in the search path of a *dsa-tree*. However, and as we will see, saving distances in this way is not for free. When the distance among the query and a traversed node is not computed, many search criteria would need to be relaxed, as we will see. Hence, it is not clear whether we will obtain an improvement or not.

To answer this question, we define a new search alternative for H-DSAT that avoids computing $d(x, q)$ whenever $\mathcal{D}(x, q) > r$ holds. Assume that, at search time, we reach the node a of the tree. For each $b_i \in N(a)$ in increasing order of timestamp, we perform the following steps:

Step 1: If $\mathcal{D}(b_i, q) > R(b_i) + r$ or $\mathcal{D}(b_i, q) > d_{min} + 2r$, prune the search at b_i since it is infeasible; otherwise, go to the next step.

Step 2: If $\mathcal{D}(b_i, q) > r$, b_i is not within the query radius. Therefore, we search inside the subtree of b_i *without* evaluating $d(b_i, q)$. Thus, all the descendants of b_i cannot use it as a pivot in the current query. We mark this fact by pushing an invalid distance into the query stack [1]. As $d(b_i, q)$ has not been computed, we cannot check whether $d(b_i, q) > d(b_{i+j}, q) + 2r$ holds. Therefore we cannot search for the timestamp of a younger sibling of b_i to search inside the subtree of b_i (step 10 of Algorithm 2). In order to use timestamp information even in this case, if we reach b_i searching for elements with timestamp older than t, then we also use t to search inside the subtree of b_i. This is a correct (although not optimal) timestamp to search inside b_i.

Step 3: On the other hand, if $\mathcal{D}(b_i, q) \leqslant r$, we compute $d(b_i, q)$, and we report b_i if it lies within the search radius. Also, we try to prune the search using the covering radius and hyperplane criterions: if $d(b_i, q) > R(b_i) + r$ or $d(b_i, q) > d_{min} + 2r$, the search can be pruned at b_i. If the search cannot be prunned at b_i, we compute the oldest timestamp t among the set $\{b_{i+j} \in F(a), \mathcal{D}(b_{i+j}, q) \leqslant r \wedge d(b_i, q) > d(b_{i+j}, q) + 2r\}$, and stop the search inside b_i at nodes whose timestamp is newer than t.

We call H-DSATP this search alternative, which is formalized in Algorithm 3. Notice that we have added an extra parameter $dist$, which is the value $d(a, q)$ in case a has not been discarded using pivots, otherwise $dist = 0$ holds (see line 10). Let us take a look also at the condition in line 1: every time $\mathcal{D}(a, q) > r$ holds, it also holds that $dist = 0$, hence condition $dist \leqslant R(a) + r$ is true in these cases. Thus, only the timestamp condition $time(a) < t$ can be used to prune the search when $\mathcal{D}(a, q) > r$ holds.

Notice that our algorithm can be regarded as a pivoting scheme that uses the spatial approximation approach to prune the search space. This has the additional advantage of reducing the overhead incurred when computing $\mathcal{D}$ (which uses to be high for pure pivoting algorithms [2]).

When an element b_i is not compared against the query q (lines 9 and 10), the descendants of b_i cannot use it as a pivot. As a result, the value of $\mathcal{D}$ for these descendants can become underestimated, which is obviously a drawback. However, this gives the data structure the potential to adapt itself to the difficulty of the metric space and "decide" the number of pivots used for each element.

Algorithm 3. H-DSATP Search(Node a, Query q, Radius r, Timestamp t, distance $dist$)

// $dist$ is $d(a,q)$ in case it has been computed, 0 otherwise.

1: **if** $time(a) < t \wedge dist \leqslant R(a) + r$ **then**
2:    **if** $\mathcal{D}(a,q) \leqslant r \wedge dist \leqslant r$ **then**
3:      report a
4:    **end if**
5:    $d_{min} \leftarrow +\infty$
6:    $cr\text{-}F(a) \leftarrow \{b \in N(a), \mathcal{D}(b,q) \leqslant R(b) + r\}$
7:    **for** $b_i \in N(a)$ in increasing timestamp order **do**
8:      **if** $b_i \in cr\text{-}F(a) \wedge \mathcal{D}(b_i,q) \leqslant d_{min} + 2r$ **then**
9:        **if** $\mathcal{D}(b_i,q) > r$ **then**
10:         H-DSATP Search$(b_i, q, r, t, 0)$
11:       **else**
12:         **if** $d(b_i,q) \leqslant d_{min} + 2r$ **then**
13:           $k \leftarrow \min\{j > i, b_j \in F(a) \wedge \mathcal{D}(b_j,q) \leqslant r \wedge d(b_i,q) > d(b_j,q) + 2r\}$
14:           H-DSATP Search$(b_i, q, r, time(b_k), d(b_i,q))$
15:         **end if**
16:       **end if**
17:     **end if**
18:     **if** $d(b_i,q)$ has been already computed **then**
19:       $d_{min} \leftarrow \min\{d_{min}, d(b_i,q)\}$
20:     **end if**
21:   **end for**
22: **end if**

If for an element $b_j \in F(a)$ it holds that $\mathcal{D}(b_j,q) > r$, hence b_j cannot be used to minimize d_{min} when searching inside the subtree of an element $b_i \in F(a)$ younger than b_j: the condition $d(b_i,q) > d(b_j,q) + 2r$ implies computing $d(b_j,q)$. Since we know that $d(b_j,q) > r$, we will prefer not to use b_j to minimize d_{min}, saving the distance computation. This is a relaxation to the original spatial approximation approach. Line 13 of Algorithm 3 shows this formally.   Also, every time $d(b_i,q)$ is computed (Step 3 above), we take full advantage of this evaluation by using the pruning criterion of the original *dsa-trees*, we use $d(b_i,q)$ to minimize d_{min}, and later, the descendants of b_i can use it as a pivot.

H-DSATP might traverse more nodes of the data structure than the original H-DSATs, because if $\mathcal{D}(b_i,q) \leqslant R(b_i) + r$, $\mathcal{D}(b_i,q) \leqslant d_{min} + 2r$, and $\mathcal{D}(b_i,q) > r$, then we have not computed $d(b_i,q)$ and the search must continue in the subtree of b_i. However, it might be that $d(b_i,q) > R(b_i) + r$ or $d(b_i,q) > d_{min} + 2r$, and the search would have stopped at b_i. That is, the cost of traversing a node is, in some cases, less expensive, but we may traverse more nodes than the original H-DSATs. The experiments of the next section will show that, despite the possible drawbacks we have remarked, in general it pays off to use $\mathcal{D}$ to exchange more traversed nodes for a smaller total cost.

4   Experimental Results

For the experiments of this paper we have considered range queries on four widely different metric spaces.

NASA images: a set of 40,700 feature vectors of dimension 20, generated from images downloaded from NASA [1]. The Euclidean distance is used. This is an easy space (sparse histogram of distances). For this space we use radii 0.605740, 0.780000 and 1.009000, which retrieve on average 0.01%, 0.1%, and 1% of the database respectively.

Words: a dictionary of 69,069 English words [2]. We use the *edit* or *Levenshtein* distance, that is, the minimum number of character insertions, deletions and replacements needed to make two strings equal. This distance is useful in text retrieval to cope with spelling, typing and optical character recognition (OCR) errors. The space turns out to be of low to medium difficulty. As the distance is discrete, we use radii 1 to 4, which retrieve on average 0.00003%, 0.00037%, 0.00326% and 0.01757% of the dataset, respectively

Color histograms: a set of 112,682 color histograms (112 dimensional vectors) from an image database [3]. Any cuadratic form can be used as a distance, so we chose Euclidean distance as the simplest meaningful alternative. The resulting space is of medium difficulty. For this space we use radii 0.051768, 0.082514 and 0.131163, which retrieve on average 0.01%, 0.1%, and 1% of the database respectively.

Documents: a set of 1,265 documents under the Cosine similarity, heavily used in Information Retrieval. In this model the space has one coordinate per term and documents are seen as vectors in this high-dimensional space. The distance we use is the angle (arccos of inner product) among the vectors. The documents are the files of the TREC-3 collection [4]. This is a space of medium to high difficulty, and the distance is expensive to compute. For this space we use radii 0.140000, 0.150000 and 0.195000, retrieving on average 1, 2, and 16 documents respectively.

For all these metric spaces, we build the indexes 10 times using 90% of the database elements, leaving the remaining 10% (randomly chosen) as queries. We test with arities 4, 8, 16 and 32 in the tree [3]. Due to lack if space, we show results only for the arity that produced the best results in each case.

We will suffix "1" the versions of H-DSAT that use the ancestors as pivots, and will use "2" for the versions that use the ancestors and their older siblings as pivots [1]. Hence, the alternative proposed in this paper will have two instances, H-DSATP1 and H-DSATP2. We will compare against the best alternative in previous work [1]: H-DSATF (the resulting instances are H-DSATF1 and H-DSATF2).

---

[1] http://www.sisap.org/library/metricSpaces/dbs/vectors/nasa.tar

[2] http://www.sisap.org/library/metricSpaces/dbs/strings/dictionaries.tar

[3] http://www.sisap.org/library/metricSpaces/dbs/vectors/colors.tar

[4] http://trec.nist.gov

Figure 1 shows the experimental query cost for search variants of H-DSAT using just ancestors as pivots. In all metric spaces we tested, H-DSATP1 performs better with arity 4. This is because the pivot information is more heavily used in this alternative, hence having small arity makes the three higher, hence each element has a bigger amount of pivots. H-DSATF1, on the other hand, uses the spatial approximation idea as much as it can. Hence, it performs better using arity 16 (except for the space of documents, where H-DSATF1 has the best performance using arity 32).

In the experiments, H-DSATP1 outperforms H-DSATF1, in many cases considerably. In the space of NASA images, we obtain about 15% (large radius) to 30% (small radius) less distance evaluations at query time. For color histograms, the improvements are from 28% to 40%. For the dictionary of English words, from 11% to 35%. Finally, for documents from 10% to 11%. The best improvements are obtained for small radii —since the problem is easier in these cases, on which pivots are more effective— and easier metric spaces —e.g., color histograms.

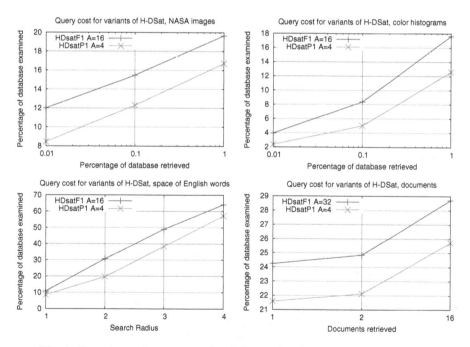

Fig. 1. Experimental query cost for different search alteratives of H-DSAT1

Figure 2 shows the experimental query cost for variants of H-DSAT2. As it can be seen, H-DSATF2 and H-DSATP2 perform better with arity 32 (except in the space of documents, where H-DSATP2 performs better using arity 16). As before, H-DSATP2 outperforms H-DSATF2, obtaining better improvements compared with the former alternatives. In the space of NASA images, we obtain about 14% (large radius) to 37% (small radius) less distance computations, for color histograms 35% to 68%, for English dictionary 20% to 77%, and for the document database 13% to 17%.

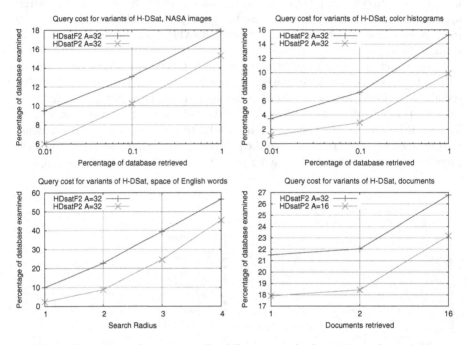

Fig. 2. Experimental query cost for different search alternatives of H-DSAT2

An important result that must be also considered is that comparing the results in Figures 1 and 2, we can conclude that HDSATP1 outperforms HDSATF2 in all cases, even though the former uses less pivots per node than the latter [1]. This reinforces the fact that our algorithm makes a better use of pivots, compared to the algorithms proposed in [1].

Finally, we obtained the following results regarding the number of traversed nodes by our algorithm. In the space of NASA images, H-DSATP1 traverses from about 23% (small radius) to 28% (large radius) of the tree nodes. This is 1.92 and 1.45 times the number of nodes traversed by H-DSATF1. For color histograms, H-DSATP1 traverses from about 17% (small radius) to 28% (large radius) of the tree nodes. This is 4.34 and 1.62 times the number of nodes traversed by H-DSATF1, respectively. For the English dictionary, H-DSATP1 traverses from about 44% ($r = 1$) to 77% ($r = 4$) of the tree nodes. This is 4.04 and 1.19 times the number of nodes traversed by H-DSATF1. For H-DSATP2, the results are similar, yet smaller than those of H-DSATP1.

Note that, even though H-DSATP traverses more nodes of the data structure than the original *dsa-tree* data structures, the total number of traversed nodes is a relatively small fraction of the whole tree. This is important in cases where one wants to reduce the overhead incurred by traversing the whole database. Tree-based pivoting schemes are specifically good for this matter. However, given a fixed amount of storage, they must encode the tree structure, hence using space that the array-based indexes could use just for pivots (hence, storing a bigger number of pivots, improving the overall search performance). Nowadays, however, trees (even dynamic ones, as in our case) can be encoded using about 2 bits per node [4].

Our results clearly indicate a trend: for small radii and easier spaces (e.g., color histograms), we obtain the best improvements over H-DSATF (in number of distance evaluations), yet the number of traversed nodes by H-DSATP is higher than for H-DSATF. This is because in such cases our algorithm behaves like a pivoting scheme in these cases. For large radii and more difficult spaces, on the other hand, the improvements over H-DSATF are moderate (yet important), and the number of traversed nodes is similar to H-DSATF. This is because in these cases the data structure tends to behave as *dsa-trees*.

5   Conclusions

From our experimental results, we conclude that it is worth to relax some spatial approximation criteria (hence probably traversing more *dsa-tree* nodes) provided pivot information is used at every tree node as we propose. The search algorithm we proposed in this paper makes a better use of the available memory space used by pivots in *dsa-trees*. Compared with previous approaches [1] that use pivots on *dsa-trees*, our range search algorithm carries out from 10% to 70% less distance evaluations at query time. Our best improvements on previous results [1] were obtained in cases of small radii and easier spaces.

Our experimental results seem to indicate that our algorithm is adaptive to the difficulty of the search: on easier cases (i.e., easier metric spaces and small query radii) the data structure tends to behave as a pivoting algorithm; on harder cases (i.e., harder metric spaces and large radii), the data structure behaves like a *dsa-tree*, which are known to be more resistant to hard spaces. This deserves future research.

References

1. Arroyuelo, D., Muñoz, F., Navarro, G., Reyes, N.: Memory-adaptative dynamic spatial approximation trees. In: Nascimento, M.A., de Moura, E.S., Oliveira, A.L. (eds.) SPIRE 2003. LNCS, vol. 2857, pp. 360–368. Springer, Heidelberg (2003)
2. Chávez, E., Navarro, G., Baeza-Yates, R., Marroquín, J.: Searching in metric spaces. ACM Computing Surveys 33(3), 273–321 (2001)
3. Navarro, G., Reyes, N.: Dynamic spatial approximation trees. ACM Journal of Experimental Algorithmics (JEA) 12:article 1.5, 68 pages (2008)
4. Navarro, G., Sadakane, K.: Fully-functional static and dynamic succinct trees. ACM Transactions on Algorithms 10(3):article 16 (2014)

Large-Scale Distributed Locality-Sensitive Hashing for General Metric Data

Eliezer Silva[2,*], Thiago Teixeira[1],
George Teodoro[1], and Eduardo Valle[2]

[1] Dep. of Computer Science, University of Brasilia, Brasilia, Brazil
`thiagotei@gmail.com`, `teodoro@cic.unb.br`
[2] RECOD Lab. / DCA / FEEC / UNICAMP, Campinas, Brazil
`{eliezers,dovalle}@dca.fee.unicamp.br`

Abstract. Locality-Sensitive Hashing (LSH) is extremely competitive for similarity search, but works under the assumption of uniform access cost to the data, and for just a handful of dissimilarities for which locality-sensitive families are available. In this work we propose Parallel Voronoi LSH, an approach that addresses those two limitations of LSH: it makes LSH efficient for distributed-memory architectures, and it works for very general dissimilarities (in particular, it works for all metric dissimilarities). Each hash table of Voronoi LSH works by selecting a sample of the dataset to be used as seeds of a Voronoi diagram. The Voronoi cells are then used to hash the data. Because Voronoi diagrams depend only on the distance, the technique is very general. Implementing LSH in distributed-memory systems is very challenging because it lacks referential locality in its access to the data: if care is not taken, excessive message-passing ruins the index performance. Therefore, another important contribution of this work is the parallel design needed to allow the scalability of the index, which we evaluate in a dataset of a thousand million multimedia features.

1 Introduction

Content-based Multimedia Information Retrieval (CMIR) is an alternative to keyword-based or metadata-based retrieval. CMIR works by extracting *features* from the multimedia content itself, and using those feature to represent the multimedia objects. The features capture perceptual characteristics (color, texture, motion, etc.) from the documents, helping to bridge the so-called "semantic gap": the disparity between the amorphous low-level multimedia coding (e.g., image pixels or audio samples) and the complex high-level tasks (e.g., classification or retrieval) performed by CMIR engines.

That way, searching for similar multimedia documents becomes the more abstract operation of finding the closest features in the feature space — an operation known as similarity search. CMIR systems are usually complex and may include several phases,

* E. Silva was supported by a CAPES scholarship. All authors thank to CAPES, CNPq (including project 2012/14: Examind) and FAPESP (including FAPESP-JP: 2010/05647-4) for financial support to this research. We also thank the XSEDE for granting us access to the Keeneland cluster, where the large-scale experiments where conducted.

A.J. Machado Traina et al. (Eds.): SISAP 2014, LNCS 8821, pp. 82–93, 2014.
DOI: 10.1007/978-3-319-11988-5_8 © Springer International Publishing Switzerland 2014

but similarity search will often be a critical step. Similarity search for multimedia descriptors is a complex problem that has been extensively studied for decades [1]. Challenges of this task include (i) the very large and increasing volume of data to be indexed/searched; (ii) the high dimensionality of the multimedia descriptors; (iii) the diversity of dissimilarity functions employed [2].

Among indexing algorithms for efficient searching in high-dimensional datasets, *Locality-Sensitive Hashing* (LSH) [3,4,5] deserves attention as one of the best performing and most cited algorithms in the literature [3,4,5]. LSH uses functions that hash together, with higher probability, points in the space that are close to each other. For architectures with uniform access cost to the data, and for a handful of dissimilarities (mainly the Hamming, Euclidean and Manhattan metrics) LSH will often be the technique with best compromise between precision and speed. However, the need to discover a new family of locality-sensitive hashing functions for each dissimilarity function precludes the generalization of LSH; in addition, the poor referential locality of LSH makes its adaptation to distributed-memory systems — essential for scalability — very challenging.

In this work, we address those two shortcomings with *Parallel Voronoi LSH*, which extends LSH for very general dissimilarities (in particular, any metric distances), and for distributed-memory architectures, enabling the index to scale-up to huge datasets. Each hash table of Voronoi LSH works by selecting a sample of the dataset to be used as the seeds of a Voronoi diagram. The Voronoi cells are then used to hash the data. Because Voronoi diagrams depend only on distances, the technique is very general.

Parallel Voronoi LSH builds upon previous works in literature that employ, implicitly or explicitly, the notion of Voronoi cells in order to perform similarity search using the principles of locality-sensitive hashing [6,7,8]. We aim at tackling a large class of dissimilarity functions (including all metric distances), and at scaling-up the index to very large corpora. As far as we know, our work is the first to tackle at once both problems of general metric spaces and very large scales (thousand of millions) for locality-sensitive hashing. An additional contribution of this work is the evaluation of adaptive locality-sensitive functions for metric data. First proposed for Euclidean data [6], in this work we adapt them for general metric data, and measure the impact of chosing adaptive functions versus random ones.

2   Background

A comprehensive review of similarity search for multimedia would include hundreds of papers and is beyond the scope of this work. For a starting point, the reader is referred to [1,9]. In this section, we focus on the key papers of Locality-Sensitive Hashing and some recent developments. Since our focus is on practical index design instead of theoretical computational geometry, we review less the recent theoretical advances of LSH, and more the papers with a practical/experimental slant.

LSH relies on the existence of families H of locality-sensitive hashing functions, to map points from the original space into a set of hash keys, such that points that are near in the space are hashed to the same value with higher probability than points that are far apart. The seminal work on LSH [3] proposed locality-sensitive hashing families for Hamming distances in Hamming spaces, and for Jacquard indexes in spaces of sets. Later, extensions for L_1-normed (Manhattan) and L_2-normed (Euclidean) spaces were proposed by embedding those spaces into Hamming spaces [4]. The practical success of LSH, however, came with E2LSH[1] (Exact Euclidean LSH) [5], which proposed a new family of locality-sensitive functions, based upon projections onto random lines, which worked "natively" for Euclidean spaces.

LSH works by boosting the locality sensitiveness of the hash functions. This is done by building from the original $\{h_i \in H\}$ locality-sensitive function family, a family $\{g_j \in H\}$, where each g_j is the concatenation of M randomly sampled h_i, i.e., each g_j has the form $g_j(\mathbf{v}) = (h_1(\mathbf{v}), ..., h_M(\mathbf{v}))$. Then, we sample L such functions g_j, each to hash an independent hash table. As M grows, the probability of a false positive (points that are far away having the same value on a given g_j) drops sharply, but so grows the probability of a false negative (points that are close having different values). But as L grows and we check all hash tables, the probability of false negatives falls, and the probability of false positives grows. LSH theory shows that it is possible to set M and L so to have a small probability of false negatives, with an acceptable number of false positives. That allows the correct points to be found among a small number of candidates, dramatically reducing the number of distance computations needed to answer the queries.

The need to maintain and query L independent hash tables is the main weak point of LSH. In the effort to keep both false positives and false negatives low, there is an "arms race" between M and L, and the technique tends to favor large values for those parameters. The large number of hash tables results in excessive storage overheads. Referential locality also suffers, due to the need to random-access a bucket in each of the large number of tables. More importantly, it becomes unfeasible to replicate the data on so many tables, so each table has to store only pointers to the data. Once the index retrieves a bucket of pointers on one hash table, a cascade of random accesses ensues to retrieve the actual data.

Multiprobe LSH [10] considerably reduces the number of hash tables of LSH, by proposing to visit many buckets on each hash table. It analyses the relative position of the query in the boundaries of the hash functions of E2LSH to estimate the likelihood of each bucket to contain relevant points. *A posteriori LSH* [11] extends that work by turning the likelihoods into probabilities using priors estimated from training data. Those works reduce the storage overhead of LSH at the cost of greatly increasing the number of random accesses to the data.

2.1 Unstructured Quantizers, General Spaces

In *K-means LSH* [6], the authors address LSH for high-dimensional Euclidean spaces, introducing the idea of hash functions adapted to the data, generated by running a K-means or hierarchical K-means on a sample, using the centroids obtained as seeds to a Voronoi diagram. The Voronoi diagram becomes the hash function (each cell induces a hash value over the points it contains). They call those hash functions *unstructured quantizers*, in contrast to the regular hash-functions (intervals on projections over

---

[1] *LSH Algorithm and Implementation (E2LSH)*. http://www.mit.edu/~andoni/LSH/

random lines, cells on lattices), which are blind to the data, that they call *structured quantizers*. Their experimental results show that the data-adapted functions perform better than the data-blind ones.

DFLSH (Distribution Free Locality-Sensitive Hashing) [7], works on the same principle of Voronoi-diagram induced hash functions, but instead of applying the K-means, randomly chooses the centroids from the dataset. The advantage of the scheme is generality: while the averaged centroids of K-means LSH imply an Euclidean (or at the very least coordinate) space, DFLSH work for any space in which Voronoi diagrams work.

Another LSH technique for general metric spaces is based on *Brief Proximity Indexing (BPI)* [8]. Similarity is inferred by the perspective on a group of points called *permutants* (if point p sees the permutants in the same order as point q, p and q are likely to be close to each other). This is similar to embedding the data into a space for which LSH is available and then applying a LSH function for that space. Indeed the method consists of those two steps: first it creates a permutation index; and then it hashes the permutation indexes using LSH for Hamming spaces [3].

M-Index [12] is a Metric Access Method for exact and approximate similarity search constructed over a universal mapping from the original metric space to scalars values. The values of the mapping are affected by the permutation order of a set of reference points and the distance to these points. In a follow-up work [13], this indexing scheme is analyzed empirically as a locality-sensitive hashing for general metric spaces.

Works on generalizing LSH for all metric spaces have focused more on proposing practical techniques and show that they work empirically rather than in proving theoretically that the scheme strictly follows the axioms of locality sensitiveness as proposed by Indyk and Motwani [3]. K-means LSH offers no theoretical analysis. In [7] a proof sketch showing that the hashing family is locality-sensitive is presented, but only for Euclidean spaces. Permutation-based index employs Hamming spaces, whose locality-sensitive family was proved to be so in the seminal LSH paper, but they offer no formal proof that the embedding they propose into Hamming spaces preserves the metric of the original space.

In this paper we combine and extend the efforts of those previous works. As in DFLSH we employ Voronoi cells over points chosen from the dataset, allowing us to tackle very general dissimilarity functions[2]. As in K-means LSH, we are interested in evaluating the impact of the adaptation to the data in the performance of the technique. Unlike existing art, we are concerned on how to scale-up the index to *very* large datasets. In the next two sections we will describe the basic technique, and the distributed implementation proposed to tackle those large collections.

3   Voronoi LSH and Parallel Voronoi LSH

Each hash table of Voronoi LSH employs a hash function induced by a Voronoi diagram over the data space. If the space is known to be Euclidean (or at least coordinate)

---

[2] A precise characterization of which dissimilarities are compatible with our technique (and with DFLSH, for that matter) seems very complex, but it is sure that it includes all metric distances (e.g. Euclidean and Manhattan distance on coordinate spaces, edit distance on strings, etc.), and all non-decreasing functions of metric distances (e.g., squared Euclidean distance).

the Voronoi diagram can use as seeds the centroids learned by a Euclidean clustering algorithm, like K-means (in which case Voronoi LSH coincides with K-means LSH). However, if nothing is known about the space, points from the dataset must be used as seeds, in order to make as few assumptions as possible about the structure of the space. In the latter case, randomly sampled points can be used (in which case Voronoi LSH coincides with DFLSH). However, it is also possible to try select the seeds by employing a distance-based clustering algorithm, like K-medoids, and using the medoids as seeds.

Neither K-means nor K-medoids are guaranteed to converge to a global optimum, due to dependencies on the initialization. For Voronoi LSH, however, obtaining *the* best clustering is not a necessity. Moreover, when several hash tables is applied, the difference on the clustering results between runs is an advantage, since it is one of the sources of diversity among the hash functions (Figure 1). We further explore the problem of optimizing clustering initialization below.

Fig. 1. Each hash table of Voronoi LSH employs a hash function induced by a Voronoi diagram over the data space. Differences between the diagrams due to the data sample used and the initialization seeds employed allow for diversity among the hash functions.

We define more formally the hash function used by Voronoi LSH in Equation 1. In summary, it computes the index of the Voronoi seed closest to a given input object. In addition, we present the indexing and querying phases of Voronoi LSH, respectively, in Algorithms 1 and 2. Indexing consists in creating L lists with k Voronoi seeds each $(C_i = \{c_{i1}, \ldots, c_{ik}\}, \forall i \in \{1, \ldots, L\})$. When using K-medoids, which is expensive, we suggest the Park and Jun fast K-medoids algorithm (apud [6]), performing this clustering over a sample of the dataset, and limiting the number of iterations to 30. Then, for each point in the dataset, the index stores a reference in the hash table T_i ($i \in \{1, \ldots, L\}$), using the hashing function defined in Equation 1 ($h_{C_i}(x)$). When using K-means, we suggest the K-means++ variation [14,15]. The seeds can also simply be chosen at random (like in DFLSH). The querying phase is conceptually similar: the same set of L hash functions ($h_{C_i}(x)$) is computed for a query point, and all hash tables are queried to retrieve the references to the candidate answer set. The actual points are retrieved from the dataset using the references, forming a candidate set (shortlist), and the best answers are selected from this shortlist.

In this work, we focus on k-nearest neighbor queries. Therefore, this final step consists of computing the dissimilarity function to all points in the shortlist and selecting the k closest ones.

We mention, in passing, that several variations of K-medoids clustering exist, from the traditional PAM (Partitioning Around Medoids) [16] to the recently proposed FAMES [17]. We chose the method of Park and Jun [18] due to its simplicity of implementation and good speed. K-medoid is expensive (even more so than K-means, which is already not cheap), but Park and Jun restrict the search for new medoid candidates to other points already assigned to cluster, making it much faster. Still, our technique can in principle be employed with the other variations, if desired.

Definition 1. *Given a metric space (U,d) (U is the domain set and d is the distance function), the set of Voronoi seeds $C = \{c_1,\ldots,c_k\} \subset U$ and an object $x \in U$:*

$$h_C : U \to \mathbb{N}$$
$$h_C(x) = \text{argmin}_{i=1,\ldots,k}\{d(x,c_1),\ldots,d(x,c_i),\ldots,d(x,c_k)\} \tag{1}$$

---

input : Set of points X, number of hash tables L, size of the sample set S and
number of Voronoi seeds k
output: list of L index tables $T_1,\ldots,T_L$ populated with all points from X and list
of L Voronoi seeds $C_i = \{c_{i1},\ldots,c_{ik}\}, \forall i \in 1,\ldots,L$
for $i \leftarrow 1$ **to** L **do**
    Draw sample set S_i from X;
    $C_i \leftarrow$ choose k seeds from sample S_i (random, K-means, K-medoids, etc.);
    for $x \in X$ **do**
        $T_i[h_{C_i}(x)] \leftarrow T_i[h_{C_i}(x)] \cup \{\text{pointer to } x\}$;
    end
end

Algorithm 1. Voronoi LSH indexing phase: a Voronoi seed list (C_i) is independently selected for each of the L hash tables used. Further, each input data point is stored in the bucket entry $(h_{C_i}(x))$ of each hash table.

---

input : Query point q, index tables $T_1,\ldots,T_L$, L lists of M Voronoi seeds each
$C_i = \{c_{i1},\ldots,c_{ik}\}$, number of nearest neighbors to be retrieved N
output: set of N nearest neighbors $\text{NN}(q,N) = \{n_1,\ldots,n_N\} \subset X$
CandidateSet $\leftarrow \emptyset$;
for $i \leftarrow 1$ **to** L **do**
    CandidateSet $\leftarrow$ CandidateSet $\cup$ $T_i[h_{C_i}(q)]$;
end
$\text{NN}(q,N) \leftarrow \{\text{k closest points to q in CandidateSet }\}$;

Algorithm 2. Voronoi LSH querying phase: the input query point is hashed using same L functions as in indexing phase, and points in colliding bucket in each hash table are used as nearest neighbors candidate set. Finally, N closest points to the query are selected from the candidate set.

Initialization. K-means and K-medoids strongly depend on the initial centroid/medoid selection. K-means++ [14,15] solve the $O(\log k)$-approximate K-means problem by carefully choosing those initial centroids. Moreover, because K-means++ initialization employs only distance information and sampling, it can be transposed to K-medoids. Park and Jun [18] also propose a special initialization for their fast K-medoids algorithm, based on a distance-weighting scheme. We evaluated both of those special initializations, as well as random initialization.

3.1   Parallelization Strategy

The parallelization strategy is based on the dataflow programming paradigm. Dataflow applications are typically represented as a set of computing *stages*, which are connected to each other using directed *streams*.

Our parallelization decomposes Voronoi LSH into five computing stages organized into two conceptual pipelines, which execute the index building and the search phases of the application. All stages may be replicated in the computing environment to create as many copies as necessary. Additionally, the streams connecting the application stages implement a special type of communication policy referred here as *labeled-stream*. Messages sent through a labeled-stream have an associated label or tag, which provides an affordable scheme to map message tags to specific copies of the receiver stage in a stream. We rely on this communication policy to partition the input dataset and to perform parallel reduction of partial results computed during a query execution. The data communication streams and processes management are built on top of Message Passing Interface (MPI).

The index building phase of the application, which includes the Input Reader (IR), Bucket Index (BI), and Data Points (DP) stages, is responsible for reading input data objects and building the distributed LSH indexes that are managed by the BI and the DP stages. In this phase, the input data objects are read in parallel using multiple IR stage copies and are sent (1) to be stored into the DP stage (message i) and (2) to be indexed by the BI stage (message ii). First, each object read is mapped to a specific DP copy, meaning that there is no replication of input data objects. The mapping of objects to DPs is carried out using the data distribution function *obj_map* (labeled-stream mapping function), which calculates the specific copy of the DP stage that should store an object as it is sent through the stream connecting IR and DP. Further, the pair <object identifier, DP copy in which it is stored> is sent to every BI copy holding buckets into which the object was hashed. The distribution of buckets among BI stage copies is carried out using another mapping function: *bucket_map*, which is calculated based on the bucket value/key. Again, there is no replication of buckets among BIs and each bucket value is stored into a single BI copy. The *obj_map* and *bucket_map* functions used in our implementation are modulo operation based on the number of copies of the receiver in a stream. We plan to evaluate other hashing strategies in the future.

The index construction is very compute-intensive, and involves many distance calculations between the input data objects and the Voronoi seeds. For Euclidean data, we implemented a vectorized code using Intel SSE/AVX intrinsics to take advantage of the wide SIMD instructions of current processors. Preliminary measurements have shown that the use of SIMD instructions sped-up the index building 8 times.

The search phase of the parallel LSH uses four stages, two of them shared with the index building phase: Query Receiver (QR), Bucket Index (BI), Data Points (DP), and Aggregator (AG). The QR stage reads the query objects and calculates the bucket values into which the query is hashed for the L hash tables. Each bucket value computed for a query is mapped to a BI copy using the *bucket_map* function. The query is then sent to those BI stage copies that store at least one bucket of interest (message iii). Each BI copy that receives a query message visits the buckets of interest, retrieves the identifier of the objects stored on those buckets, aggregates all object identifiers to be sent to the same DP copy (list(obj_id)), and sends a single message to each DP stage that stores at least one of the retrieved objects (message iv). For each message received by a DP copy, it calculates the distance from the query to the objects of interest, selects the k-nearest neighbors objects to the query, and sends those local NN objects to the AG stage. Finally, the AG stage receives the message containing the DPs local NN objects from all DPs involved in that query computation and performs a reduction operation to compute the global NN objects. The DP copies (message v) use the query_id as a label to the message, guaranteeing that the same AG copy will process all messages related to a specific query. As a consequence, multiple AG copies may be created to execute different queries in parallel. Although we have presented the index building and the search as sequential phases for sake of simplicity, their executions may overlap.

The parallelization approach we have proposed exploits task, pipeline, replicated and intra-stage parallelism. Task parallelism results from concurrent execution that allows indexing and searching phases to overlap, e.g. during an update of the index. Pipeline parallelism occurs as the search stages, for instance, execute different queries in parallel in a pipeline fashion. Replicated parallelism is available in all stages of the application, which may have an arbitrary number of copies. Finally, intra-stage parallelism results of the application's ability to use multiple cores within a stage copy. This parallelism has the advantages of sharing the same memory space among computing cores in a stage copy, and a reduced number of messages exchanged, since a smaller number of state partitions may be used.

4   Experimental Evaluation

Datasets: two datasets were used: the English dictionary of strings with Levenshtein distance from SISAP Metric Library [19]; and BigANN [20]. The English dataset has 69,069 strings, 500 strings are randomly removed from the set to serve as query dataset. The BigANN contains a thousand million (10^9) 128-dimensional SIFT local feature vectors extracted from images, and 10,000 query feature vectors. Euclidean distance is used for SIFT. We perform k-NN searches, with k=5, 10, and 30 for the Dictionary dataset, and k=10 for the BigAnn dataset.

Metrics: We employ the *recall* as a metric of quality, and the *extensiveness* as metric of cost for the techniques. The recall is defined as usual for information retrieval, as the fraction of relevant answers that was effectively retrieved. The *extensiveness* metric is the fraction of the dataset selected into the shortlist for linear scan. As the indexes work by selecting a (hopefully small) fraction of the dataset as candidates, and then computing the actual distance to the query for those candidates, the size of the shortlist corresponds to the number of distances computed for that query. The related *selectivity* metric ($selectivity = 1 - extensivity$) has distinct (and sometimes conflicting) use in the database and image retrieval research communities: some authors used it as a synonymous for extensiveness, while others and we use it as a complementary notion (as selectivity grows, extensivity drops). We also employ the *query runtime* as metric of cost. Runtimes are difficult to compare across the literature, due to differences in code optimization, programming language, and execution environment, but in controlled experiments like ours they can offer a perspective on the cost of different techniques.

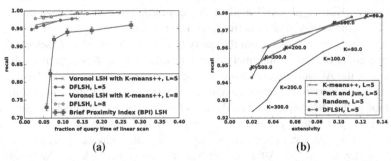

Fig. 2. (a) Voronoi LSH recall–cost compromises are competitive with those of BPI (error bars are the imprecision of *our interpretation* of BPI original numbers). To make the results commensurable, time is reported as a fraction of brute-force linear scan. (b) Different choices for the seeds of Voronoi LSH: K-medoids with different initializations (K-means++, Park & Jun, random), and using random seeds (DFLSH). Experiments on the English dictionary dataset.

4.1 Comparison of Voronoi LSH and BPI-LSH

We compared our Voronoi LSH with Brief Permutation Indexing (BPI) LSH [8] in a sequential (non-distributed) environment. We implemented our Voronoi LSH on Java with The Apache Commons Mathematics Library[3]. Also included is Distribution-Free LSH (DFLSH) [7], which we evaluate as a specific configuration of our implementation of Voronoi LSH with the seeds of the Voronoi diagram chosen at random.

In order to allow the comparison with BPI LSH, we followed an experiment protocol as close as possible to the one described in their paper, and used the recall and query runtimes reported there. In order to remove the effects of implementation details and execution environment differences, we normalize both our times and theirs by the runtime of the brute-force linear scan. Because the authors report their numbers graphically, we added error bars to indicate the imprecision of *our* reading of their results.

The experiments used the English Dictionary dataset. Figure 2 summarizes the results. We compare the impact of the Clustering algorithm initialization to the search quality of the K-medoids nearest neighbors results. For sake of this analysis, the initialization strategy proposed in K-means++ and in the work of "Park and Jun" and the naive random initialization are considered. Figure 2b presents the recall and query time for varying number of centroids. As shown, the K-means++ initialization and the random initialization performance are not very distinguishable. On the other hand, the initialization of Park and Jun is clearly poorer than the others.

4.2 Large-Scale Distributed Voronoi LSH

The large-scale evaluation used a distributed-memory machine with 70 nodes interconnected through a FDR Infiniband switch. Each computation node was equipped with a dual-socket Intel E5 2.60 GHz Sandy Bridge processor with a total of 16 CPU cores,

---

[3] *Commons Math: The Apache Commons Mathematics Library.*
http://commons.apache.org/proper/commons-math/

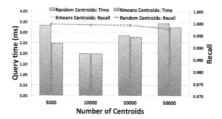

Fig. 3. Random seeds vs. K-means++ centroids on Parallel Voronoi LSH (BigAnn). Results reported are averaged over 10,000 queries. Well-chosen seeds have little effect on recall, but occasionally lower query times.

32 GB of DDR3 RAM, running Linux OS kernel version 2.6.32. Parallel Voronoi LSH was implemented on C++ and MPI.

The Figure 3 shows the query phase execution time and recall of parallel Voronoi LSH using BigANN dataset with Voronoi seeds chosen at random, and by using K-means++ centroids. Although the difference in recall is neglectable, using K-means++ centroids in general resulted in better query times than using random seeds. The best execution time in both cases was with 10,000 Voronoi seeds. For minimizing the distance computations, there is an expected compromise between too few seeds (cheap hash functions, large shortlists due to more points in buckets) and too many (costly hash functions, smaller shortlists). The theoretical sweet spot is obtained with $\sqrt{n}$ seeds, where n is the dataset size (around 30,000 for BigAnn). However, in our tests, the empirical best value was always much smaller than that, favoring, thus, the retrieval of fewer buckets with many points in them, instead of many buckets with fewer points. That suggests that the cost of accessing the data is overcoming the cost of computing the distances.

Our parallelism efficiency analysis employs a scale-up (weak scaling) experiment in which the reference dataset and the number of computing cores used increase proportionally. A scale-up evaluation was selected because we expect to obtain an abundant volume of data for indexing, which would only fit in a distributed system. For each CPU core allocated to the BI stage 4 CPU cores are assigned to the DP stage, resulting in a ratio of computing cores by BI:DP of 1:4. A single core is used for the AG stage.

The efficiency of the parallel Voronoi LSH is presented in Figure 4. As the number of CPU cores and nodes used increase, the application achieves a very good parallel efficiency of 0.9 when 801 computing cores are used (10 nodes for BI and 40 nodes for DP), indicating very modest parallelism overheads. The high efficiency attained is a result of (i) the application asynchronous design that decouples communication from computation tasks and of (ii) the intra-stage parallelization that allows for a single multi-threaded copy of the DP stage to be instantiated per computing node. As a consequence, a smaller number of partitions of the reference dataset are created, which reduces the number of messages exchanged by the parallel version (using 51 nodes) in more than $6\times$ as compared to an application version that instantiates a single process per CPU computing core.

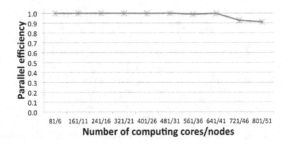

Fig. 4. Efficiency of Parallel Voronoi LSH parallelization as the number of nodes used and the reference dataset size increase proportionally. The results show that the parllelism scheme scales-up well even for a very large dataset, with modest overhead.

5 Conclusion

Efficient large-scale similarity search is a crucial operation for Content-based Multimedia Information Retrieval (CMIR) systems. But because those systems employ high-dimensional feature vectors, or other complex representations in metric spaces, providing fast similarity search for them has been a persistent research challenge. LSH, a very successful family of methods, has been advanced as a solution to the problem, but it is available only for a few distance functions. In this article we propose to address that limitation, by extending LSH to general metric spaces, using a Voronoi diagram as basis for a LSH family of functions. Our experiments show that employing Voronoi diagrams to index the data works well both for metric and for Euclidean data. The experiments do not show any clear advantage in learning the seeds of the Voronoi diagram by clustering: a random choice seems to work just as well. The lack of effect of clustering on recall is somewhat disappointing, and must be confirmed by evaluating a more diverse selection of datasets. However, if confirmed, it will also be an important hint for scalability, since learning the seeds by clustering is expensive. On the other hand, clustering might in some cases affect *query times*, which is surprising. This seems to be due to a more uniform partition of the data, since random seeds tend to create an unbalanced distribution of the dataset on the buckets of the hash tables. The large-scale experiments show that our proposed parallelization has very modest overhead and scales-up well even for a very large collection.

As a future work, we would like to explore very large collections of non-Euclidean metric data. This is currently a challenge because creating an exact ground truth for such corpora is very expensive.

References

1. Chávez, E., Navarro, G., Baeza-Yates, R., Marroquín, J.L.: Searching in metric spaces 33(3), 273–321 (September 2001)
2. Akune, F., Valle, E., Torres, R.: MONORAIL: A Disk-Friendly Index for Huge Descriptor Databases. In: 20th Int. Conf. on Pattern Recognition, pp. 4145–4148. IEEE (August 2010)

3. Indyk, P., Motwani, R.: Approximate nearest neighbors: towards removing the curse of dimensionality. In: Proc. of 13th Ann. ACM Symp. on Theory of Comp., pp. 604–613 (1998)

4. Gionis, A., Indyk, P., Motwani, R.: Similarity search in high dimensions via hashing. In: Proc. of the 25th Int. Conf. on Very Large Data Bases, pp. 518–529 (1999)

5. Datar, M., Immorlica, N., Indyk, P., Mirrokni, V.S.: Locality-sensitive hashing scheme based on p-stable distributions. In: Proc. of the 20th Ann. Symp. on Computational Geometry, p. 253 (2004)

6. Paulevé, L., Jégou, H., Amsaleg, L.: Locality sensitive hashing: A comparison of hash function types and querying mechanisms 31(11), 1348–1358 (August 2010)

7. Kang, B., Jung, K.: Robust and Efficient Locality Sensitive Hashing for Nearest Neighbor Search in Large Data Sets. In: NIPS Workshop on Big Learning (BigLearn), Lake Tahoe, Nevada, pp. 1–8 (2012)

8. Tellez, E.S., Chavez, E.: On locality sensitive hashing in metric spaces. In: Proc. of the Third Int. Conf. on Similarity Search and Applications, SISAP 2010, pp. 67–74. ACM, New York (2010)

9. Zezula, P., Amato, G., Dohnal, V., Batko, M.: Similarity Search: The Metric Space Approach. Advances in Database Systems, vol. 32. Springer (2006)

10. Lv, Q., Josephson, W., Wang, Z., Charikar, M., Li, K.: Multi-probe LSH: efficient indexing for high-dimensional similarity search. In: Proc. of the 33rd Int. Conf. on Very large data bases. VLDB 2007, pp. 950–961. VLDB Endowment (2007)

11. Joly, A., Buisson, O.: A posteriori multi-probe locality sensitive hashing. In: Proc. of the 16th ACM Int. Conf. on Multimedia, MM 2008, pp. 209–218. ACM, New York (2008)

12. Novak, D., Batko, M.: Metric Index: An Efficient and Scalable Solution for Similarity Search. In: 2009 Second Int. Workshop on Similarity Search and Applications, pp. 65–73. IEEE Computer Society (August 2009)

13. Novak, D., Kyselak, M., Zezula, P.: On locality-sensitive indexing in generic metric spaces. In: Proc. of the Third Int. Conf. on Similarity Search and Applications, SISAP 2010, pp. 59–66. ACM Press, New York (2010)

14. Ostrovsky, R., Rabani, Y., Schulman, L., Swamy, C.: The Effectiveness of Lloyd-Type Methods for the k-Means Problem. In: Focs, pp. 165–176. IEEE (December 2006)

15. Arthur, D., Vassilvitskii, S.: K-means++: the advantages of careful seeding. In: Proc. of the 18th Annual ACM-SIAM Symp. on Discrete Algorithms, SODA 2007, Philadelphia, PA, USA, pp. 1027–1035 (2007)

16. Kaufman, L., Rousseeuw, P.J.: Finding Groups in Data: An Introduction to Cluster Analysis, 9th edn. Wiley-Interscience, New York (1990)

17. Paterlini, A.A., Nascimento, M.A., Junior, C.T.: Using Pivots to Speed-Up k-Medoids Clustering 2(2), 221–236 (June 2011)

18. Park, H.S., Jun, C.H.: A simple and fast algorithm for K-medoids clustering 36(2), 3336–3341 (2009)

19. Figueroa, K., Navarro, G., Chávez, E.: Metric spaces library (2007),
   http://www.sisap.org/Metric_Space_Library.html

20. Jegou, H., Tavenard, R., Douze, M., Amsaleg, L.: Searching in one billion vectors: Re-rank with source coding. In: ICASSP, pp. 861–864. IEEE (2011)

Dynamic List of Clusters in Secondary Memory

Gonzalo Navarro[1,*] and Nora Reyes[2]

[1] Center of Biotechnology and Bioengineering, Department of Computer Science,
University of Chile, Chile
gnavarro@dcc.uchile.cl
[2] Departamento de Informática, Universidad Nacional de San Luis, Argentina
nreyes@unsl.edu.ar

Abstract. We introduce a dynamic and secondary-memory-based variant of the *List of Clusters*, which is shown to be competitive with the literature, especially on higher-dimensional spaces, where it outperforms the *M-tree* in searches and I/Os used for insertions. The basic principles of our design are applicable to other secondary-memory structures.

1 Introduction

The metric space approach has become popular in recent years [2,14,16,6] and a large number of indexing methods have flourished. Most of the research, however, is still in the stage of static solutions that work in main memory. Static indexes have to be rebuilt from scracth when the set of indexed objects undergoes insertions or deletions. In-memory indexes can handle only small datasets, suffering serious performance degradations when the objects reside on disk. Most real-life database applications require indexes able to work on disk and to support insertions and deletions of objects interleaved with the queries.

To date, there exist only a few indexing structures supporting dynamism and designed for secondary memory. Some are based on so-called pivots [5,8,13], some on hierarchical clustering [3,11,12], and some on combinations [4,15].

A further challenge is that the metric spaces arising in many applications are intrinsically high-dimensional, that is, the histogram of distances is concentrated. Pivot-based indexes are known to perform well on low-dimensional spaces, whereas hierarchical clustering indexes handle medium dimensions better. A simple structure that has shown to perform well on higher-dimensional spaces is the *List of Clusters (LC)* [1], but it is a static in-memory structure. There is a dynamic version of LC, named *Recursive List of Clusters (RLC)* [9], but it is also designed to work in main memory.

In this paper we introduce a dynamic and secondary-memory variant of the List of Clusters, aiming at higher-dimensional spaces. Our secondary memory version, *DLC*, retains the good features of the *LC*, and in addition performs well on secondary memory. In this paper we focus on handling searches and insertions (thus enabling incremental construction), leaving deletions for future work

---

* Funded with Basal Funds FB0001, Conicyt, Chile.

A.J. Machado Traina et al. (Eds.): SISAP 2014, LNCS 8821, pp. 94–105, 2014.
DOI: 10.1007/978-3-319-11988-5_9 © Springer International Publishing Switzerland 2014

(these are usually handled with lazy deletion mechanisms). Our experimental comparisons show that our structures need little extra space, achieve very good disk page utilization, and are competitive with state-of-the-art alternatives. For example, compared to the *M-tree* [3], the best known alternative structure, the *DLC* is more efficient at searches. For insertions, the *DLC* performs fewer I/Os, but more distance computations. Overall, the *DLC* turns out to be a practical and easy-to-implement index that fits several practical scenarios.

2 Basic Concepts

Let $\mathbb{U}$ be a universe of *objects*, with a nonnegative *distance function* $d : \mathbb{U} \times \mathbb{U} \longrightarrow \mathbb{R}^+$ defined among them. This distance function satisfies the three axioms that make $(\mathbb{U}, d)$ a *metric space: strict positiveness, symmetry,* and *triangle inequality*. We handle a finite *dataset* $S \subseteq \mathbb{U}$, which is a subset of the universe of objects and can be preprocessed (to build an index). Later, given a new object from the universe (a *query* $q \in \mathbb{U}$), we must retrieve all similar elements found in the dataset. There are two basic kinds of queries: *range query* and *k-nearest neighbor queries*. We focus this work on range queries, where given $q \in U$ and $r > 0$, we need to retrieve all elements of S within distance r to q.

In a dynamic scenario, the set S may undergo insertions and deletions, and the index must be updated accordingly for the subsequent queries. It is also possible to start with an empty index and build it by successive insertions.

The distance is assumed to be expensive to compute. However, when we work in secondary memory, the complexity of the search must also consider the I/O time; other components such as CPU time for side computations can usually be disregarded. The I/O time is composed of the number of disk pages read and written; we call B the size of the disk page.

In terms of memory usage, one considers the extra memory required by the index on top of the data, and in the case of secondary memory, the disk page utilization, that is, the average fraction of the disk pages that is used.

3 List of Clusters

We briefly recall the list of clusters (LC) [1]. The LC splits the space into zones (or "clusters"). Each zone has a center c and a radius r_c, and it stores the *internal* objects $I = \{x \in S, \ d(x, c) \le r_c\}$, which are at distance at most r_c from c.

The construction proceeds by choosing c and r_c, computing I, and then building the rest of the list with the remaining elements, $E = S - I$. Many alternatives to select centers and radii are considered [1], finding experimentally that the best performance is achieved when the zones have a fixed number of elements m (and r_c is defined accordingly for each c), and when the next center c is selected as the element that maximizes the distance sum to the centers previously chosen. The brute force algorithm for constructing the list takes $O(n^2/m)$ time.

A range query (q, r) visits the list zone by zone. We first compute $d(q, c)$, and report c if $d(q, c) \le r$. Then, if $d(q, c) - r_c \le r$, we search exhaustively the set of internal elements I. The rest of the list is processed only if $r_c \le d(q, c) + r$.

4 Our Proposal

In this section we introduce the DLC. We base our index on the LC [1], and also use some ideas from the M-$tree$ [3]. The challenge is to maintain a disk layout that minimizes both distance computations and I/Os, and achieves a good disk page utilization.

4.1 Structure

We store the objects I of a cluster in a single disk page, so that the retrieval of the cluster incurs only one disk page read. Therefore, we use clusters of fixed size m, which is chosen according to the disk page size B.[1]

For each cluster C the index stores (1) the center object $c = center(C)$; (2) its covering radius $r_c = cr(C)$ (the maximum distance between c and any object in the cluster); (3) the number of elements in the cluster, $|I| = \#(C)$; and (4) the objects in the cluster, $I = cluster(C)$, together with the distances $d(x, c)$ for each $x \in I$. In order to reduce I/Os, we will maintain components (1), (2) and (3) in main memory, that is, one object and a few numbers per cluster. The cluster objects and their distances to the center (component (4)) will be maintained in the corresponding disk page.

Unlike in the static LC, the dynamic structure will not guarantee that I contains *all* the objects that are within distance r_c to c, but only that all the objects in I are within distance r_c to c. This makes maintenance much simpler, at the cost of having to consider, in principle, all the zones in each query.

The structure starts empty and is built by successive insertions. The first arrived element becomes the center of the first cluster, and from then on we apply a general insertion mechanism described next.

4.2 Insertions

To insert a new object x we must locate the most suitable cluster for accommodating it. The structure of the cluster might be improved by the insertion of x. Finally, if the cluster overflows upon the insertion, it must be split somehow.

Two orthogonal criteria determine which is the "most suitable" cluster. On one hand, choosing the cluster whose center is closest to x yields more compact zones, which are then less likely to be read from disk and scanned at query time. On the other hand, choosing clusters with lower disk page occupancy yields better disk usage, fewer clusters overall, and a better value for the cost of a disk page read. We consider the two following policies to choose the insertion point:

Compactness: the cluster C whose $center(C)$ is nearest to x is chosen. If there is a tie, we choose the one whose covering radius will increase the least. If there is still a tie, we choose the one with least elements.

---

[1] In some applications, the objects are large compared to disk pages, so we must relax this assumption and assume that a cluster spans a constant number of disk pages.

Occupancy: the cluster C with lowest $\#(C)$ is chosen. If there is a tie, we choose the cluster whose $center(C)$ is nearest to x, and if there is still a tie, we choose the one whose covering radius will increase the least.

As it can be noticed, to determine the cluster where the new element will be inserted it suffices with the information maintained in main memory, thus no I/Os are incurred, only distance computations between x and the cluster centers. Once the cluster C that will receive the insertion is determined, we increase $\#(C)$ in main memory and read the corresponding page from secondary memory.

Before updating the page on disk, we consider whether x would be a better center of C than $c = center(C)$: We compute $cr_x = \max\{d(x,y),\ y \in I \cup \{c\}\}$, the covering radius C would have if x were its center. If $cr_x < \max(cr(C), d(x,c))$, we set $center(C) \leftarrow x$ and $cr(C) \leftarrow cr_x$ in main memory, and write back $I \cup \{c\}$ to disk, with all the distances between elements and the (new) center recomputed. Otherwise, we leave the current $center(C)$ as is, set $cr(C) \leftarrow \max(cr(C), d(x,c))$, and write back $I \cup \{x\}$ to disk, associating distance $d(x,c)$ to x

This improvement of cluster qualities justifies our "compactness" choice of minimizing the distance $d(x, center(C))$ against, for example, choosing the center C with smallest $cr(C)$ resulting after the insertion of x: The insertion of elements into the clusters of their smallest centers will, in the long term, reduce the covering radii of the clusters.

On large databases, a sequential scan for the center most appropriate for insertion can be too expensive in terms of distance evaluations. To reduce this time, the centers stored in memory are organized in a *Dynamic Spatial Approximation Tree (DSAT)* [10], a fully-dynamic in-memory metric index that uses little extra space per element. Any change involving a center is then reflected in the DSAT. For insertions, we determine K candidate centers with a K-NN query in the DSAT and then select one of them according to the policy to choose the insertion point. We use K to be 10% of the centers.

When the cluster chosen for insertion is full, the procedure is different. We must split it into two clusters, the current one (C) and a new one (N), choose centers for both (according to a so-called "selection method") and choose which elements in the current set $\{c\} \cup cluster(C) \cup \{x\}$ stay in C and which go to N (according to a so-called "partition method"). Finally, we must update C and add N in the list of clusters (and in the DSAT) maintained in memory, and write C and N to disk. The combination of a selection and a partition method yields a *split policy*, several of which have been proposed for the *M-tree* [3].

Split Policies. The *M-tree* [3] considers various requirements for split policies: *minimum volume* refers to minimizing $cr(C)$; *minimum overlap* to minimizing the amount of overlap between two clusters (and hence the chance that a query must visit both); and *maximum balance* to minimizing the difference in number of elements. The latter is less relevant to our structure, because the LC is not a tree, but still it is important to maintain a minimum occupancy of disk pages.

The selection method may maintain the old center c and just choose a new one c' (the so-called "confirmed" strategy [3]) or it may choose two fresh centers (the "non-confirmed" strategy). The confirmed strategy reduces the splitting cost in terms of distance computations, but the non-confirmed one usually yields clusters of better quality. We use their same notation [3], adding _1 or _2 to the strategy names depending on whether the partition strategy is confirmed or not.

Random: The center(s) are chosen at random, with zero distance evaluations.
Sampling: A random sample of s objects is chosen. For each of the $\binom{s}{2}$ pairs of centers, the m elements are assigned to the closest of the two. Then, the new centers are the pair with least sum of the two covering radii. It requires $O(s^2m)$ distance computations ($O(sm)$ for the confirmed variant, where one center is always c). In our experiments we use $s = 0.1m$.
M_LB_DIST: Only for the confirmed case. The new center is the farthest one from c. As we store those distances, this requires no distance computations.
mM_RAD: Only for the non-confirmed case. It is equivalent to sampling with $s = m$, so it costs $O(m^2)$ distance computations.
M_DIST: Only for the non-confirmed case, and not used for the M-tree. It aims to choose as new centers a pair of elements whose distance approximates that of the farthest pair. It selects one random cluster element x, determines the farthest element y from x, and repeats the process from y, for a constant number of iterations or until the farthest distance does not increase. The last two elements considered are the centers. The cost of this method is $O(m)$ distance calculations.

Once the centers c and c' are choosen, the M-tree proposes two partition methods to determine the new contents of the clusters C and $C' = N$. The first yields unbalanced splits, whereas the second does not.

Hyperplane Partition: It assigns each object to its nearest center.
Balanced Partition: It starts from the current cluster elements (except the new centers) and, until assigning them all, (1) moves to C the element nearest to c, (2) moves to C' the elment nearest to C'.

A third strategy ensures a minimum occupancy fraction αm, for $0 < \alpha < 1/2$:

Mixed Partition: Use balanced partitioning for the first $2\alpha m$ elements, and then continue with hyperplane partitioning.

4.3   Range Search

Upon a search for (q, r), we determine the candidate clusters as those whose zone intersects the query ball, using the data maintained in memory. More precisely, for each C, we compute $d = d(q, center(C))$, and if $d \leq r$ we immediately report $c = center(C)$. Independently, if $d - cr(C) \leq r$, we read the cluster elements from disk and scan them. Note that, in the dynamic case, the traversal of the list cannot be stopped when $cr(C) \leq d + r$, as explained.

The scanning of a cluster also has a filtering stage: Since we store $d(x, c)$ for all $x \in cluster(C)$, we compute $d(x, q)$ explicitly only when $|d(x, q) - d(q, c)| \leq r$. Otherwise, we already know that $d(x, q) > r$ by the triangle inequality.

Finally, in order to perform a sequential pass on the disk when reading the candidate clusters, and avoid unnecessary seeks, we first sort all the candidate clusters by their disk page number before starting reading them one by one.

For lack of space we have focused on range search. Nearest neighbor search algorithms can be systematically built over range searches in an optimal way [7]. To find the k objects nearest to q, the main difference is that the set of candidate clusters must be traversed ordered by the lower-bound distances $d(q, center(C)) - cr(C)$, in order to shrink the current search radius as soon as possible, and the process stops when the currently known kth nearest neighbor is closer than the least $d(q, center(C)) - cr(C)$ value of an unexplored cluster.

5   Experimental Results

In order to give a broad picture of the performance of our index, we have selected three widely different metric spaces, all from the SISAP Metric Library (www.sisap.org). The disk page size used in this experiments is 4KB.

Words: a dictionary of 69,069 English words. The distance is the *edit distance*, that is, the minimum number of character insertions, deletions and substitutions needed to make two strings equal.

Images: 40,700 20-dimensional feature vectors, generated from NASA images, using Euclidean distance.

Histograms: 112,682 8-D color histograms (112-dimensional vectors) from an image database. Euclidean distance is used.

5.1   Search Performance

For the search experiments, we built the indexes with 90% of the elements and used the other 10% (randomly chosen) as queries. All our results are averaged over 10 index constructions using different permutations of the datasets. We have considered range queries retrieving on average 0.01%, 0.1% and 1% of the dataset. This corresponds to radii 0.605740, 0.780000 and 1.009000 for the images, and 0.051768, 0.082514 and 0.131163 for the histograms. Words have a discrete distance, so we used radii 1 to 4, which retrieved on average 0.00003%, 0.00037%, 0.00326% and 0.01757% of the dataset, respectively. The same queries were used for all the experiments on the same datasets.

For lack of space, we show the results of the best alternatives considering mainly search costs. From the point of view of searches, the best alternatives are: compactness (COMP) for the search of the insertion point, mM_RAD_2, Sampling_1 (SAMP_1), and M_LB_DIST_1 for center selection, and pure (HYPERPL) or combined with balancing (MIXED) hyperplane distribution for partitioning. As expected, the balanced partitioning obtains worse search costs than the others, because it prioritizes occupancy over compactness. The same occurs with the insertion strategy that looks for improved occupancy. Fig. 1 shows the search costs in terms of distance evaluations (1(a)) and pages read (1(b)).

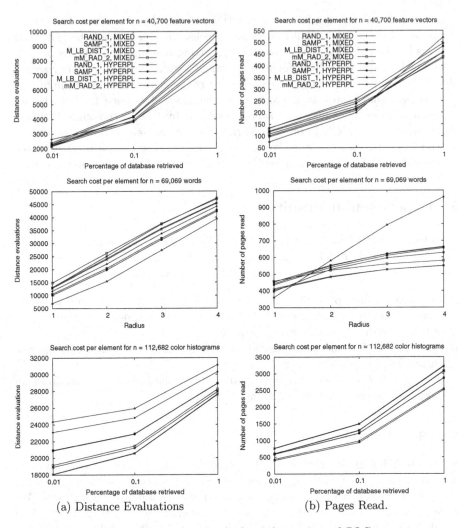

(a) Distance Evaluations   (b) Pages Read.

Fig. 1. Search costs for the best alternatives of *DLC*

As it can be seen, with respect to distance evaluations, in general better costs are obtained with hyperplane distributions. For the NASA images, the best strategy of center selection depends on the radius, but a good global alternative is M_LB_DIST_1. For Words, the best alternative of center selection for all radii is M_LB_DIST_1. If we consider the number of pages read during searches for NASA images, better costs are obtained with the two versions of M_LB_DIST_1, with the distribution that ensures a minimum occupancy of disk pages (MIXED) and with hyperplane distribution. For Words, the best results are achieved with MIXED distribution and the center selection strategies SAMP_1 and Random_1 (RAND_1). Notably, the confirmed center selection policies are better than non-confirmed ones, except for mM_RAD_2. This fact suggests that if we are not

Table 1. Average space usage for the different datasets

Dataset	Fill ratio		Total pages used		
	DLC	DSA+-tree	DLC	DSA+-tree	M-tree
Words	34%	66%	1,288	1,536	1,608
Images	54%	67%	1,431	1,726	1,973
Histograms	45%	67%	24,922	21,136	31,791

willing to spend the necessary number of distance evaluations to test all pairs of elements as centers, we should leave the old center and choose only a new one. Finally, on Histograms, the MIXED alternative is better than the HYPERPL one regarding distances, and conversely considering the number of pages read.

5.2   Comparison with Other Indexes

The *M-tree* [3] is the best-known dynamic and secondary-memory index, and its code is freely available[2]. We have used the parameter setting suggested by the authors: SPLIT_FUNCTION = G_HYPERPL, PROMOTE_PART_FUNCTION = MIN_RAD, SECONDARY_PART_FUNCTION = MIN_RAD, RADIUS_FUNCTION = LB, MIN_UTIL = 0.2.

Another suitable index is the *DSA+-tree* [11]. Its only parameter is the maximum arity, for which we use the best values reported before [11] for each metric space: 4 for all the spaces except Words, where it is 32.

There are other suitable metric indexes [5,8,13,4,15], not all of which have available code. For this conference version we compare our structure with the two indexes described above, plus the static *LC*, using the same bucket size used in *DLC*, as a reference.

Table 1 shows the average disk page occupancy achieved, considering the best search alternative for the different spaces: M_LB_DIST_1 HYPERPL for Words, SAMP_1 HYPERPL for NASA images, and mM_RAD_2 HYPERPL for Histograms. The table also shows the total number of disk pages used, compared to the M-tree and the *DSA+-tree*. Our fill ratios vary depending of the space, but they are always over 30%. Although 30% occupancy is not good, even then the *DLC* is more compact than the other indexes. We remind that, by using the MIXED partition strategy, we can guarantee a minumum disk page occupancy, if desired.

Fig. 2 compares the search costs, considering distance computations (2(a)) and pages read (2(b)). In terms of distance computations, the *DSA+-tree* always takes over as the search radius grows, even outperforming the static *LC*. A larger query range makes the problem harder, equivalently to a higher dimension. For smaller radii, however, the *DSA+-tree* or the *LC* are significantly faster. In terms of disk pages read, however, the *DLC* is significantly better than the *M-tree* and the *DSA+-tree*. Only the latter gets close for small search radii on NASA images.

---

[2] At http://www-db.deis.unibo.it/research/Mtree/

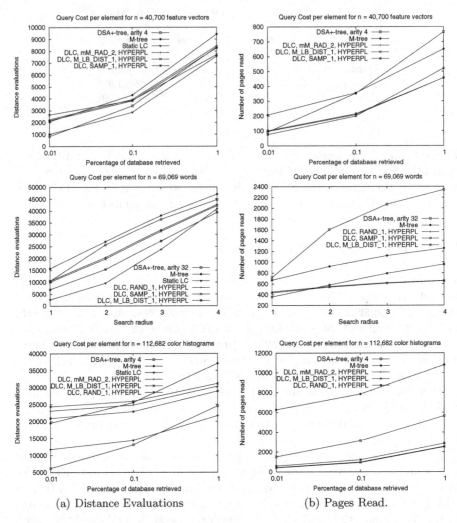

(a) Distance Evaluations   (b) Pages Read.

Fig. 2. Comparison of search costs of *DLC*, *LC*, *DSA+-tree*, and *M-tree*

5.3   Insertion Performance

Now we analyze the insertion costs of our alternatives, and compare the best ones with previous indexes. Fig. 3(a) shows the insertion cost per element as the database grows, measured in number of distance computations. All the methods have basically the same I/O cost, 1 read and 1 write per insertion, plus a very small number equal to the average number of page splits produced, which is the inverse of the average number of objects per disk page.

Fig. 4 compares our best alternatives with previous methods, both in distance computations and I/Os. In general, *DLC* pays more distance computations for insertions than the other indexes, but it outperforms them in number of I/Os.

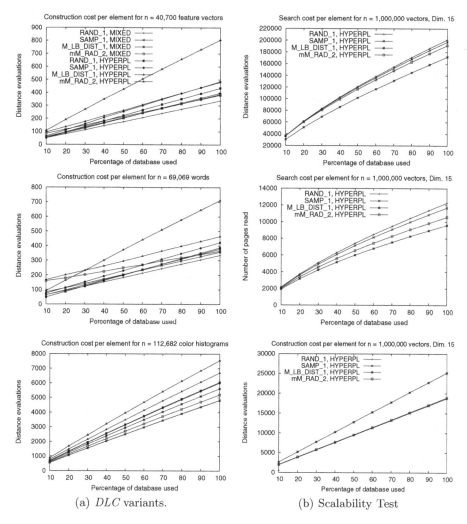

Fig. 3. Construction costs for best *DLC* alternatives (left) and scalability test (right)

5.4 Scalability

Fig. 3(b) shows the search costs in terms of distance evaluations, number of pages read, and construction costs (in terms of distance evaluations) on a larger synthetic dataset composed of 1,000,000 random vectors on dimension 15, uniformly distributed on the unitary hypercube.

The conclusions obtained for the smaller datasets are roughly maintained for this larger one. The fill ratio for the best searching strategy is over 30%. A more thorough study of the performance of the index on more massive scenarios is left for future work.

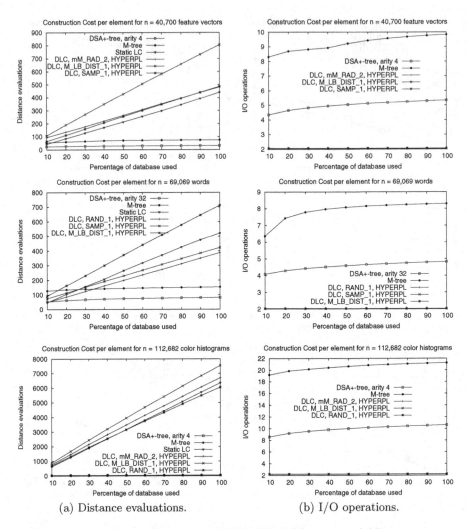

(a) Distance evaluations.   (b) I/O operations.

Fig. 4. Construction costs of *DLC*, *LC*, *DSA+-tree*, and *M-tree*

6   Conclusions

We have presented the *Dynamic List of Clusters (DLC)*, a dynamic and
secondary-memory variant of the *List of Clusters* [1], which maintains its simplicity, low space overhead, and a good search performance in high dimensions. The
DLC, in addition, supports efficient insertions and works in secondary memory.
It achieves a reasonable disk page utilization (30% to 54%) and is competitive
in both distance computations and I/Os. For the journal version we plan to add
experimental results over larger real datasets and measure the evolution of the
search performance as a function of n.

The weakest point of our structure is its high cost for insertions in terms of distance computations (whereas its number of I/Os is outstanding). We plan to study ways to optimize our idea of using an in-memory index to lower the cost of insertions. A variant of this structure can also be used to discard clusters at query time, without comparing their centers against the query.

Another important remaining work is to handle deletions, which is likely to work well with a lazy deletion mechanism that reconstructs clusters when they reach a fraction of marked elements. Adapting the original construction algorithm for the LC as a bulk-loading mechanisms for the DLC seems promising as well.

References

1. Chávez, E., Navarro, G.: A compact space decomposition for effective metric indexing. Pattern Recognition Letters 26(9), 1363–1376 (2005)
2. Chávez, E., Navarro, G., Baeza-Yates, R., Marroquín, J.: Searching in metric spaces. ACM Computing Surveys 33(3), 273–321 (2001)
3. Ciaccia, P., Patella, M., Zezula, P.: M-tree: an efficient access method for similarity search in metric spaces. In: Proc. 23rd VLDB, pp. 426–435 (1997)
4. Dohnal, V., Gennaro, C., Savino, P., Zezula, P.: D-index: Distance searching index for metric data sets. Multimedia Tools and Applications 21(1), 9–33 (2003)
5. Santos Filho, R.F., Traina, A.J.M., Traina Jr., C., Faloutsos, C.: Similarity search without tears: The OMNI family of all-purpose access methods. In: Proc. 17th ICDE, pp. 623–630 (2001)
6. Hetland, M.L.: The basic principles of metric indexing. In: Coello, C.A.C., Dehuri, S., Ghosh, S. (eds.) Swarm Intelligence for Multi-objective Problems in Data Mining. SCI, vol. 242, pp. 199–232. Springer, Heidelberg (2009)
7. Hjaltason, G., Samet, H.: Index-driven similarity search in metric spaces. ACM Transactions on Database Systems 28(4), 517–580 (2003)
8. Jagadish, H.V., Ooi, B.C., Tan, K.-L., Yu, C., Zhang, R.: iDistance: An adaptive B+-tree based indexing method for nearest neighbor search. ACM Transactions on Database Systems 30(2), 364–397 (2005)
9. Mamede, M.: Recursive lists of clusters: A dynamic data structure for range queries in metric spaces. In: Yolum, p., Güngör, T., Gürgen, F., Özturan, C. (eds.) ISCIS 2005. LNCS, vol. 3733, pp. 843–853. Springer, Heidelberg (2005)
10. Navarro, G., Reyes, N.: Dynamic spatial approximation trees. ACM Journal of Experimental Algorithmics 12:article 1.5 (2009)
11. Navarro, G., Reyes, N.: Dynamic spatial approximation trees for massive data. In: Proc. 2nd SISAP, pp. 81–88 (2009)
12. Navarro, G., Uribe, R.: Fully dynamic metric access methods based on hyperplane partitioning. Information Systems 36(4), 734–747 (2011)
13. Ruiz, G., Santoyo, F., Chávez, E., Figueroa, K., Tellez, E.S.: Extreme pivots for faster metric indexes. In: Brisaboa, N., Pedreira, O., Zezula, P. (eds.) SISAP 2013. LNCS, vol. 8199, pp. 115–126. Springer, Heidelberg (2013)
14. Samet, H.: Foundations of Multidimensional and Metric Data Structures. Morgan Kaufmann Publishers Inc. (2005)
15. Skopal, T., Pokorný, J., Snásel, V.: PM-tree: Pivoting metric tree for similarity search in multimedia databases. In: ADBIS (Local Proceedings) (2004)
16. Zezula, P., Amato, G., Dohnal, V., Batko, M.: Similarity Search: The Metric Space Approach. Advances in Database Systems, vol. 32. Springer (2006)

Index-Based R-S Similarity Joins

Spencer S. Pearson and Yasin N. Silva

Arizona State University, Glendale, AZ, USA
{sspearso,ysilva}@asu.edu

Abstract. Similarity Joins are some of the most useful and powerful data processing operations. They retrieve all the pairs of data points between different data sets that are considered similar within a certain threshold. This operation is useful in many situations, such as record linkage, data cleaning, and many other applications. An important method to implement efficient Similarity Joins is the use of indexing structures. The previous work, however, only supports self joins or requires the joint indexing of every pair of relations that participate in a Similarity Join. We present an algorithm that extends a previously proposed index-based algorithm (eD-Index) to support Similarity Joins over two relations. Our approach operates over individual indices. We evaluate the performance of this algorithm, contrast it with an alternative approach, and investigate the configuration of parameters that maximize performance. Our results show that our algorithm significantly outperforms the alternative one in terms of distance computations, and reveal interesting properties when comparing execution time.

1 Introduction

The Similarity Join (SJ) is one of the most useful and studied data processing operators. It has applications in many different situations or domains, such as multimedia applications, sensor networks, marketing analysis, and many others. Many different implementations and algorithms for SJ have been proposed, ranging from on-the-fly algorithms to index-based techniques. Index-based algorithms have the potential to significantly reduce execution time since they store pre-computed information that can be used during query execution. One such technique is the eD-Index [1]. This index enables efficient similarity-aware operations such as similarity search and Self-SJ. In this paper, we present an algorithm that significantly extends this technique to support generic SJ queries over two relations. The main contributions of our work are:

- We implemented the Range Query Similarity Join (RQ-SJ) algorithm using successive similarity search operations for the case of SJ with two relations. This technique was previously proposed in [2] for the case of Self-SJ only.
- We designed and implemented an efficient algorithm, i-SimJoin to extend eD-Index to support SJ operations over two relations using only the individual indices.
- We evaluated the performance of i-SimJoin and RQ-SJ. Our preliminary results show that i-SimJoin significantly outperforms the alternative one in terms of distance computations and interesting properties when comparing execution time.
- We explore ways to tune the eD-Index parameters to improve performance.

A.J. Machado Traina et al. (Eds.): SISAP 2014, LNCS 8821, pp. 106–112, 2014.
DOI: 10.1007/978-3-319-11988-5_10   © Springer International Publishing Switzerland 2014

The remaining part of the paper is organized as follows. Section 2 presents the related work. Section 3 gives a brief overview of the indexing structures we used in our algorithms, the RQ-SJ algorithm, and a detailed explanation of our i-SimJoin algorithm. Section 4 presents the performance evaluation of the i-SimJoin and RQ-SJ algorithms. Section 5 presents the conclusions and future work directions.

2   Related Work

Significant work has been carried out on the study of Similarity Joins. Much of this work has focused on standalone operators – both index-based and dynamic (on-the-fly) – while some has focused on implementing Similarity Join operators inside of database systems. The distance range join (retrieves all pairs whose distances are smaller than or equal to μ) is one of the most studied Similarity Join types [1,2,3,4,5,6,7,8]. This is the Similarity Join type focused on in this paper. Of the non-index-based approaches, some of the most relevant algorithms are Epsilon Grid Order (EGO) [4], Generic External Space Sweep (GESS) [5], and Quickjoin [6]. These algorithms dynamically partition and cluster the data into smaller, easier to process subsets in such a way that all similar pairs are still captured by the algorithm. The index-based approaches include such algorithms as Pass-Join [7], an algorithm proposed for string data, and the D-Index [2, 3], eD-Index [1] and List of Twin Clusters (LTC) [8], which are indices that can apply to any metric space. Pass-join [7] partitions strings into substrings and used inverted indices in order to efficiently prune dissimilar pairs. LTC [8] is an indexing approach that constructs a combined index for both datasets involved in the Similarity Join. This indexing structure consists of clusters of data points within a fixed radius of given reference points. This structure allows Similarity Join queries with a μ less than or equal to the radius of the clusters to be easily computed. An important disadvantage of this approach is the need to build joint or combined indices for every pair of datasets that can be joined. The D-Index [3] and eD-Index [1] construct an index structure based around separate buckets arranged in a hierarchical structure of levels. This index-structure allows for efficient similarity search and Self-Similarity Join queries. Our work extends on the D-Index and eD-Index, and focuses on algorithms to utilize the eD-Index functionality to efficiently perform Similarity Joins between two relations in metric spaces.

3   The i-SimJoin Algorithm

3.1   The eD-Index

The structure of the D-Index and its extension, the eD-Index, are detailed in [3] and [1] respectively. In brief, the eD-Index makes use of multiple levels where each level is organized into separable buckets and an exclusion set. The top level partitions the initial dataset. Each subsequent level after the first is created by partitioning the exclusion set of the previous level. Separable buckets are constructed by picking n pivots and a radius d from each pivot. d can be different for each pivot and is calculated when constructing the index so as to attempt to balance the number of tuples in each separable bucket. n can also vary between levels. Objects are placed in the appropriate separable bucket or the exclusion set based on their distance from the pivots and the global parameters ρ and ε, which determine the maximum query radius the eD-Index can be used to answer efficiently. Objects with a distance between $d + \rho$ and $d - \rho$ from a pivot are placed into the exclusion set. All other objects are placed into a separable bucket determined by the objects' distances from all pivots on that level. All objects with a distance of between $d \pm \rho$ and $d \pm (\rho + \varepsilon)$ from a pivot are duplicated into the exclusion set in addition to being placed in a separable bucket.

3.2   RQ-SJ: Range Query Similarity Join

The Range Query Similarity Join is an algorithm proposed in [2] for the case of Self-SJ only. As part of our work, we implemented and evaluated the performance of this algorithm for the case of SJ for two relations. This algorithm applies successive similarity search operations over the indexed dataset R, using all elements of the dataset S as the targets of the similarity searches. For each object s in S, the output is the collection of all objects in R that are within μ of s.

3.3   i-SimJoin: Index-Based Similarity Join

i-SimJoin is an algorithm for performing Similarity Join operations over two datasets indexed using the D-Index (individual indices). The indices are constructed so that they share the same index structure – that is, that the index for relation S uses the same number of levels and the same pivots for each level as relation R does. This allows the indices to be treated as the same logical index containing two separate relations while maintaining the index of each relation as a separate structure. On this logical index, we can apply an extension of a Self SJ operation such as the sliding window algorithm proposed in [1] with the added modification of awareness of which relation the tuples originally came from. This last modification ensures that only matches of pairs between both relations R and S will be returned.

To create the indices for the relations, an index is first created for relation R. The index structure generated from this is then used to create the index for relation S. This allows for the index of relation R to be used for Similarity Join queries with relation S, while still allowing the index on relation R to be independently used for other similarity-aware queries. Other approaches to create logical indices over the two relations while maintaining independent physical indices is a task for future work.

The i-SimJoin algorithm consists of two routines. First, we process the indices simultaneously, treating the corresponding buckets as a logical combined bucket as shown in Algorithm 1. Algorithm 2 is the algorithm that is run on each combined bucket. The getNextObject() function returns the next object in the combined bucket, ordered by the pre-computed distance from the object to the pivot. *upObject* and *loObject* are pointers to the current objects from each relation being compared – the

```
iSimJoin(indices, mu)
```

Input: indices (logical combined indices from relations), mu (query
radius)

Output: all the results of the Similarity Join operation R $\bowtie_{\theta\mu(r,s)}$ S

```
1 for each CombinedLevel L in indices
2  for each CombinedBucket b in L
3   b.iSimJoin_bucket(mu)
4  end for
5 end for
6 indices.exclusionSet.iSimJoin_bucket(mu)
```

Alg. 1. iSimJoin

upObject is the highest-ordered of the two objects, while the *loObject* iterates through
the current sliding window. A marking system is used to correctly slide the window
through the combined bucket.

```
iSimJoin_bucket(mu)
```

Input: mu (query radius)

Output: all the results of the Similarity Join operation R $\bowtie_{\theta\mu(r,s)}$ S
for one logical CombinedBucket

```
1  Object loObject = getNextOb-       19  else
   ject()                             20   markS = loObject
2  Object upObject = loObject          21  end if
3  while(upObject.relation ==          22  while(loObject.distance <=
   loObject.relation)                      upObject.distance)
4   upObject = getNextObject()         23   if(loObject.distance ==
5  end while                               upObject.distance && loOb-
6  if(loObject.relation == R)              ject.relation == S)
7   markR = loObject                   24    break
8   markS = upObject                   25   end if
9  else                               26   if(dist(loObject, upObject)
10  markR = upObject                       <= mu
11  markS = loObject                   27    report (loObject, upObject)
12 end if                              28   end if
13 while(upObject != NULL)             29  end while
14  while(upObject.distance -          30  upObject = getNextObject()
    loObject.distance > mu)            31  if(upObject.relation == R)
15   loObject = loObject.next()        32   loObject = markS
16  end while                          33  else
17  if(loObject.relation == R)         34   loObject = markR
18   markR = loObject                  35  end if
                                       36 end while
```

Alg. 2. iSimJoin_bucket

Lines 1-12 of the iSimJoin_bucket algorithm are the initial setup of the data structures and the marking system. Lines 14-21 advance the rear of the window as it slides through the bucket and marking the new *loObject* appropriately. Lines 22-29 report all similar matches in the current window. Additional checks are done here to prune out dissimilar pairs before performing the final distance calculation on the candidate matches. The check at line 23 ensures that no match will be added twice if the current *upObject* and *loObject* have the same pre-computed distance. Lines 30-36 advance the *upObject* to the next element in the bucket and set *loObject* to the correct marked position for its relation.

4   Performance Evaluation

We implemented the i-SimJoin algorithm as a stand-alone application written in C++. In this section, we present preliminary results comparing the i-SimJoin algorithm to the Range Query (RQ-SJ) algorithm over two relations in terms of number of distance computations and execution time.

All experiments are performed on an Intel Core I-5 1.70 GHz 4-core machine with 6GB of RAM running Linux (OpenSUSE 12.3 64-bit) as the operating system. The dataset used for this experiment is a synthetically generated, 10-dimensional vector dataset with randomly-generated values for each dimension ranging from 0 to 100. This dataset contains 100K tuples per relation, or 200K in total. The Euclidean distance function is used to calculate distances between objects.

The strategy taken for constructing each index was to choose a number of pivots for each level such that the number of objects in the largest separable bucket in that level fell within the range of 5,000 to 10,000 objects, resulting in an index structure with 3 levels and 9 pivots. The value of ρ was 0.5% of the maximum distance and the value of ε was 1.0% of the maximum distance.

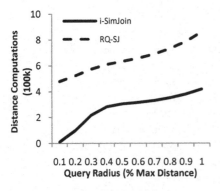

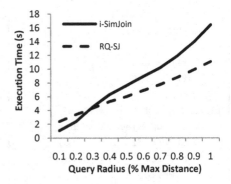

Fig. 1. Comparing Distance Computations while Increasing Query Radius

Fig. 2. Comparing Execution Time while Increasing Query Radius

Fig. 1 presents the number of distance calculations needed for each algorithm as the query radius increases. i-SimJoin requires a significantly lower amount of distance computations than the RQ-SJ algorithm, ranging from only 2.1% of the distance computations of the RQ-SJ algorithm at a query radius of 0.1% of the maximum distance, to 48.7% at a query radius of 1.0%.

Fig. 2 presents the execution time of the algorithms, with the execution time of i-SimJoin being comparable to that of the Range Query algorithm. The execution time of i-SimJoin ranges from 44.8% when the query radius is low to 148.3% of that of the Range Query algorithm. Note that i-SimJoin performs better than RQ-SJ in terms of execution time when the number of distance computations required for i-SimJoin is very low compared to that of RQ-SJ.

As part of our initial tests, we also compared the original Self-SJ algorithms proposed in [1] (range query and sliding window algorithms) and extended in this paper. We used the described 10D vector dataset and Euclidean distance function. The results were very similar to the ones reported in this paper for the case of SJ over two relations, i.e., while the sliding window technique performs significantly less distance computations, both algorithms have similar execution times. Note that the work in [1] only reports the number of distance computations and not the execution times.

The comparison of these approaches contrasting both distance computations and execution time is actually quite revealing. These results highlight the fact that the overhead required in processing these algorithms is a significant factor in the execution time. For instance, the i-SimJoin algorithm performs many tests to prune out pairs that are not in the result set. Although these checks do not necessarily involve distance computations, they still contribute to the processing that needs to be done. Since the results reported in this paper were obtained from a 10-dimensional dataset, the distance computations involved were not highly expensive. More complicated distance functions over more complex data types can significantly increase the complexity of the distance computations, and this would be expected to result in the number of distance computations being more significant in terms of the execution time. While i-SimJoin is expected to outperform RQ-SJ for complex data types and distance functions, it is also important to observe that the simple RQ-SJ algorithm can be the most efficient approach for simple data types and distance functions.

5   Conclusions

This paper presents i-SimJoin, an algorithm to perform Similarity Joins on two relations using physically independent indexing structures. Our performance evaluation shows that i-SimJoin requires far fewer distance calculations than an alternative SJ algorithm, and has a comparable execution time. Our future work will include: (1) extensive performance evaluations of more complex data types and distance functions to investigate how this affects the execution times of i-SimJoin and RQ-SJ, and (2) generalization of the i-SimJoin algorithm to the case of multiple SJ predicates.

References

1. Dohnal, V., Gennaro, C., Rabitti, F., Zezula, P.: Similarity join in metric spaces. In: Sebastiani, F. (ed.) ECIR 2003. LNCS, vol. 2633, pp. 452–467. Springer, Heidelberg (2003)
2. Dohnal, V., Gennaro, C., Rabitti, F., Zezula, P.: Similarity join in metric spaces. In: Sebastiani, F. (ed.) ECIR 2003. LNCS, vol. 2633, pp. 452–467. Springer, Heidelberg (2003)
3. Dohnal, V., Gennaro, C., Savino, P., Zezula, P.: D-Index: Distance searching index for metric data sets. Multimeda Tools and Applications 21, 9–33 (2003)
4. Böhm, C., Braunmüller, B., Krebs, F., Kriegel, H.-P.: Epsilon grid order: An algorithm for the similarity join on massive high-dimensional data. In: Proceedings of the 2001 ACM SIGMOD International Conference on Management of Data, SIGMOD 2001, pp. 379–388. ACM, New York (2001)
5. Dittrich, J.-P., Seeger, B.: Gess: A scalable similarity-join algorithm for mining large data sets in high-dimensional spaces. In: Proceedings of the 7th ACM SIGKDD International Conference on Knowledge Discovery and Data Mining, KDD 2001, pp. 47–56. ACM, New York (2001)
6. Jacox, E.H., Samet, H.: Metric space similarity joins. ACM Trans. Database Syst. 33, 7:1–7:38 (2008)
7. Li, G., Deng, D., Wang, J., Feng, J.: Pass-join: a partition-based method for similarity joins. Proc. VLDB Endow. 5(3), 253–264 (2011)
8. Paredes, R., Reyes, N.: Solving similarity joins and range queries in metric spaces with the list of twin clusters. J. of Discrete Algorithms 7, 18–35 (2009)

A Compressed Index for Hamming Distances

Francisco Santoyo[1], Edgar Chávez[2], and Eric S. Téllez[3]

[1] School of Electrical Engineering
Universidad Michoacana, México
psantoyo@dep.fie.umich.mx
[2] Department of Computer Science
CICESE, México
elchavez@cicese.mx
[3] INFOTEC
Aguascalientes, México
eric.tellez@infotec.com.mx

Abstract. Some instances of multimedia data can be represented as high dimensional binary vectors under the hamming distance. The standard index used to handle queries is Locality Sensitive Hashing (LSH), reducing approximate queries to a set of exact searches. When the queries are not selective and multiple families of hashing functions are employed, or when the collection is large, LSH indexes should be stored in secondary memory, slowing down the query time.

In this paper we present a compressed LSH index, queryable without decompression and with negligible impact in query speed. This compressed representation enables larger collections to be handled in main memory with the corresponding speedup with respect to fetching data from secondary memory.

We tested the index with a real world example, indexing songs to detect near duplicates. Songs are represented using an entropy based audio-fingerprint (AFP), of independent interest.

The combination of compressed LSH and the AFP enables the retrieval of lossy compressed audio with near perfect recall at bit-rates as low as 32 kbps, packing the representation of 30+ million music tracks of standard length (which is about the total number of unique tracks of music available worldwide) in half a gigabyte of space. A sequential search for matches would take about 15 minutes; while using our compressed index, of size roughly one gigabyte, searching for a song would take a fraction of a second.

Keywords: Audio indexing, Succinct Audio-Fingerprint, Succinct LSH Indexes.

1 Introduction

High dimensional binary vectors under the hamming distance can represent many interesting objects for applications. The standard index used to handle queries in this setup is Locality Sensitive Hashing (LSH), reducing approximate queries to a set of exact searches. When the queries are not selective and multiple families of hashing functions are employed, or when the collection is large, LSH indexes should be stored in secondary memory, slowing down the query time.

A.J. Machado Traina et al. (Eds.): SISAP 2014, LNCS 8821, pp. 113–126, 2014.
DOI: 10.1007/978-3-319-11988-5_11 © Springer International Publishing Switzerland 2014

Compressing the data is an option to avoid overflow to secondary memory as long as the compressed representation is usable without decompressing. In this paper we present a compressed LSH index, which can be queried without decompression and with negligible impact in the query speed. This defer the use of secondary memory for larger collections, implying a non-trivial speedup with respect to a secondary memory index.

We performed a real world test for our algorithms. We selected the problem of indexing music tracks. The total amount of music tracks worldwide is in the order of 30 million. The *Apple iTunes* music-store lists less than 30 million songs in its catalog. Other on-line music-stores like *Amazon MP3* offer a 22 million song catalog to choose from and *Deezer* or *Spotify* just advertise more than 30 million songs. With the increasing number of records, the creation of high performance music search algorithms becomes a basic requirement for any music on demand application. The industry should respond with systems being able to discover, navigate, and recommend music. One basic tool for music retrieval is the simple matching of a track in a collection.

The task of matching whole songs in audio collections has been tackled by fingerprinting the audio, and then comparing the corresponding fingerprints. This method serves multiple purposes, on the one hand the fingerprinting procedure masks subtle differences between audio objects and conflates near duplicates. On the other hand, having a succinct representation of the audio avoids a lengthy comparison in the original domain.

While audio fingerprints (AFP) can be made very robust to ambient noise and other severe degradations, there is a tension between robustness and the memory footprint of the representation [1]. Other commercial approaches use a time-frequency representation of the audio (as in [2] and [3]) with limitations in both, the processing power required to obtain the AFP, and the type of index to be engineered to obtain fast answers.

We did focus on whole-song identification with the only expected degradation transcoding (e.g. lossy compression), which induces very mild distortions to the songs. We will also assume that both the query song and the song in the database have the same length and that they are correctly aligned. This is the case, for example, of an audio labeling service; where the user rips the audio from a CD and wants it automatically labeled.

Our second contribution consists in a lightweight AFP using just a few bits per minute of audio (precisely one bit every two seconds). Every song will produce a string of bits of the same size (the strings are cyclically completed to a fixed size). To compare two songs we use the hamming distance between the corresponding AFPs. With this procedure near duplicates are conflated (they have small Hamming distance) and non corresponding songs have large distances. These two facts allow extremely fast searches with no false positives. The unique combination of speed, precision and small memory footprint is unparalleled in the literature. We can pack about 300 million minutes of audio in about one gigabyte, and query a database of this size in a fraction of a second.

2   Related Work

A variant of the classical KD-tree algorithm which efficiently indexes high-dimensional data by recursive spatial partitioning is presented by McFee and Lanckriet [4]. They perform experiments on the One Million Song Dataset [5] to demonstrate that content-based similarity search can be significantly accelerated by the use of spatial partitioning structures. However, KD-tree suffers (as any exact spatial and metric index) of the so called *curse of dimensionality* (Samet [6] and Chavez et al. [7]); as any spatial method working with the explicit dimensionality, it becomes suboptimal on high dimensional datasets [7].

The interested reader on a more general point of view of Music Information Retrieval is referred to the surveying works of Lu [8], which provides a comprehensive survey of audio indexing and retrieval techniques; Stober and Nürnberger [9] present a structured view on the last decade of Music Information Retrieval research; and Yan et al. [10] present a rich review of large-scale multimedia analysis techniques.

3   Computing the Fingerprint

Our approach is derived from [11]. The signal is framed and for each frame we measure the information content, directly in the time domain. The Information content or self information $I(p_i)$ of a value v_i, depends only on its probability $p_i = P(v_i)$ to occur, the less likely a value to appear, the more information it will bring when it shows up. Therefore, the self information must be a monotonically decreasing function of the probability, usually it is defined as $I(p_i) = \ln(\frac{1}{p_i}) = -\ln(p_i)$

Let $X = \{x_1, x_2, \cdots, x_n\}$ a sequence of values, with f_i denoting the frequency of x_i. The entropy $H(X)$ is the average of all the information contents weighted by their probabilities to occur

$$H(X) = \sum_{i=1}^{m} \frac{f_i}{n} \log(\frac{x_i}{n}) = -\sum_{i=1}^{n} p_i \log(p_i)$$

Transcoded versions of the same song will differ in the amount of information packed, producing a vertical shift. To avoid this shift we keep only the sign of the derivative,

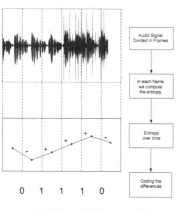

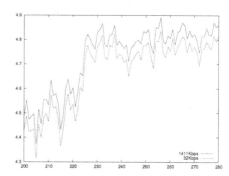

(b) The entropy curve for the song *Chilanga Banda - Café Tacuba* for versions @1411Kbps and @32Kbps. The vertical axis shows the magnitude of the entropy as a function of time (horizontal axis).

(a) The audio fingerprinting process

Fig. 1. The process for obtaining the fingerprint and an example of two versions of the same song

which can be computed measuring only the relative change between frames, with 1 if the change is positive and 0 otherwise, see Figure (1a). The frames can be overlapped to smooth the changes in the sequence, because transcoding induces a time shift. In other words, two versions of a song will be almost aligned. The frame overlapping smooths the time shift, however this increases the size of the fingerprint and also modifies the distance distribution between songs. The optimal frame size and overlap amount can be found experimentally, the optimization goal is to minimize the distance between near duplicates while maximizing the distance between unrelated songs. Figure 1a illustrates the process of obtaining the fingerprint of a song. One of the nice characteristics of this approach is the trivial parallelization in obtaining the fingerprint, fitting well in modern hardware. Furthermore, the operations needed are simple enough to be implemented in low-end processors, such as mobile devices.

We computed the AFP with frames of sizes half, 1 and 2 seconds with overlaps of 0, 50, 75, 90 and 95%. The experiments were performed in a sample of 4000 songs with mp3 encoding at different bit-rates mp3@$\{128, 96, 64, 32\}$ Kbps, also, we denote the original as wav@1411Kbps. We observed two things in this experiment, the first is that overlap increases the distance between near duplicates; the minimum distance (and variance) is obtained with no overlap. The second observation was that the frame size was not critical, changes in the distance were not significant. Due to the later fact, we selected a 2 second frame with no overlap because it gives the smallest memory footprint.

Table 1 shows both the average distance and the standard deviation matrix for a 2-second frame with no overlap. For near duplicates the average distance goes from 1.8% to 4.2% with a small variance.

Table 1. Average normalized distances from one song to all its versions. (μ, ρ)

	@1411Kbps	@128Kbps	@96Kbps	@64Kbps	@32Kbps
@1411Kbps	(0, 0)	(0.018, 0.076)	(0.022, 0.076)	(0.026, 0.076)	(0.054, 0.075)
@128Kbps		(0, 0)	(0.008, 0.010)	(0.012, 0.013)	(0.042, 0.028)
@96Kbps			(0, 0)	(0.013, 0.013)	(0.042, 0.028)
@64Kbps				(0, 0)	(0.042, 0.029)
@32Kbps					(0, 0)

Complementarily, for the same setup, the average distance between unrelated songs is one order of magnitude larger as shown in Table 2. The average distance is 42.7%, with standard deviation depending on the bit-rate.

Table 2. Average normalized distances from one song to all the other songs. (μ, ρ)

	@1411Kbps	@128Kbps	@96Kbps	@64Kbps	@32Kbps
@1411Kbps	(0.426, 0.061)	(0.427, 0.061)	(0.427, 0.061)	(0.427, 0.060)	(0.429, 0.058)
@128Kbps		(0.427, 0.061)	(0.427, 0.061)	(0.427, 0.060)	(0.429, 0.058)
@96Kbps			(0.427, 0.061)	(0.427, 0.060)	(0.429, 0.058)
@64Kbps				(0.427, 0.060)	(0.429, 0.058)
@32Kbps					(0.429, 0.058)

Since we do not have overlap, frames are independent of each other which gives us the capability to parallelize the fingerprint computation.

We ended up with a succinct representation of a song, fast to compute, and with nice conflation properties. Near duplicates are one order of magnitude closer than unrelated songs, respectively shown in Tables 1 and 2. This fact avoids the retrieving of false positives when querying by content, as discussed in the next section.

4   Matching Songs

Since we are using the Hamming distance with the AFP, it is natural to use *locality sensitive hashing* (LSH) (Gionis et al. [12]) as the base of our index.

LSH is a fast approximate proximity searching technique giving probabilistic guarantees on the quality of the result. The general idea of an LSH index is to find hashing functions that applied to close objects give the same bucket with high probability. This technique is prone to two types of errors, namely: 1) False positives, when two non related objects fall in the same bucket, and 2) False negatives, when two near duplicates end in a different bucket. Those errors can be alleviated by using more than one LSH function.

In general, the process of finding hashing functions g_i can be tricky; however Hamming spaces are the most studied and hash functions are very simple, they are just random samples of the bit strings.

Definition 1 (Locality Sensitive Hashing, Gionis et al. [12]). *A family of hashing functions* $\mathcal{H} = \{g_1, g_2, \cdots, g_h\}$, $g_i : U \to \{0,1\}$ *is called* (p_1, p_2, r_1, r_2)-*sensitive, if for any* p, q:

— *If* $d(p, q) < r_1$ *then* $Pr[hash(p) = hash(q)] > p_1$
— *If* $d(p, q) > r_2$ *then* $Pr[hash(p) = hash(q)] < p_2$

Where $hash(u)$ *is the concatenation of the output of individual hashing functions* g_i, *following a fixed order, i.e.* $hash(u) = g_1(u)g_2(u) \cdots g_h(u)$.

Let d_{max} be the maximum possible distance between objects in the metric space; the probability that some g_i computes the same hash for $u, v \in U$ is determined as $Pr[g_i(u) = g_i(v)] = 1 - d(u, v)/d_{max}$.

If hashing functions are selected independently, with replacement, and equally probably to fail, we obtain $Pr[hash(u) = hash(q)] = 1 - (d(u, v)/d_{max})^h$. In order to have a sound LSH scheme, the above formula should comply with definition 1.

Other data models and distance functions, like vectors measured with Minkowski norms or sets with Jaccard distances are also studied in the literature, Gionis et al. [12], and Andoni & Indyk [13].

4.1   Normalizing AFP Size

Our AFP is a bit string of variable length, proportional to the length of the song. The LSH based index needs a fixed size representation. Hence each fingerprint is *conceptually* expanded to be of the size of the largest fingerprint on our database, or any fixed large size if that length is unknown beforehand. Let ℓ_{max} be such length, and ℓ_i be the length of the i-th fingerprint s_i. Let $s[j]$ be the j-th bit in the fingerprint s.

Our database S of size n is denoted as $S = \{\hat{s}_1, \hat{s}_2, \cdots, \hat{s}_n\}$, where $\hat{s}_i[j] = s_i[1 + (j \mod \ell_i)]$ for all $1 \leq j \leq \ell_{max}$. The same method is applied to obtain any valid object (e.g. queries $q \in U$), using ℓ_{max} obtained from S. The distance between any two database objects is D (Hamming distance).

A single index would be enough if the length of the objects are not very different, all of them will have the same length using the above normalization. However in a database with disparate lengths, the normalization would be unfair since LSH captures (relatively) very large hashes for smaller fingerprints. In this case the database should be partitioned into several sets, and indexed separately.

4.2   Indexing the AFP Database

We have seen before that all versions of the same song are quite close to each other as shown in Table 1, and complementarily unrelated songs are distant from each other as shown in Table 2. There is a balance between the size of the sample in an LSH index, the implicit searching radius, and the time to retrieve the near duplicates. We cascade a set of indexes of non decreasing selectivity, all of them using different samplings. We apply the hashing in order until reaching the last index, and give up when we reach it.

We compute the minimum and maximum sampling sizes using Table 1 and the size of the AFPs. We can use a single parameter α (where $0 \leq \alpha \leq 1$) establishing what is the selectivity of each index in the cascade, i.e, anything with larger distance than α will be discarded. The maximum α would be fixed to 0.15 because this will capture most of the near duplicates.

The average searching cost will be very low since most queries are solved using only the first indexes, only a fraction of the queries would require all the indexes. The searching steps are described in Algorithm 1.

Algorithm 1: Near-duplicate searching of songs

Input: The query song Q, the index set $\mathcal{L}$ corresponding to the size of Q, the maximum distance α.
Output: The set R of near-duplicate objects of Q.

1: Process Q and obtain its fingerprint q
2: Normalize the audio-fingerprint as the object $\hat{q}$
3: Initialize $R \leftarrow \emptyset$
4: **for all** $I \in \mathcal{L}$ **do**
5:     Lookup I to match similar objects $\hat{q}$, put candidates in C
6:     Remove from C objects not matching the length of q
7:     $R \leftarrow R \cup \{\hat{u} \in C \mid d(q, u) \leq \alpha\}$.
8:     **Stop** the iteration if $|R| \geq 1$ (or the minimum desired cardinality)
9: **end for**

---

5 Compressing LSH

To reduce false positives and false negatives multiple hashing functions are needed; thus, the search index will require more memory and probably will resort to secondary memory for large instances. Our goal is to produce a representation of the LSH index with close to optimal storage.

The general idea is to represent the hashing tables as inverted indexes, and in turn to represent those inverted indexes as an indexed sequence. This type of representation can be compressed and can be queried without decompression.

Let $T = s_1 s_2 \cdots s_n$ be a sequence of symbols on the alphabet Σ of size σ, i.e. $s_i \in \Sigma$. Without loss of generality, let Σ be a set of integers, that is $\Sigma = \{1, 2, \cdots, \sigma\}$. The i-th symbol in T is denoted as T_i.

An index of sequences (IoS) provides three basic operations:

- $\mathsf{Rank}_c(T, pos)$ counts how many c's occurs in T until pos, $c \in \Sigma$.
- $\mathsf{Select}_c(T, r)$ returns the smaller position pos such that $\mathsf{Rank}_c(T, pos) = r$.
- $\mathsf{Access}(T, pos)$ retrieves the symbol stored at the position pos in T, i.e., T_{pos}.

Notice that an IoS replaces T, since we can reconstruct it using Access, but our notation requires to put T in the arguments even when it is not necessarily stored.

5.1 A Brief Survey for Indexes of Sequences

There are several indexes achieving near optimal space bounds, we briefly review some of them. We start establishing the memory costs for any representation.

Memory usage. Let n_c be the number of symbols c in T, then from information theory we can obtain the following formulation, using a fixed code word for each symbol, we require at least $nH_0(T) \leq n \log \sigma$ bits, here, $H_0(T)$ is the order zero empirical entropy of T, i.e., $nH_0(T) = n \sum_{c \in \Sigma} p_c \log \frac{1}{p_c} = \sum_{c \in \Sigma} n_c \log \frac{n_c}{n}$ bits. Here p_c is the probability of occurrence of c in T, empirically, $p_c = n_c/n$.

Binary alphabets (Bitmaps). Most IoS use as building block the binary case, an alphabet of two symbols $c \in \{0, 1\}$ without loss of generality. Consider a bitmap B with n bits, let n_0 and n_1 the number of 0's and 1's respectively in the bitmap.

Gonzalez et al. [14] developed a fast practical approach. It consists on a directory structure of absolute Rank_c samples every $\log^2 n$ bits. This structure solves Rank_c and Select_c in $O(\log n)$ time, and Access in constant time. It stores the plain bitmap using n bits and used $o(n)$ bits to store the absolute samples.

Several indexes achieve near-optimal space for binary alphabets. For example, Raman et al. [15] based on classifying bit blocks and then codifying blocks using tuples (c_i, offset) where c_i describes the class of the block (the number of bits set to 1) and an offset to distinguish a block inside the class c_i. These tuples are cleverly codified such that classes with few members will produce smaller offset's codes. This approach uses $nH_0(B) + O(\log \log n) + o(n)$ bits and solves the three basic operations in constant time; however, the constants are too large in practice. Claude and Navarro [16] improved the practical, sample based implementation introduced by Gonzales in [14], achieving better performance in practice; however, the space space complexity is similar and can be a waste of resources when $n_1 \ll n$. On the other side, Okanohara and Sadakane [17] presented the *sparse array* (SArray) which achieves $n_1 \log n/n_1 + O(n_1)$ space with $O(1)$ time for Select_1.

In addition to the above, there exists specialized indexes achieving near optimal space. One example is presented in Tellez et al. [18,19] they key idea consist in storing differences with variable length integer codifications along the necessary directory structures to accelerate operations. Tellez introduced Diffset, which is basically the representation of the bitmap as a compressed sorted list with directory structures to provide fast Rank_c, Select_c, and Access performances. Diffset achieves $nH_0(B) + o(n)$ bits. Similarly, Diffset-RL is defined adding run-length compression for large consecutive runs of ones; depending of the distribution it could produce much better compression and times or add n bits in the worst case.

Larger alphabets. For $\sigma > 2$ there are several canonical techniques to index a sequence T of length n as described below.

Grossi et al. [20] introduce the Wavelet Tree (WT), it uses $n \log \sigma + O(\sigma \log n)$ bits solving all operations in $O(\log \sigma)$ time. There exists several variants of the WT. For example, the WT with Huffman shape or with internal bitmaps compressed to nH_0, like surveyed by Navarro and Mäkinen [21]. Very large alphabets are problematic with this scheme since the time complexity of all operations depend on σ.

Golinsky et al. [22] introduce a fast index, robust to large σ. It uses $n \log \sigma + o(n \log \sigma)$ bits, it solves Select_c in constant time, and both Rank_c and Access on $O(\log \log \sigma)$ time. Claude and Navarro [16] show an implementation of Rank_c and Access on $O(\log \sigma)$ time performing better in practice for most instances.

Tellez [18] introduces the Extra Large Bitmap (XLB) family of indexes for large alphabets achieving both compression and fast operations, specially on sequences with low local entropy. The main idea is to codify a sequence using a permutation of $[1 \ldots n]$; the trick is to store the inverse in $o(n)$ bits extra, while the direct is represented with a large bitmap that takes advantage of the sparseness of the resulting bitmap. The sequence $T = T_1 T_2 \cdots T_n$ is represented with a bitmap $P[1, \sigma n]$ where the i-th bit is 1 if $T_{i \bmod \sigma} = \frac{i}{\sigma}$, and 0 otherwise. Then, P is a large bitmap, with regions of length n corresponding to each symbol. The basic algorithm solves Rank_c and Select_c on T performing Rank_1 and Select_1 on P. Access is solved using Π^{-1} where $\Pi(i) = \mathsf{Select}_1(P, i) \bmod n$. Also, Π^{-1} is stored with the cyclic representation of Munro et al. [23] using $\frac{1}{t} \log n$ bits; it solves Π^{-1} in t time (Select_1 operations on P), where $t \geq 1$. Since all operations are delegated to the P bitmap, the efficiency is tightly linked to P. Since we need to represent a very large bitmap of $n\sigma$ bits with n bits set to 1, then we need an underlying bitmap taking advantage of the sparseness of the represented bitmap.

5.2   The Sequence Representation of LSH

Consider the database $S \subseteq U$, $S = \{u_1, u_2, \cdots, u_n\}$, and a family of hashing functions $\mathcal{H} = \{g_1, g_2, \cdots, g_h\}$, where $h = |\mathcal{H}|$ and $g_i : U \to \{0, 1\}$. A tag of an object is defined as $\mathsf{tag}(u) = g_1(u)g_2(u)\cdots g_h(u)$. The set of all possible values of $\mathsf{tag}(\cdot)$ is called the alphabet, $\Sigma = \{0, 1, 2, \cdots, \sigma - 1\}$, where $\sigma = |\Sigma| \leq 2^h$. Even when $\mathsf{tag}(u) = \mathsf{hash}(u)$, conceptually tag is an atomic item (indivisible and recognized as a unit), and defines a sequence's symbol. Let us define $T = \mathsf{tag}(u_1)\,\mathsf{tag}(u_2)\cdots\mathsf{tag}(u_n)$. We can store T using $\log \binom{n}{n_1, n_2, \cdots, n_\sigma}$ bits, where n_i is the number of occurrences of the tag i in T.

Recall that high quality results with LSH require several LSH tables, which increase the memory cost and hence the need of a memory efficient representation. The alphabet derived from the LSH representation is large. One option of index is WT, described in the previous section, but the operations for the simulation of LSH make heavy use of the Select_c operation. The performance of WT and most of its variants is poor for our needs. Hence we focus on the Golynski and the XLB approaches. In particular, we use XLB with SArray, Diffset and Diffset-RL; XML-SArray will use $n \log \sigma + o(n)$ bits, while XML-Diffset and XML-DiffsetRL can achieve better compression under particular entries with low local entropy; however, the latter two will introduce a minor term of $O(n \log \log n)$ bits which can impact on sequences with high local entropy since they will be added to the resulting worst case.

5.3 Solving Approximate Nearest Neighbors with T

The abstract data structure for LSH needs access to the buckets. To solve a query, the structure needs to count the number of items in a bucket, and retrieve all items on it. Figure 2b shows a hash table of an example database of 16 objects. Each row is a bucket, represented by some hash value. Figure 2a shows the sequence T of the hash table.

As an abstract data structure T solves similarity queries using the same proximity properties than LSH tables. Algorithm 2 solves the approximate $\mathsf{nn}_{d,S,U}(q)$ queries. The idea is to retrieve all items using $\mathsf{Select}_{\mathsf{tag}(q)}$.

LSH is essentially an indexed table, we can emulated its functionality as follows. i) The number of items with the same hash c is computed with $\mathsf{Rank}_c(T, n)$ (Figure 2c); ii) all items with the same tag c are retrieved as $\mathsf{Select}_c(T, i)$ for $i = 1, 2, \cdots,$ $\mathsf{Rank}_c(T, n)$.

Algorithm 2: Searching for the approximate $\mathsf{nn}_{d,S,U}(q)$

Input: The query q, the distance function d, and T.
Output: The approximate nearest neighbor $\mathsf{nn}^*(q)$

1: Let $c = \mathsf{tag}(q)$
2: Let $\mathsf{nn}^*(q) \leftarrow \mathsf{undefined}$
3: **for** $i = 1$ to $\mathsf{Rank}_c(T, n)$ **do**
4:     Define p as $\mathsf{Select}_c(T, i)$-th object in T
5:     $\mathsf{nn}^*(q) \leftarrow p$ **if** $\mathsf{nn}^*(q)$ is $\mathsf{undefined}$ or p is closer to q than the previous $\mathsf{nn}^*(q)$
6: **end for**

---

i   1 2 3 4 5 6 7 8 9 10 11 12 13 14 15 16
tag(u_i) 8 7 4 6 2 1 3 8 8 4 0 1 1 1 5 9

(a) The sequence T representing the LSH table of Figure 2b

hash tag	occurrences list
0000 0	$\rightarrow$ 11
0001 1	$\rightarrow$ 6, 12, 13, 14
0010 2	$\rightarrow$ 5
0011 3	$\rightarrow$ 7
0100 4	$\rightarrow$ 3, 10
0101 5	$\rightarrow$ 15
0110 6	$\rightarrow$ 4
0111 7	$\rightarrow$ 2
1000 8	$\rightarrow$ 1, 8, 9
1001 9	$\rightarrow$ 16

(b) An example of the LSH hash table representation

	$\text{Select}_c(T, i)$			
	1	2	3	4
$\text{Rank}_0(T, n) = 1$	11			
$\text{Rank}_1(T, n) = 4$	6	12	13	14
$\text{Rank}_2(T, n) = 1$	5			
$\text{Rank}_3(T, n) = 1$	7			
$\text{Rank}_4(T, n) = 2$	3	10		
$\text{Rank}_5(T, n) = 1$	15			
$\text{Rank}_6(T, n) = 1$	4			
$\text{Rank}_7(T, n) = 1$	2			
$\text{Rank}_8(T, n) = 3$	1	8	9	
$\text{Rank}_9(T, n) = 1$	16			

(c) Reconstructing the LSH table

Fig. 2. An example of the LSH sequence representation LSH, and its operations

6   Experimental Results

All experiments were performed in a 16 core Intel Xeon 2.40 GHz workstation with 32GiB of RAM, running CentOS. All tasks were restricted to run into a single core, we did not exploited the parallel capabilities of our workstation.

6.1   Case of Study

For the real world example, we collected 3.7 million songs (about 1.5 Tb) and finger-printed them with the techniques described, using 1 bit every two seconds of music. A collection of this size is necessarily diverse. Our database of fingerprints requires 54 MiB (15 bytes per song), that is always maintained in main memory. Proximity between fingerprints is measured with the Hamming distance. Each fingerprint requires 0.31 seconds in average to be computed (reading PCM files with 16 bits per sample, and 44100 Hz, 1411kbps). We randomly selected 400 songs from the songs database and compressed them to @128kbps, @96kbps, @64kbps, and @32kbps. These versions of the song are similar to the versions in personal music libraries.

Table 3 contains the average size of the LSH indexes for our setups (over our 3.7 million song audio-fingerprint database). It is interesting to notice that the compression ratio (smaller is better) decreases as h does. Using the compact sequence representation of LSH we expect to use from 54% to 77% of the original space of an LSH index. This improvement is important because we are trying to maintain our data structures in the higher places of the memory hierarchy, asymptotically obtaining faster indexes.

Table 3. Average memory requirements

method	h		
	10	15	20
LSH	14.0 MiB	14.3 MiB	22.0 MiB
compressed LSH	7.6 MiB	9.9 MiB	17.0 MiB
compression ratio	0.54	0.69	0.77

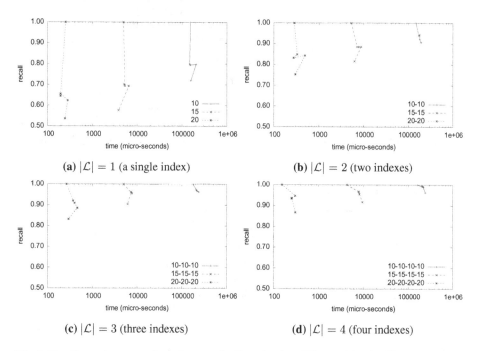

(a) $|\mathcal{L}| = 1$ (a single index) **(b)** $|\mathcal{L}| = 2$ (two indexes)

(c) $|\mathcal{L}| = 3$ (three indexes) **(d)** $|\mathcal{L}| = 4$ (four indexes)

Fig. 3. Recall vs. time to retrieve the near-duplicated using different LSH families and several indexes. The points in the curves are generated searching for different versions of the songs, i.e., perfect recall is achieved for 1411kbps, and from it each point matches with versions with decreasing quality 128, 96, 64, and 32 kbps.

The *recall* increases as h decreases. However, as h decreases we expect higher searching times since the database is partitioned in fewer buckets. The searching and recall compromise is shown in Figure 3. For instance, Figure 3a shows the performance of a single index. Decreasing h increases the searching time exponentially, while the recall is only moderately improved. An option to improve the performance is using more than one index, a set of them $\mathcal{L}$ (as described in Algorithm 1). Using two indexes, Figure 3b, performs better than reducing h, at the cost of using twice the memory. The improvements with three and four indexes, Figures 3c and 3d respectively, are more notorious. Note that the average searching time is smaller than just multiplying the searching time for the number of indexes, this is because we only advance to the next index if the

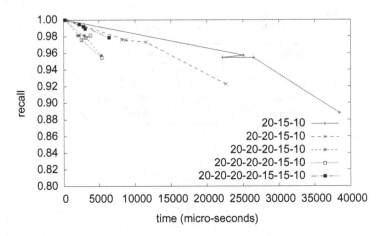

Fig. 4. Our mixed setups that optimize both recall and time performances. The points in the curves are generated searching for different versions of the songs, i.e., perfect recall is achieved for 1411kbps, and from it each point matches with versions with decreasing quality 128, 96, 64, and 32 kbps.

Table 4. Average memory requirements for our mixed setups

setup	memory		
	compressed LSH	LSH	compression ratio
20, 15, 10	34.5 MiB	50.3 MiB	0.686
20, 20, 15, 10	51.5 MiB	72.3 MiB	0.712
20, 20, 20, 15, 10	68.5 MiB	94.3 MiB	0.726
20, 20, 20, 20, 15, 10	85.0 MiB	116.3 MiB	0.735
20, 20, 20, 20, 15, 15, 10	95.4 MiB	130.6 MiB	0.730

current one fails to retrieve the near-duplicates (Algorithm 1). It is clear that the better time-recall tradeoff is found for several indexes with large h values, also the memory cost is reduced using the compressed LSH index (Table 3).

As seen on Figure 3d, we obtain at least 85% of recall for all versions when $h = 20$. However we always obtain at least 90% for $h = 15$, and 95% for $h = 10$. Based on the above facts, we tuned the strategy improving the recall with a small impact in both average searching time and memory cost. The idea is to filter by solving most queries very fast, with only a few queries passing the filter and using a more expensive procedure. This is illustrated in Figure 4. Here we can see that most configurations perform better than 95% of recall with a moderate searching time. For example those instances with at least 5 indexes achieve more than 98% for high quality songs, and less than 7 milliseconds searches. Our setup with seven indexes gives more than 99% of recall and 3 millisecond searches for songs of quality @64, @96, @128 and @1411 kbps.

Table 4 shows the cost of storage of the setups of Figure 4. The memory usage is maintained below 100 MiB, since compressed indexes require close to 70% of the uncompressed LSH. Please notice that the compression is important, since it can be central for running a standalone version of the indexes in mobile devices.

7 Conclusions

We presented a compressed index for LSH. The index can be queried without decompression and with negligible impact in the query time. We tested the index for the problem of near duplicate detection in whole-song querying. We made experiments with 3.7 million songs, obtaining near perfect recall and searching times of 5 milliseconds. The index fits well under 100MiB of RAM, and requires only simple operations, easily cacheable. The fingerprint of a 4 minute song is computed in 0.3 seconds in a standard CPU without parallelization. This allows to think in applications running standalone in small devices. Making a simple linear extrapolation of our index, which is a pessimistic assumption, we can fingerprint about 37 million songs in half gigabyte of RAM; being able to query the collection in a fraction of a second. The assumption is pessimistic because the compression ratio and searching times scale sub-linearly.

In future work we will try to estimate with very few parameters the audio quality of a song using our fingerprinting technique. We believe we only need the area under the curve of the time entropy profile. This feature can act as filter in a third party storage and streaming service, for example, the service provider may reject to stream low quality audio found in the users folders.

From the searching point of view we will investigate generalizations of the $\mathcal{L}$ set of indexes using metric indexes, more robust than LSH with higher error rates.

Acknowledgements. We want to thank the anonymous referees who helped us to improve the presentation with insightful observations. This work was partially supported by CONACyT and CICESE grants. The third author was a postdoc in Universidad Michoacana under the CONACyT's grant 179795 (project "Bases de Datos Multimedia Superescalables").

References

1. Chandrasekhar, V., Sharifi, M., Ross, D.: Survey and evaluation of audio fingerprinting schemes for mobile query-by-example applications. In: Proceedings of ISMIR (2011)
2. LTD, S.: (2006), http://www.shazam.com/
3. SoundHound (2008), http://www.soundhound.com/
4. McFee, B., Lanckriet, G.R.G.: Large-scale music similarity search with spatial trees. In: Klapuri, A., Leider, C. (eds.) ISMIR, pp. 55–60. University of Miami (2011)
5. Bertin-Mahieux, T., Ellis, D.P.W., Whitman, B., Lamere, P.: The million song dataset. In: Klapuri, A., Leider, C. (eds.) ISMIR, University of Miami, pp. 591–596. University of Miami (2011)
6. Samet, H.: Foundations of Multidimensional and Metric Data Structures, 1st edn. The morgan Kaufman Series in Computer Graphics and Geometic Modeling. Morgan Kaufmann Publishers, University of Maryland at College Park (2006)

7. Chávez, E., Navarro, G., Baeza-Yates, R., Marroquín, J.L.: Searching in metric spaces. ACM Comput. Surv. 33(3), 273–321 (2001)
8. Lu, G.: Indexing and retrieval of audio: A survey. Multimedia Tools Appl. 15(3), 269–290 (2001)
9. Stober, S., Nürnberger, A.: Adaptive music retrieval - a state of the art. Multimedia Tools and Applications (2012), 'Online First' article
10. Yan, R., Huet, B., Sukthankar, R.: Large-scale multimedia retrieval and mining. IEEE Multimedia 18, 11–13 (2011)
11. Camarena-Ibarrola, A., Chavez, E.: A robust entropy-based audio-fingerprint. In: Proceedings of the International Conference on Multimedia and Expo, ICME 2006, pp. 1729–1732 (2006)
12. Gionis, A., Indyk, P., Motwani, R.: Similarity search in high dimensions via hashing. In: Proceedings of the 25th International Conference on Very Large Data Bases, VLDB 1999, pp. 518–529. Morgan Kaufmann Publishers Inc., San Francisco (1999)
13. Andoni, A., Indyk, P.: Near-optimal hashing algorithms for approximate nearest neighbor in high dimensions. Communications ACM 51, 117–122 (2008)
14. González, R., Grabowski, S., Mäkinen, V., Navarro, G.: Practical implementation of rank and select queries. In: Poster Proc. Volume of 4th Workshop on Efficient and Experimental Algorithms (WEA), pp. 27–38. CTI Press and Ellinika Grammata, Greece (2005)
15. Raman, R., Raman, V., Rao, S.S.: Succinct indexable dictionaries with applications to encoding k-ary trees and multisets. In: Proceedings of the Thirteenth Annual ACM-SIAM Symposium on Discrete Algorithms (SODA), pp. 233–242. ACM/SIAM, San Francisco (2002)
16. Claude, F., Navarro, G.: Practical rank/select queries over arbitrary sequences. In: Amir, A., Turpin, A., Moffat, A. (eds.) SPIRE 2008. LNCS, vol. 5280, pp. 176–187. Springer, Heidelberg (2008)
17. Okanohara, D., Sadakane, K.: Practical entropy-compressed rank/select dictionary. In: Proceedings of the Workshop on Algorithm Engineering and Experiments, ALENEX 2007. SIAM, New Orleans (2007)
18. Tellez, E.S.: Practical Proximity Searching in Large Metric Databases. PhD thesis, Universidad Michoacana, Morelia, Michoacán, México (July 2012)
19. Tellez, E.S., Chavez, E., Navarro, G.: Succinct nearest neighbor search. Information Systems 38(7), 1019–1030 (2013)
20. Grossi, R., Gupta, A., Vitter, J.S.: High-order entropy-compressed text indexes. In: Proceedings of the Fourteenth Annual ACM-SIAM Symposium on Discrete Algorithms, SODA 2003, pp. 841–850. Society for Industrial and Applied Mathematics, Philadelphia (2003)
21. Navarro, G., Mäkinen, V.: Compressed full-text indexes. ACM Computing Surveys 39(1) (2007)
22. Golynski, A., Munro, J.I., Rao, S.S.: Rank/select operations on large alphabets: a tool for text indexing. In: Proceedings of the Seventeenth Annual ACM-SIAM Symposium on Discrete Algorithm, SODA 2006, pp. 368–373. ACM, New York (2006)
23. Munro, J.I., Raman, R., Raman, V., Rao, S.S.: Succinct representations of permutations. In: Baeten, J.C.M., Lenstra, J.K., Parrow, J., Woeginger, G.J. (eds.) ICALP 2003. LNCS, vol. 2719, pp. 345–356. Springer, Heidelberg (2003)

Perils of Combining Parallel Distance Computations with Metric and Ptolemaic Indexing in kNN Queries

Martin Kruliš[1], Steffen Kirchhoff[2], and Jakub Yaghob[1]

[1] Parallel Architectures/Applications/Algorithms Research Group
Faculty of Mathematics and Physics, Charles University in Prague, Czech Republic
{krulis,yaghob}@ksi.mff.cuni.cz
[2] School of Engineering and Applied Sciences, Harvard University, Cambridge, USA
skirchhoff@seas.harvard.edu

Abstract. Similarity search methods face serious performance issues since similarity functions are rather expensive to compute. Many optimization techniques were designed to reduce the number of similarity computations, when a query is being resolved. Indexing methods, like pivot table prefiltering, based on the metric properties of feature space, are one of the most popular methods. They can increase the speed of query evaluation even by orders of magnitude. Another approach is to employ highly parallel architectures like GPUs to accelerate evaluation by unleashing their raw computational power. Unfortunately, resolving the k nearest neighbors (kNN) queries optimized with metric indexing is a problem that is serial in nature. In this paper, we explore the perils of kNN parallelization and we propose a new algorithm that basically converts kNN queries into range queries, which are perfectly parallelizable. We have experimentally evaluated all approaches using a highly parallel environment comprised of multiple GPUs. The new algorithm demonstrates more than $2\times$ speedup to the naïve parallel implementation of kNN queries.

Keywords: parallel, GPU, kNN queries, similarity search, indexing.

1 Introduction

Similarity search presents a specific concept of information retrieval where complex objects are being looked up based on their contents. It is being used in a wide variety of applications including computer vision, pattern recognition, data mining, content-based image retrieval, or bioinformatics. It introduces a query-by-example paradigm, where the user provides an object as a query and expects to receive the most similar objects from the database in return.

Unfortunately, similarity search suffers from serious performance issues, since the functions that determine similarity of two objects are quite computationally demanding in most models [5,16]. There are various optimization techniques that address the performance issue. In this work, we will focus on combining two approaches:

A.J. Machado Traina et al. (Eds.): SISAP 2014, LNCS 8821, pp. 127–138, 2014.
DOI: 10.1007/978-3-319-11988-5_12 © Springer International Publishing Switzerland 2014

- Metric indexing techniques which can be employed to reduce the number of similarity computations and which have been shown to increase the overall performance by orders of magnitude [3,14].
- Utilizing the raw computational power of current parallel hardware, especially the parallel accelerators such as GPGPUs or Xeon Phi devices. In the previous work [12,10], we have shown that GPUs can be utilized to accelerate SQFD adaptive distance measure by two orders of magnitude with respect to common CPU.

Our main objective is to investigate the problems that arise when these two methods are combined in one system. We have discovered that when most of the metric indexing methods are applied for standard k nearest neighbour (kNN) queries, any attempt to parallelize the algorithm will lead to a suboptimal solution (i.e., a solution that has to evaluate the similarity function more times than the serial solution).

In the remainder of this paper, we will narrow our focus to similarity functions, with nontrivial computational demands, like Signature Quadratic Form Distance (SQFD) [5,4] or Earth Mover's Distance (EMD) [16]. Very cheap functions (e.g., a simple Euclidean distance in low-dimensional space) can utilize only limited variety of metric indexing methods, which can efficiently prune the object candidates, so that the overhead of the indexing does exceed the computational time of the similarity function. Analogically, extremely expensive functions (e.g., protein alignment models from bioinformatics [8]) require best possible metric access methods and their internal parallelization (i.e., computing each similarity in parallel) is becoming more important than computing multiple distances concurrently.

We have selected multimedial database as our referential dataset for experimental evaluation. Feature signatures [16] that reflect localized image properties regarding color, contrast, and coarseness were used as the object descriptors and SQFD was used as the distance (i.e., inverse of similarity) function. It has been established in previous work [12,10] that SQFD can be accelerated by GPUs, so we have used this parallel platform in our evaluation. Finally, we have selected pivot table prefiltering (which is based on LAESA [15] method) for indexing. The regular structure of pivot table makes it a very good candidate for parallelization. Furthermore, the prefiltering can use not only metric axioms (triangular inequality), but also Ptolemaic inequality [14], since the SQFD conforms to Ptolemaic axioms. Despite the fact we have selected a specific setup for our experiments, our observations and proposed solutions may be extended to other similarity problems and indexing methods and to other parallel hardware.

The paper is organized as follows. Section 2 revises the fundamentals of similarity search and querying principles. In Section 3, we address the issues that rise when parallel processing is introduced in query evaluation. Possible strategies and our proposed solution are described in Section 4 and empirically evaluated in Section 5. Section 6 concludes our work.

2 Similarity Search

Similarity model consists of two parts – object descriptors and a distance (dissimilarity) function. The object descriptors represent the object features that are essential for similarity comparisons. The distance function measures the dissimilarity of two object descriptors.

In the similarity search paradigm, the user provides an example object as a query. A query descriptor is extracted from the query object and it is compared with descriptors in the database using the distance function. The closest (least dissimilar) objects to the query object are then returned to the user.

There are two basic types of queries. The *range query* selects all objects with distance to the query lower than given range r. The k *nearest neighbours* query (kNN) returns k the most similar objects from the database. The kNN query is the prefered choice in many systems, since its concept is obvious to the user and the size of the result is predetermined by parameter k. The range query applications are limited to cases when the distance function produce values that can be somehow related to object attributes (e.g., in case of Levenshtein edit distance [13]) or when the filtering range can be determined by other means (e.g., from the results of previous queries).

2.1 Sequential Scan

When no indexing method is applied, both queries are evaluated using a *sequential scan* algorithm. It sequentially computes distances from the query object to all objects in the database and filter out the results based on these distances. In case of the range query, the filtering step simply compares computed distances with the range r and all objects closer than r are included into the result.

The kNN query evaluation usually maintains an intermediate set of k closest objects, which is (possibly) updated with each computed distance. When all distances are computed the intermediate set becomes the result. From this point of view, we can perceive the kNN query as a range query with dynamic filtering range that is decreasing as the intermediate results gets more refined.

2.2 Metric and Ptolemaic Indexing

The indexing methods are designed to reduce the number of distance computations made during the query evaluation. The idea is to select object candidates, which can be possibly included into the result, thus it is worth computing their exact distance. In another words, the indexing method prunes out objects which cannot pass the final filtering. In the remainder of our work, we will address the indexing step as *prefiltering* since it pre-filters the candidates from the database.

We have focused on pivot table prefiltering (which is based on the LAESA method [15], and is often denoted 2-phase LAESA), as it has little requirements and it can be easily parallelized. The method selects some objects from the database called *pivots* or *vantage points* and precomputes a matrix (*pivot table*) containing distances between the pivots and all objects in the database. The pre-computed values can be used to determine *lower bounds* for the object distances if the distance function conforms to metric or even Ptolemaic axioms.

The prefiltering algorithm works in two phases. In the first phase, it computes distances between the query object and all pivot objects. In the second phase, it sequentially computes lower bounds for all non-pivot objects by triangular inequality [3] or Ptolemaic inequality [14] and compare them with the filtering range – a constant value r in case of range queries or maximal distance in the intermediate top-k set in case of kNN queries. If the lower bound is greater than the filtering range, the object may be pruned out without computing its distance, which must be greater than or equal to its lower bound. The schema of kNN algorithm that employs pivot table prefiltering is depicted in Figure 1.

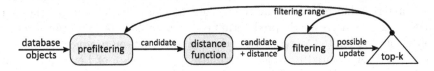

Fig. 1. Schema of a kNN algorithm that employs pivot table prefiltering

Let us emphasize that the intermediate top-k result may be refined after each computed distance. This refinement is particularly important when the indexing is employed, since the prefiltering step uses the same range (i.e., the greatest distance from the intermediate top-k set) as the final filtering step.

3 Introducing Parallel Processing

The parallelization efforts of the SQFD presented in previous work [12,10] were twofold. Each distance has to be computed in parallel manner and multiple distances has to be computed concurrently in order to utilize the hardware to its full potential. Since the host-GPU communication and data transfers are bound with nontrivial overhead, the SQFD distances need to be computed in batches.

3.1 Parallel Query Evaluation

Despite the word "sequential" in the name, the sequential scan is an embarrass-ingly parallel algorithm for both range queries and kNN queries. If we assume that the distance function is computationally demanding, it is reasonable to compute only the distances on the parallel hardware (GPUs in our case). The filtering step can be performed sequentially by the CPU and overlapped with GPU workload, since the time required for filtering is negligible in comparison to the time consumed by the distance computations.

The situation gets slightly more difficult when an indexing technique is em-ployed in the query processing. The first complication is risen by the compu-tational demands of the prefiltering itself. Computing the lower bounds and dispatching batches of the candidate objects to the GPUs may become a bottleneck, especially when multiple state-of-the-art GPUs are used in combination with Ptolemaic inequality[1]. Fortunately, this problem can be solved by precomputing the lower bounds on GPUs [10] and by dispatching multiple CPU threads to feed the GPUs [11].

Much more serious issue, which we address in this work, present the updates of kNN filtering range as illustrated in Figure 1. When a candidate object passes the filtering step, it is added into the intermediate top-k set, which ultimately changes the value of the filtering range. The filtering range is not employed only in filtering, but also in the prefiltering step. In order to achieve optimal prefiltering effectivity, each candidate yielded by the prefiltering has to be processed completely (since it may lower the filtering range) before another database object is prefiltered.

When multiple distances are computed concurrently, the prefiltering step must yield multiple candidates at once, so they can be dispatched to a GPU in a batch. Therefore, any parallel algorithm will lead to a suboptimal solution, since it may (and usually will) compute more distances than its serial counterpart. On the other hand, the parallel version could still outperform the serial algorithm significantly if we compensate the inefficiency of prefiltering with the raw computational power of the GPUs.

3.2   Related Work

The kNN query problem has been extensively studied and reaches well beyond the realms of similarity search. One of the first parallel approaches to nearest neighbour search [6] was proposed by a research team from Munich university in 1997. In their work, the authors assumed that the object descriptors are mapped to high dimensionality spaces and compared by rather cheap distance functions. They proposed to cluster the feature space, so the clusters may be searched concurrently.

One of the first implementations of kNN query on GPUs was presented by Bustos et al. [7] in 2006. Their implementation was restricted to compute k nearest neighbours using Manhattan distance in $\mathbb{R}^d$ spaces, where d varied up to 256. The work proposed a GPU-specific data representation, which allowed better utilization of the texture cache (shared memory) of its symmetric multiprocessors.

A similar approach was taken by Garcia [9] in 2008. Their experiments tested Euclidean and Manhattan distances in $\mathbb{R}^d$ spaces for dimensions between 8 to 96. The brute force implementation outperformed not only naïve serial algorithm, but also the version which used kd-tree to index the space [1].

The most recent work on the topic was done by Barrientos et al. [2]. They improved the performance by replacing parallel sorting with standard 2-regular heap that holds the intermediate top-k result. The GPU parallel implementation

---

[1] Computing Ptolemaic lower bounds have time complexity $O(P^2)$, where P is the number of pivots.

was based on heap reduction. The heap is kept replicated, so each thread in a warp has its own copy and the data from multiple heaps are then combined by a parallel reduction algorithm.

To the best of our knowledge, all papers that address the parallel processing of kNN queries focus on concurrent distance computations and on combining the top-k results efficiently. Our work integrates the prefiltering step into kNN query processing. This modification affects the parallelism in a specific way, which has not been investigated yet.

4   Parallel Top-k Queries with Pivot Table Prefiltering

We have established that the metric indexing employed in kNN queries is optimal only when the evaluation is conducted serially. If the query is processed in parallel, we sacrifice optimality, thus we compute more distances than would be required by the sequential method. Our goal is to compute as few unnecessary distances as possible whilst exploiting the parallelism to its full potential.

4.1   Computing Distances on GPU

We assume that the distance function is not expensive (or not parallelizable) enough so that one distance computation would occupy all available cores. Hence, the data parallel approach is employed and multiple distances are computed concurrently to fully utilize the hardware. Blocks of signatures are dispatched to the GPUs and blocks of corresponding distances are returned back. The details of this process are thoroughly described in the previous work [12,10].

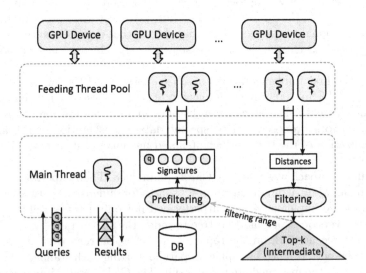

Fig. 2. Schema of our GPU framework

The framework for computing SQFDs on GPUs is depicted in Figure 2. The feeding threads are used to serve data to the GPUs and fetch the results back. The main thread is responsible for prefiltering the candidates, dispatching work to feeding threads, gathering distances returned by these threads, and refining the intermediate top-k set. Several different strategies can be implemented in the main thread. These strategies have significant impact on the performance.

4.2 The Naïve Approach

The simplest way of parallelizing the kNN query with pivot table prefiltering has been suggested along with the parallelization of the SQFD [10]. We denote this method *the naïve approach*, as it is a direct extension of the serial algorithm. The main thread prefilters candidates, puts them in blocks of constant size S, and dispatches them to GPUs. There can be at most B blocks simultaneously dispatched. When this limit is reached, the main thread waits for the first block of the distances to return from one of the GPUs, incorporates these distances into the intermediate top-k set, and prepares another block of candidates which are immediately dispatched.

We have empirically observed that the system works at its peak performance if the limit is set to $B = 2G$, where G is the number of GPUs. This number corresponds with the overlapping strategy, when each GPU is simultaneously processing one block and transferring another block from the host memory.

The block size S tunes the ratio between effectiveness of the prefiltering and the efficiency of the parallel hardware. Lower values causes that the intermediate top-k result is updated more often, so the prefiltering works better. Higher values present larger workloads for the GPUs, thus the parallel hardware is better utilized and the overhead is reduced. The best block size depends on the computational costs of the distance function, on properties of the parallel hardware, and the indexability of the database (as presented in Section 5).

Initial Phase Optimization. We have made a simple observation about the indexing effectiveness. At the beginning of the query evaluation, the candidate prefiltering is not very effective since the intermediate top-k set is produced from a very small portion of the database. As the evaluation advances, the filtering range is getting lower, thus the prefiltering pruning ratio is getting higher.

Our first attempt to improve the naïve algorithm is based on this observation. We divide the query processing into two phases. The *initial phase* does not use the candidate prefiltering, since it is not efficient at the beginning of the query evaluation anyway. Instead, the distances are computed in a highly parallel manner on the GPUs. When the distances for the initial part are computed, the intermediate top-k set is updated and the remaining objects are processed by the naïve algorithm. We have empirically observed that this optimization has measurable impact on the performance if the size of the initial part is selected properly.

4.3   Converting kNN Query to Range Query

Introducing parallelism into kNN queries with pivot table prefiltering caused a serious complication. On the other hand, the range queries remain to be an embarrassingly parallel problem even if we use prefiltering, since the filtering (and prefiltering) range is constant for the whole time.

Let us define an optimal range r_{opt} for a given k as a filtering range for the range query algorithm that produces a result set $R_{r_{opt}}$ equal to the result set R_{kNN} produced by the kNN algorithm for the same k. If we knew the r_{opt} in advance, we could easily convert the kNN algorithm to the range algorithm, which would compute significantly less distances, and which would be perfectly parallelizable.

Since we do not know r_{opt} in advance, we propose *a range estimation* algorithm. It computes *a range estimate* $r_{est} \geq r_{opt}$, which is then used for processing of the range query. The results of the range query $R_{r_{est}}$ are obviously a superset of the kNN result since $r_{est} \geq r_{opt} \Rightarrow R_{r_{est}} \supseteq R_{r_{opt}} = R_{kNN}$. The final kNN result can be quickly extracted from the range query result as the distances have been already computed.

The Range Estimation The most essential part of the range estimation algorithm is its initial part that estimates the range. Our approach is based on the observation that the filtering works significantly better, if the database objects are sorted by their lower bounds and processed in an ascending order. However, the sorting process takes a significant amount of time (even when performed on the GPU), so we just take a database subset of the objects with the smallest lower bounds[2]. We can summarize the algorithm into the following steps.

1. Object lower bounds are computed on GPUs (based on a combination of the triangular inequality and Ptolemaic inequality).
2. *An estimate set* $E \subseteq D$ of a given size is computed, so that $\forall e \in E, o \in D \setminus E : LB(e) \leq LB(o)$.
3. SQFD distances for all objects in E are computed on GPUs and an intermediate top-k set is constructed from E.
4. Let r_{est} be the greatest distance from the intermediate top-k set.
5. A set of candidate objects $c \in D \setminus E : LB(c) \leq r_{est}$ is constructed on CPU (using all available cores).
6. SQFD distances for the candidates are computed on GPUs and the top-k set is refined.

In both SQFD steps (3 and 6), the distances are computed for a predefined set of objects. Therefore, we can dispatch the SQFD batches in a way that optimally utilizes all available GPUs and minimize the overhead. Furthermore, the estimate set of objects is picked based on their lower bounds, thus it is very likely that the range estimate algorithm would require less distance computations than the naïve algorithm.

---

[2] We assume that the size of the estimate set is significantly smaller than the size of the entire database, so it is much efficient to select its items using a d-regular heap data structure for instance instead of sorting the whole database.

Combining Range and Naïve Algorithm. The range estimation algorithm is designed to perform better in common cases. However, if the range set consists of objects that are in fact very far from the query, the estimate range r_{est} will be quite high. A poor range estimation could affect the performance significantly as the number of computed distances depends on it strongly and the subsequent range query algorithm does not refine the filtering range. We can alter the original range estimation algorithm and replace its last two steps by the naïve algorithm. This combination is less suitable for the parallelism, but it is significantly more resilient to range misestimation.

The choice between the original range estimation version and the range-naïve version could be made at runtime. The first four steps of both algorithms are exactly the same so the decision can be postponed until the distances of the estimate set are computed. These distances can be used help with this decision and we are planning to investigate this possibility in our future work.

5   Experimental Results

This section summarizes the intensive empirical research we have conducted. We have selected only the most intriguing data due to the limited space. All times were measured by the real-time clock of the operating system. The experiments were performed multiple times to verify the measured values.

Our experiments were performed on a GPU server equipped with a Xeon E5645 processor that contains 12 logical cores running at 2.4 GHz, 96 GB of DDR3-1333 RAM, and four NVIDIA Tesla M2090 GPU cards based on Fermi architecture. Each GPU card has 512-core chip and 6 GB of memory.

We have used 5 million photographs as testing dataset and 100 randomly picked queries. Three signature sets with different signature sizes were created in order to simulate different computational costs of distance functions[3]. We denote these sets *small*, *medium*, and *large*, and they have average signature sizes of 14.3, 90.5, and 286, respectively. We have also tested various values of the SQFD parameter α in the range 0.1 to 2, which tunes the ratio between indexability and precision.

5.1   The Naïve Algorithm

The naïve algorithm has only one important parameter – the block size. We keep the block size constant during the whole evaluation time since we have observed that varying the block size disrupts the pipeline data flow to the GPUs and the workload imbalance causes significant drop in performance. The number of blocks which are *on fly* was empirically determined as 2× the number of GPU devices.

The results presented in Figure 3 support our original assumptions. The more expensive the distance function is, the smaller blocks are required to achieve

---

[3] The SQFD time complexity is quadratically proportional to the size of signatures.

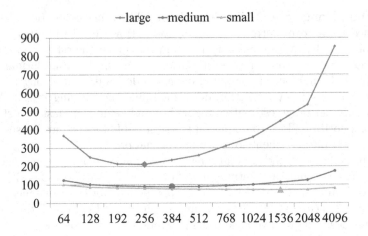

Fig. 3. Query times (in ms) of the naïve algorithm for different block sizes ($\alpha = 0.5$)

optimal performance. The expensive functions benefit from the prefiltering more, thus smaller blocks (which refine the filtering range more often) are a better choice. Extremely expensive distance functions cannot be efficiently computed by this data parallel model and the only option is to parallelize their internal evaluation.

On the other hand, very cheap distance functions do not benefit from the candidate prefiltering, since the cost of the prefiltering and the cost of the distance computations are tending to be similar. Therefore, it is better to compute all the distances concurrently without any disruptions.

5.2 The Range Estimation Algorithm

The Figure 4 depicts the comparison of all tested approaches. The *optimal* algorithm is a theoretical baseline created by modified range algorithm which has been given the optimal filtering range by an oracle in advance. Hence, the optimal algorithm computes the fewest distances in an optimally parallel manner.

The range estimation algorithm is quite close to the optimal algorithm in every case and significantly outperforms the naïve algorithm for selected alpha[4]. The range-naïve combination does not perform better than range algorithm. However, this difference gets smaller as we increase the alpha value and for $\alpha = 2$ (the prefiltering gets less effective), the 58% of queries perform better with the range-naïve algorithm than the original range algorithm (on the large signature set). We will focus on more detailed analysis in the future work.

We did not provide comparison with serial kNN and parallel kNN without pivot table prefiltering (i.e., the sequential algorithm) as both of these approaches are slower by more than an order of magnitude. This observation supports our

---

[4] The $\alpha = 0.5$ has been selected as a good compromise between indexability and precision.

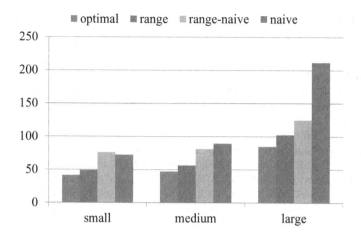

Fig. 4. Comparison of query evaluation times (in ms) for small, medium, and large signature sizes ($\alpha = 0.5$)

original assumption that we need the combination of parallel processing and metric indexing to achieve peak performance.

6   Conclusions

We have performed an analysis of various strategies to evaluate kNN queries that combine parallel distance computations with metric and Ptolemaic indexing. The naïve algorithm, which is a direct extension of the serial algorithm, performs adequately if we select a block size that is appropriate for given situation. The block size can be determined by a statical analysis of the similarity model computational costs for instance.

Furthermore, we have proposed a new algorithm that basically converts kNN queries into range queries, which are embarrassingly parallel and require less distance computations in most cases. The combination of the range estimation and the naïve algorithm has been proven as a fall back solution for outlier queries.

In the future work, we will focus on a more detailed cost analysis. We are currently in the process of designing a model that will help us select the best algorithm and its parameters based simple cost estimations and on the query properties at runtime. We also wish to test different types of parallel hardware, to determine the scalability of our approach.

Acknowledgements. This paper was supported by Czech Science Foundation (GAČR), projects no. P103/14/14292P and P103/13/08195, and by a training grant from the NIH Blueprint for Neuroscience Research (T90DA022759/ R90DA023427). Its contents are solely the responsibility of the authors and do not necessarily represent the official views of the NIH.

References

1. Arya, S., Mount, D.M., Netanyahu, N.S., Silverman, R., Wu, A.Y.: An optimal algorithm for approximate nearest neighbor searching fixed dimensions. Journal of the ACM (JACM) 45(6), 891–923 (1998)
2. Barrientos, R., Gómez, J., Tenllado, C., Prieto, M.: Heap based k-nearest neighbor search on gpus. In: Congreso Espanol de Informática (CEDI), pp. 559–566 (2010)
3. Beecks, C., Lokoč, J., Seidl, T., Skopal, T.: Indexing the Signature Quadratic Form Distance for Efficient Content-Based Multimedia Retrieval. In: Proc. ACM Int. Conf. on Multimedia Retrieval, pp. 24:1–24:8 (2011)
4. Beecks, C., Uysal, M.S., Seidl, T.: Signature Quadratic Form Distances for Content-Based Similarity. In: Proc. 17th ACM Int. Conference on Multimedia (2009)
5. Beecks, C., Uysal, M.S., Seidl, T.: Signature Quadratic Form Distance. In: Proc. ACM International Conference on Image and Video Retrieval, pp. 438–445 (2010)
6. Berchtold, S., Böhm, C., Braunmüller, B., Keim, D.A., Kriegel, H.P.: Fast parallel similarity search in multimedia databases, vol. 26. ACM (1997)
7. Bustos, B., Deussen, O., Hiller, S., Keim, D.: A graphics hardware accelerated algorithm for nearest neighbor search. In: Alexandrov, V.N., van Albada, G.D., Sloot, P.M.A., Dongarra, J. (eds.) ICCS 2006, Part IV. LNCS, vol. 3994, pp. 196–199. Springer, Heidelberg (2006)
8. Galgonek, J., Kruliš, M., Hoksza, D.: On the parallelization of the sprot measure and the tm-score algorithm. In: Caragiannis, I., et al. (eds.) Euro-Par Workshops 2012. LNCS, vol. 7640, pp. 238–247. Springer, Heidelberg (2013)
9. Garcia, V., Debreuve, E., Barlaud, M.: Fast k nearest neighbor search using gpu. In: IEEE Computer Society Conference on Computer Vision and Pattern Recognition Workshops, CVPRW 2008, pp. 1–6. IEEE (2008)
10. Krulis, M., Skopal, T., Lokoc, J., Beecks, C.: Combining cpu and gpu architectures for fast similarity search. Distributed and Parallel Databases (2012)
11. Krulis, M., Falt, Z., Bednárek, D., Yaghob, J.: Task Scheduling in Hybrid CPU-GPU Systems. In: ITAT, pp. 17–24 (2012)
12. Kruliš, M., Lokoč, J., Beecks, C., Skopal, T., Seidl, T.: Processing the signature quadratic form distance on many-core gpu architectures. In: CIKM, pp. 2373–2376 (2011)
13. Levenshtein, V.I.: Binary codes capable of correcting deletions, insertions and reversals. Soviet Physics Doklady 10, 707 (1966)
14. Lokoč, J., Hetland, M., Skopal, T., Beecks, C.: Ptolemaic indexing of the signature quadratic form distance. In: Proceedings of the Fourth International Conference on SImilarity Search and APplications, pp. 9–16. ACM (2011)
15. Moreno-Seco, F., Micó, L., Oncina, J.: Extending LAESA fast nearest neighbour algorithm to find the k nearest neighbours. In: Caelli, T.M., Amin, A., Duin, R.P.W., Kamel, M.S., de Ridder, D. (eds.) SPR 2002 and SSPR 2002. LNCS, vol. 2396, pp. 718–724. Springer, Heidelberg (2002)
16. Rubner, Y., Tomasi, C., Guibas, L.J.: The Earth Mover's Distance as a Metric for Image Retrieval. International Journal of Computer Vision 40(2), 99–121 (2000)

Transition-Sensitive Distances

Kaoru Yoshida

Sony Computer Science Laboratories, Inc.
3-14-13 Higashigotanda, Shinagawa-ku, Tokyo, 141-0022 Japan
kaoru@csl.sony.co.jp

Abstract. In information retrieval and classification, the relevance of the obtained result and the efficiency of the computational process are strongly influenced by the distance measure used for data comparison. Conventional distance measures, including Hamming distance (HD) and Levenshtein distance (LD), count merely the number of mismatches (or modifications). Given a query, samples mapped at the same distance have the same number of mismatches, but the distribution of the mismatches might be different, either disperse or blocked, so that other measures must be cascaded for further differentiation of the samples. Here we present a new type of distances, called transition-sensitive distances, which count, in addition to the number of mismatches, the cost of transitions between positionally adjacent match-mismatch pairs, as part of the distance. The cost of transitions that reflects the dispersion of mismatches can be integrated into conventional distance measures. We introduce transition-sensitive variants of LD and HD, referred to as TLD and THD. It is shown that while TLD and THD hold properties of the metric similarly as LD and HD, they function as more strict distance measures in similarity search applications than LD and HD, respectively.

Keywords: Transition-sensitive Distance, Transition-sensitive Levenshtein Distance, Transition-sensitive Hamming Distance, distance measure, metric, string matching, pattern matching, dynamic programming.

1 Introduction

Recently, a variety of information has been accumulated to a growing scale. Highly demanded is a simple and efficient method to retrieve relevant information of interest out of the accumulated source. In the core of information retrieval systems is data comparison or pattern matching. Numerous methods and strategies have been developed for comparison of various kinds of symbolic data, including text, voice, music, image, and video [1–7]. The relevance of the retrieved result and the efficiency of the computational process are both strongly influenced by the distance measure used for comparison.

When two patterns of equal size in arbitrary dimensions, such as multi-dimensional bitmap image data, to be compared, Hamming distance (HD) [8] has been widely used, which is defined as the minimum number of substitutions required to transform one pattern into the other. For comparison of two strings,

A.J. Machado Traina et al. (Eds.): SISAP 2014, LNCS 8821, pp. 139–150, 2014.
DOI: 10.1007/978-3-319-11988-5_13 © Springer International Publishing Switzerland 2014

whose lengths may be different, Levenshtein (or edit) distance (LD) [9] has been used instead, which is defined as the minimum number of insertions, deletions, and substitutions required to transform one string to the other.

Given a query string 'form', for example, two strings, 'forms' and 'forum', are mapped by LD at the same distance, 1, both for one insertion: the last character 's' of 'forms' and the fourth character 'u' of 'forum'. Mismatches, including deletions, insertions, and substitutions, break up one string into fragments, each composed of all matching (or mismatched) characters. The former 'forms' is split into two fragments of lengths 4 and 1, such as form-s, while the latter 'forum' is broken into three fragments of lengths 3, 1 and 1, such as for-u-m. Thus the LD measure gives the same distance as long as the number of mismatches is the same no matter whether the mismatches are blocked or distributed, whether they are on the edge or in the middle.

In many different applications, such as linguistic analysis, it is often presumed that strings with mismatches at the head or tail may be related objects, while those with mismatches found in the middle or distributed throughout could be independent objects. In the previous example, 'forms' (form-s) is a variant of 'form', while 'forum' (for-u-m) is an independent word. Simple methods for segregating variants from others are highly demanded for natural language processing systems as in [10].

To further differentiate those two strings (or arrays) which are mapped at the same LD (or HD), other measures need to be cascaded as additional steps. To capture the locations of mismatches, the N-gram method that is to conduct pattern matching locally in a window of length N sliding along each string has been used as in [11, 12]. To assess the degree of fragmentation of mismatches, Shannon entropy, which is defined: $H = -\Sigma p_i \cdot \log p_i$ with the occupancy p_i of a fragment of length i, has been used as in [13–15].

In this paper, we present a new type of distances, called transition-sensitive distances, that reflect not only the sum but also the distribution of mismatches between the subjects of comparison.

2   Transition-Sensitive Distances

2.1   Transition-Sensitivity

Suppose that the subjects of comparison are arrays of arbitrary number of dimensions and size. Each array is composed of symbolic elements that are either atomic or composite. Atomic elements are quantitatively comparable symbols. Composite elements are those consisting of two or more atomic elements.

The difference between two corresponding elements is referred to as the element dissimilarity. Depending on the comparison method used, whether discrete or fuzzy, the element dissimilarity may be a discrete binary integer (0 for match and 1 for mismatch) or a real number in the range [0,1] in fuzzy indicating a degree of mismatch.

The difference between two element dissimilarities at adjacent positions, which is referred to as the transition, is represented as a real number in the range

[-1, 1], where negative numbers, positive numbers, and zero represent ascents, descents and none, respectively. The positional adjacency is defined by the spatial properties of the arrays. Transition-sensitivity is the property that the distance is variable depending on the transitions in the element dissimilarities. In the following, we extend two conventional metrics, Levenshtein distance and Hamming distance, to be transition-sensitive.

2.2   Transition-Sensitive Levenshtein Distance

Transition-sensitive Levenshtein distance (TLD) is a distance between two strings, which is formulated in the dynamic programming manner similarly as Levenshtein distance (LD) is.

Definition 1 (Transition-sensitive Levenshtein Distance, TLD). *Given a string X of length m and a string Y of length n, the transition-sensitive Levenshtein distance (TLD) between the two strings is:*

$$TLD(X, Y) = D[m, n]$$

where

1. *$D[i, j]$ $(0 \leq i \leq m, 0 \leq j \leq n)$ is a string distance defined as:*

$$D[i, 0] = i, \ (0 \leq i \leq m); \quad D[0, j] = j, \ (0 \leq j \leq n);$$

$$D[i,j] = min \begin{cases} D[i-1,j] + 1 + t(d[i-1,j], d[i,j], A, B) & \text{(deletion)} \\ D[i,j-1] + 1 + t(d[i,j-1], d[i,j], A, B) & \text{(insertion)} \\ D[i-1,j-1] + d[i,j] + t(d[i-1,j-1], d[i,j], A, B) & \text{(substitution)} \end{cases},$$

$$(0 \leq i \leq m, 0 \leq j \leq n).$$

2. *$d[i, j]$ $(0 \leq i \leq m, 0 \leq j \leq n)$ is an element dissimilarity defined as:*

$$d[0, 0] = -1; \quad d[i, 0] = 1, \ (1 \leq i \leq m); \quad d[0, j] = 1, \ (1 \leq j \leq n);$$

$$d[i, j] = c(X_i, Y_j), \ (1 \leq i \leq m, 1 \leq j \leq n);$$

X_i and Y_j are the i-th element of X and the j-th element of Y, respectively.

3. *$c(x, y)$ is a function that returns a real number within the range $[0, 1]$ representing the dissimilarity (or normalized distance) between two elements, x and y as follows:*

$$0 \leq c(x, y) = |x - y| \leq 1$$

Note that 0 indicates a complete match and 1 a complete mismatch. The element dissimilarity may be either taken as it is or further binarized with a given threshold γ, called the dissimilarity threshold, as:

$$c(x, y) = \begin{cases} 0 \text{ if } |x - y| \leq \gamma & \text{(match)} \\ 1 \text{ otherwise} & \text{(mismatch)} \end{cases}$$

4. $t(d1, d2, A, B)$ *is a function that returns a non-negative real number representing the cost for a transition from one element dissimilarity $d1$ to the other $d2$ as follows:*

$$t(d1, d2, A, B) = \begin{cases} A \cdot (d2 - d1) & \text{if } 0 \leq d1 < d2 \quad \text{(ascent)} \\ B \cdot (d1 - d2) & \text{if } 0 \leq d2 < d1 \quad \text{(descent)} \\ 0 & \text{otherwise} \quad \text{(no transition)} \end{cases}$$

A and B are the cost coefficient for the ascent or descent, respectively, under the following constraint:

$$0 \leq A + B \leq 1.$$

□

Proposition 1. *Given two strings x and y and transition cost coefficients A and B, $TLD(x, y)$ satisfies the following conditions:*

1. *$TLD(x, y) \geq 0$ (non-negativity)*
2. *$|\, |x| - |y| \,| \leq TLD(x, y) \leq max(|x|, |y|)$ (lower and upper bounds)*
3. *$TLD(x, y) = 0$ if and only if $x = y$ and $|x| = |y|$ (identity)*
4. *$TLD(x, y) = TLD(x, y)$ (symmetry)*
5. *Given another string z, $TLD(x, z) \leq TLD(x, y) + TLD(y, z)$ (triangle inequality)*

where $|x|$ denotes the length of string x.

Proof.

1. $TLD(x, y)$ is defined with addition, multiplication and minimum operators on non-negative numbers, thereby resulting in a non-negative number.
2. If x and y do not match in any elements, the element dissimilarity matrix is filled up with 1s, yielding the maximum value of $TLD(x, y)$ through the shortest path of the matrix, which is equal to the larger one of the string lengths. For x and y of different lengths, the case where the shorter string fully matches either the beginning or ending part of the longer string yields the minimum value of $TLD(x, y)$, $|\, |x| - |y| \,| + min(A, B)$, where $min(A, B) = 0$ if $A = B = 0$.
3. if and only if x and y are of the same length and fully match, the element dissimilarity matrix is filled up with 0s, yielding $TLD(x, y) = 0$.
4. In computing $TLD(x, y)$ and $TLD(y, x)$, element dissimilarities and transition costs are similarly maintained, except that their coordinates are transposed. Since the computation algorithm is uniform for each axis of the coordinates, the same distance is yielded.
5. It is trivial for special cases involving identity, including $x = y = z$, $x = y \neq z$, $x \neq y = z$, and $x = z \neq y$. In other cases, there is at least one mismatch in all three paths: $x \to y$, $y \to z$ and $x \to z$. If element dissimilarities and accompanied transitions on the path $x \to y$ and those on the path $y \to z$ do not positionally overlap in y, $TLD(x, z) = TLD(x, y) + TLD(y, z)$ can hold. Otherwise, $TLD(x, z) \leq TLD(x, y) + TLD(y, z)$ holds, since positional overlaps on the sequential path $x \to y \to z$ may be reduced on the direct path $x \to z$, such as double substitutions to a single substitution or zero for the reversion. □

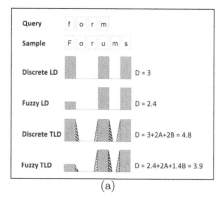
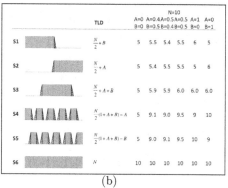

Fig. 1. Levenshtein distance (LD) and Transition-sensitive Levenshtein distance (TLD). (a) Comparison of strings, 'form' and 'Forums', with LD and TLD in discrete and fuzzy matching modes. Gray rectangles indicate element dissimilarities (or mismatches), each of which is either 0 or 1 in the discrete mode or a real number in the range [0,1] in the fuzzy mode. such as c(*Uppercase, Lowercase*)=0.4. Striped triangles indicate ascending or descending transitions, whose cost coefficients,A=0.4 and B=0.5, are assumed. (b) TLD for different patterns of mismatches and various cost coefficients. Note that TLD with A=0 and B=0 is equivalent to LD.

Figure 1(a) illustrates comparison of two strings, 'form' and 'Forums', using four different distance measures: Discrete LD, Fuzzy LD, Discrete TLD and Fuzzy TLD. While LD counts merely mismatches, TLD counts not only mismatches but also their transition costs. 'Discrete' or 'Fuzzy' implies whether the mismatch is represented with a binary or real number, respectively. The value of TLD varies depending on the distribution of mismatches and also on the cost coefficients for ascending and descending transitions, as shown in Figure 1(b). In the first five cases (S1-S5), there are five mismatches in common between the compared strings. However, the mismatches are distributed differently, as blocked in S1-S3 and distributed in S4-S5, so that they are given different TLDs (see the second or later columns). In the last string S6, all elements mismatched, so that LD and TLD are both 10. S4 and S5 contain only 50% mismatches, but the mismatches are evenly distributed, resulting in the highest transition cost. When the cost coefficients are set high, the TLDs of S4 and S5 get close or equal to the maximum 10 of S6 that contains 100% mismatches.

2.3 Transition-Sensitive Hamming Distance

Transition-sensitive Hamming distance (THD) is a distance between two matrices of equal size in arbitrary dimensions. Unlike LD and TLD, insertions and deletions are not allowed in HD and THD. For n-dimensional matrices, Transition-sensitive Hamming distances are formulated below.

Definition 2 (Transition-Sensitive Hamming Distance, THD). *Given two matrices X and Y of equal size $m_1 m_2 \ldots m_n$ in n dimensions, n-dimensional*

Hamming distance (HDn), n-dimensional transition cost (TCn), and Transition-sensitive n-dimensional Hamming distance (THDn) between the two matrices are respectively defined as:

$$HDn(X,Y) = \sum_{i_1=1}^{m_1} \cdots \sum_{i_n=1}^{m_n} d[i_1, \ldots, i_n]$$

$$TCn(X,Y) = \sum_{i_1=1}^{m_1} \cdots \sum_{i_n=1}^{m_n} \sum_{h=1}^{n} t(d[i_1, \ldots, i_{h-1}, \ldots i_n], d[i_1, \ldots, i_h, \ldots, i_n], A_h, B_h)$$

$$THDn(X,Y) = HDn(X,Y) + TCn(X,Y)$$

where

1. $d[i_1, \ldots, i_n]$ $(1 \leq h \leq n,\ 0 \leq i_h \leq m_k)$ *is an element dissimilarity defined as:*

$$d[0, \ldots, 0] = -1;$$

$$d[i_1, \ldots, i_h = 0, \ldots, i_n] = 1,\ (1 \leq h \leq n);$$

$$d[i_1, \ldots, i_n] = c(X_{i_1,\ldots,i_n}, Y_{i_1,\ldots,i_n}),\ (1 \leq h \leq n,\ 1 \leq i_h \leq m_h);$$

$X_{i_1,\ldots,i_n}$ *and* $Y_{i_1,\ldots,i_n}$ *are the elements of X and Y at the corresponding position,* $i_1, \ldots, i_n$, *respectively.*

2. $c(x,y)$ *and* $t(d1, d2, A, B)$ *are functions previously defined.*   □

Proposition 2. *Given two matrices x and y of the same dimension N and size L_i and transition cost coefficients A_i and B_i for each dimension $(1 \leq i \leq N)$, $THD(x,y)$ satisfies the following conditions:*

1. $THD(x,y) \geq 0$ *(non-negativity)*
2. $0 \leq THD(x,y) \leq L$ *(lower and upper bounds)*
3. $THD(x,y) = 0$ *if and only if $x = y$ (identity)*
4. $THD(x,y) = THD(x,y)$ *(symmetry)*
5. *Given another matrix z of the same dimension and size as x and y,*
   $THD(x,z) \leq THD(x,y) + THD(y,z)$ *(triangle inequality)*

where $L = \prod_{i=1}^{n} L_i$.

Proof.   Similarly proved as in Proposition 1.   □

Figures 2(a) and 2(b) illustrate THD used for comparison of two-dimensional (2D) and three-dimensional (3D) data, respectively. In individual figures, pattern a shows a complete match and pattern f a complete mismatch. In pattens b and c, there is one mismatch in common, but the mismatch in pattern c is in the center, costing more for the transitions than in pattern b where the mismatch is in the corner. In patterns d and e of Figure 2(a), there are five mismatches in common, but the mismatches in pattern e are more distributed, costing more for the transitions than the rather blocked mismatches in pattern d.

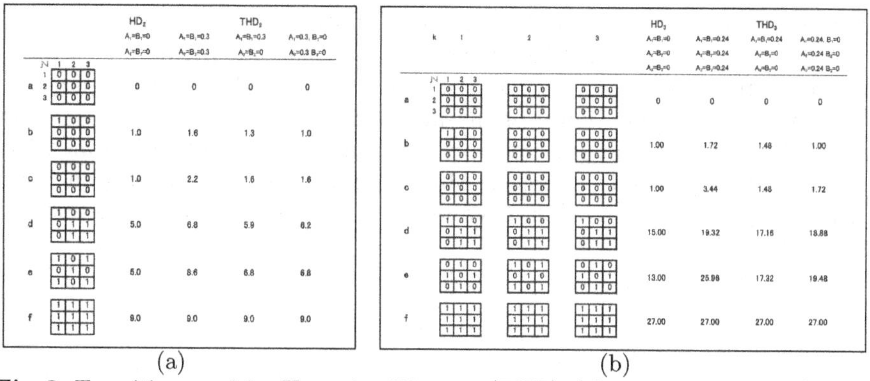

Fig. 2. Transition-sensitive Hamming Distance (THD). (a) THD for 2D data (THD2). Transitions are counted along two different axes, i and j. (b) THD for 3D data (THD3). Transitions are counted along three different axes, i, j, and k.

3 Applications

3.1 Application of TLD: Approximate Name Search

Given two strings, it is simple to compare the whole strings in the flat form, that is, in the manner of *whole-string matching*. When strings are physically large or contain semantic components, however, it is desirable to reflect the structures of the strings in the comparison. In the case of natural language texts, for example, statements can be split into coarse-grain elements (e.g., words) and further into fine-grain elements (e.g., characters). TLD is applicable to the comparison at such different grain levels.

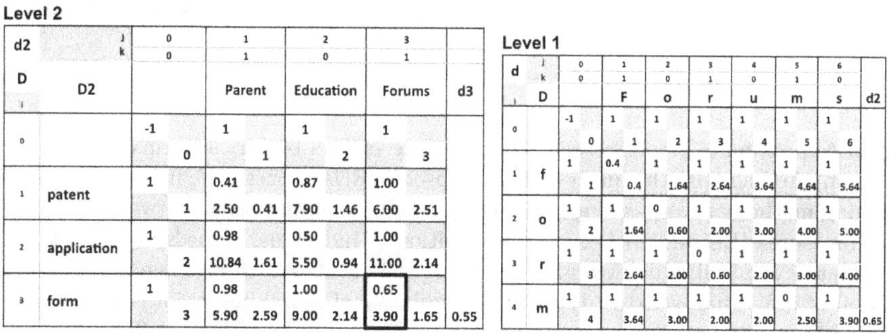

Fig. 3. Hierarchical application of TLD. Strings, 'patent application form' and 'Patent Education Forums', are split into words (Level 2), where a single space is used as the separator. The word distance D (at the left bottom) obtained from word comparison (Level 1) is normalized to the dissimilarity $d2$ (at the left top) to be used for phrase comparison (Level 2), i.e., for words u and v, $d2(u, v) = D(u, v)/max(|u|, |v|)$. The phrase distance $D2$ (at the right bottom) is similarly normalized to the dissimilarity $d3$. Note that the transition cost coefficients, $A=A2=0.4$ and $B=B2=0.5$, and the element dissimilarity, $c(Uppercase, Lowercase)=0.4$, are assumed.

Figure 3 introduces a hierarchical method of string comparison, referred to as the *word-wise matching*. After strings, 'patent application form' and 'Parent Education Forums', are individually split into words, TLD is similarly applied to both word comparison (Level 1) and phrase comparison (Level 2). The word distance D obtained from word comparison is normalized to the dissimilarity $d2$ to be used for phrase comparison. After conducting phrase comparison similarly as word comparison, the obtained phrase distance $D2$ is normalized to the dissimilarity $d3$.

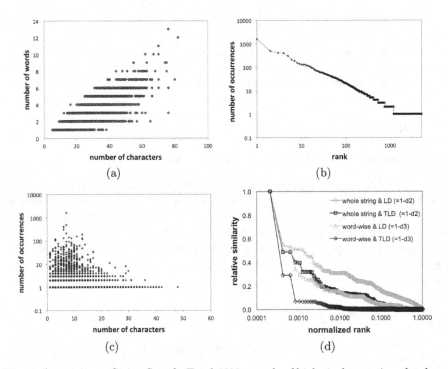

Fig. 4. Approximate String Search. Total 4688 records of biological terms in a database were compared against the query string, 'ribosomal RNA processing', using two different matching methods (*whole-string* and *word-wise*) in combination with two different distance measures (LD and TLD). (a) Composition of individual records in the database. (b) Frequency of individual words constituting the database. (c) Frequency versus composition of individual words. (d) Relative similarity of individual records to the query, evaluated through four different measures, where the transition cost coefficients, A=0.4 and B=0.5, and the element dissimilarity, c(*Uppercae, Lowercase*)=0.4, are assumed.

Using two different matching methods (*whole string* and *word-wise*) in combination with two different distance measures (LD and TLD), we conducted approximate name search on a database containing 4688 different biological terms, including gene and protein names. The content of the database is statistically characterized as shown in Figures 4(a) - 4(c). Individual string records were composed of 5-82 (mean 31) characters and of 1-13 (mean 3) words as shown in Figure 4(a). Total 5115 different words were contained in the database. As shown in Figure 4(b), the frequency of words made a power-law distribution, obeying Zipf's law [16] that is empirically known to hold for natural language texts rather than for artificial language texts. Frequently appearing words were general terms of 3-13 characters, such as 'protein', 'subunit and 'domain' ranked in the top, as shown in Figure 4(c). A string, 'ribosomal RNA processing', composed of 24 characters containing three words, was used as the query. While the query string had a complete match with only one record, individual words frequently appeared in the database, such that 'ribosomal' occurred in 129 records, 'RNA' in 117 records and 'processing' in 33 records.

After comparing individual string records against the query string, the dissimilarity obtained for each string record i was converted to the similarity and normalized against the maximum of similarity to the relative similarity: $S_i = 1 - d*_i$; $s_i = S_i/max(S)$. The relative similarity was ranked and normalized against the total number of records (N) to the relative rank: $r_i = rank(S_i)/N$. Functions of the relative similarity to the normalized rank, evaluated through the four different measures, are plotted in Figure 4(d). The inner the function curve lies, less records will chosen above the given threshold, so that the more strict the evaluation is meant to be. For each matching method, the curve of TLD was found inner than that of LD. For each distance measure, the curve of the word-wise matching was found inner than that of the whole-string matching. Given 0.4 relative similarity as the threshold, for example, only one record was fished by the word-wise matching with TLD, while 4, 3 and 10 records were found by the whole string matching with TLD, the word-wise matching with LD and the whole string matching with LD, respectively. Thus, the combination of the word-wise matching and TLD is suggested to be the most strict evaluation measure that would retrieve the least number of records as those above the given threshold of similarity.

3.2   Application of THD: Image Clustering

THD is applicable to comparison of multi-dimensional data, such as images and volumes, similarly as HD is. To see how differently THD and HD may behave, we conducted pair-wise comparison on 20 different images shown in Figure 5(a), using each of the two distance measures. Individual images are 150x150 black&white pixels in resolution and statistically characterized with their darkness (the ratio of black pixels to the entire image) and normalized standard deviation (the standard deviation of the distances of black pixels from their center, normalized against the maximum of standard deviation), as shown in Figure 5(b).

For the 20 different images, total 190 pair-wise distances were computed with HD or THD as the distance measure, and normalized to dissimilarities to produce a distance matrix. The 190 pair-wise dissimilarities (d) were converted to similarities ($s = 1 - d$) and plotted against their ranks in Figure 5(c). The curve for THD lies inner and more sharply declines than the one for HD does,

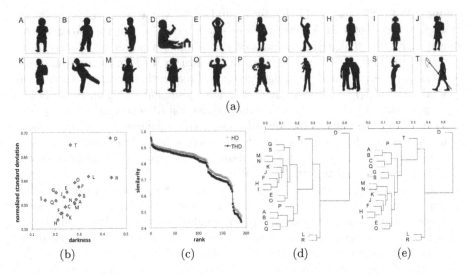

(a)

(b)   (c)   (d)   (e)

Fig. 5. Image Clustering. (a) 20 images (A-T) used in the study. (b) Darkness and normalized standard deviation of individual images. (c) Similarities ($s = 1 - d$) of 190 different pairs of images, where HD (gray circles) and THD (black squares) were used for computation of dissimilarities (d). (d-e) Dendrograms obtained through hierarchical complete-linkage clustering conducted on the distance matrix of the images, using (d) HD or (e) THD as the distance measure. Note that the transition cost coefficients, A=0.5 and B=0.5 were used in the computation of THD.

suggesting that less candidates would be retrieved with THD than with HD, given a certain similarity as the threshold.

Merely the values of similarities decreased with THD? To see if there is any change in the relationship of proximity among the images, we conducted hierarchical complete-linkage clustering [17] on individual distance matrices produced with HD and THD. Figures 5(d) and 5(e) show the resulting dendrograms for HD and THD, respectively. Images D, L, R and T, which are clearly separated from the rest in Figure 5(b), are segregated from the rest similarly in both dendrograms. The difference is in the rest. While image P is grouped together with images A, B, C and Q in one clade and the rest forms another clade in the dendrogram with HD (Figure 5(d)), image P is segregated from images A, B, C and Q and the whole rest are placed in one clade in the dendrogram with THD (Figure 5(e)). Other than image P, the relationship of proximity is maintained in both HD and THD. Thus, the cost of transitions introduced in THD seemingly made the overall evaluation of similarity more strict and contributed to the differentiation of a rather dense clade.

4   Concluding Remarks

In this paper, we presented a new type of distances, called transition-sensitive distances, which count, in addition to the number of mismatches between the compared data, the cost of transitions between positionally adjacent match-mismatch pairs, as part of the distance. By integrating the cost of transitions, conventional distance measures can be extended to be transition-sensitive.

We introduced transition-sensitive variants of LD and HD, referred to as TLD and THD. Compared with LD (or HD), each unit operation of TLD (or THD) needs one additional step of computation for the cost of transitions, but the computational order remains the same. TLD for strings of length m and n is of the order of $O(mn)$. 1D (or 2D) THD for strings of length n (or objects of size $m \times n$) is of the order of $O(n)$ (or $O(mn)$). Also, one additional matrix to store element dissimilarities is required for TLD and 2D THD, but it is possible to re-cycle a matrix of two rows in the actual implementation since only adjacent rows are used in the computation. We showed that while TLD and THD hold prop-erties of the metric as well as LD and HD, they function as more strict distance measures in similarity search applications than LD and HD, respectively.

Properties of the cost of transitions are similar to those of entropy that has been widely used as a measure of randomness (or the amount of information) in various applications, including encoding [18], music analysis [19], linguistic analysis [20], and bioinformatics [21]. The more disperse the mismatches are, the higher the cost of transitions will be. If the transition cost coefficients are set to occupy one mismatch, the cost of transitions is highest when the matches and a mismatches are alternated. When the ratio of mismatches is either smaller or larger than 0.5, the chance of getting transitions is less. Unlike entropy that is defined uniformly throughout the data space, the cost of transitions is de-fined with two separate coefficients, one for ascends and the other for descends. The separation of cost coefficients makes it possible to express the locational allowance on mismatches, i.e., whether mismatches are more acceptable in the leading or tailing part. Behaviors of transition-sensitive distances depending on various cost coefficients remain to be elucidated in the future work.

In summary, the essence of transition-sensitive distances is that the cost of transitions is embedded as part of a distance, rather than regarded as an orthog-onal measure that should be independently applied. At the cost of precision as a measure due to the dimensionality reduction, gained is computational simplic-ity that is required for a large scale of data mining, classification and machine learning. Transition-sensitive distances enable one to segregate data based on the number and dispersion of mismatches in a single pass of computation, so that they are useful for screening large datasets or information streams to retrieve and classify those objects that are highly similar to some part of the target. As more information accumulates and flows, more needs for transition-sensitive distances will rise in various fields of similarity search and applications.

References

1. Navarro, G.: A guided tour to approximate string matching. ACM Computing Surveys 33(1), 31–88 (2001)
2. Cohen, W.W., Ravikumar, P., Fienberg, S.E.: A comparison of string distance metrics for name-matching tasks. In: Proceedings of the ACM Workshop on Data Clearning, Record Linkage and Object Identification (2003)

3. Liu, C.-C., Hsu, J.-L., Chen, A.L.P.: An approximate string matching algorithm for content-based music data retrieval. In: Proceedings of the IEEE International Conference on Multimedia Computing and Systems, vol. 2, p. 9451 (1999)
4. Clifford, R., Iliopoulos, C.: Approximate string matching for music analysis. Soft Computing - A Fusion of Foundatios, Methodologies and Applications 8(9), 597–603 (2004)
5. Yeh, M.-C., Cheng, K.-T.: A string matching approach for visual retrieval and classification. In: Proceeding of the 1st ACM Conference on Multimedia Information Retrieval, pp. 52–58 (2008)
6. Adjeroh, D.A., Lee, M.C., King, I.: A distance measure for video sequences. Computer Vison and Image Understanding 75(1/2), 25–45 (1999)
7. Bezerra, F.N., Leite, N.J.: Using string matching to detect video transitions. Pattern Analysis & Applications 10(10), 45–54 (2007)
8. Hamming, R.W.: Error detecting and error correcting codes. Bell System Technical Journal 29(2), 147–160 (1950)
9. Levenshtein, V.I.: Binary codes capable of correcting deletions, insertions, and reversals. Soviet Physics=Doklady, Cybernetics and Control Theory 10(8), 707–710 (1966)
10. Zelenko, D.: System and method for variant string matching. World Intellectual Property, WO/2009/094649, PCT/US2009/032034 (2009)
11. Gravano, L., Ipeirotis, P.G., Jagadish, H.V., Koudas, N., Murthukrishnan, S., Pietarinen, L., Srivastava, D.: Using q-grams in a DBMS for approximate string processing. IEEE Data Engineering Bulletin 24, 28–34 (2001)
12. Wang, C., Li, J., Shi, S.: N-gram inverted index structures on music data for theme mining and content-basd information retrieval. Pattern Recognition Letters 27(5), 492–503 (2006)
13. Shannon, C.E.: A mathematical theory of communication. Bell System Technical Journal 27, 379–423, 623–656 (1948)
14. Shannon, C.E.: Prediction and entropy of printed english. Bell System Technical Journal 30, 50–64 (1951)
15. Lin, J.: Divergence measures based on the Shannon entropy. IEEE Transactions on Information Theory 37(1), 145–151 (1991)
16. Zipf, G.K.: Human behavior and the principle of least effort. Addison-Wesley, Cambridge (1949)
17. Defays, D.: The efficient algorithm for a complete link method. The Computer 20(4), 364–366 (1977)
18. Yang, S.: Entropy distance. Computing Research Repository, 1303.0070 (2013)
19. Camarena-Ibarrola, A., Chávez, E.: On musical performances identification, entropy and string matching. In: Gelbukh, A., Reyes-Garcia, C.A. (eds.) MICAI 2006. LNCS (LNAI), vol. 4293, pp. 952–962. Springer, Heidelberg (2006)
20. Juola, P.: Cross-entropy and linguistic typology. In: Powers, D.M.W. (ed.) NeMLaP3CoNLL98: New Methods in Language Processing and Computational Natural Language Learning, pp. 141–149. ACL
21. Benson, G.: A new distance measure for comparing sequence profiles based on path lengths along an entropy surface. Bioinformatics 18(suppl. 2), S44–S53 (2002)

Retrieval of Binary Features in Image Databases: A Study

Johannes Niedermayer and Peer Kröger

Ludwig-Maximilians-University Munich, Germany
{niedermayer,kroegerp}@dbs.ifi.lmu.de

Abstract. Many state-of-the art object recognition systems rely on local image features, sometimes hundreds per image, that describe the surroundings of detected interest points by a high-dimensional feature vector. To recognize objects, these systems have to match features detected in a query image against the features stored in a database containing millions or even billions of feature vectors. Hence, efficient matching is crucial for real applications. In the past, feature vectors were often real-valued, and therefore research focused on such feature representations. Present techniques, however, involve binary features to reduce memory consumption and to speed up the feature extraction stage. Matching such binary features received relatively little attention in the computer vision community. Often, either Locality Sensitive Hashing (LSH) or quantization-based techniques, that are known from real-valued features, are used. However, an in-depth evaluation of the involved parameters in binary space has, to the best of our knowledge, not yet been performed. In this paper, we aim at closing this research gap, providing valuable insights for application-oriented research.

1 Introduction

The emergence of sophisticated keypoint detection and description techniques such as SIFT [13] together with the development of efficient indexing techniques [25] have started a revolution in the field of object recognition that is pushed forward by the vast amount of image data available, recorded by smartphones and traditional cameras, and published in social networks, on photo sharing websites, and on the world wide web in general. The availability of head-up-displays such as Google glass[1] will surely further enforce the development of even more efficient and effective solutions.

Many state-of-the-art solutions follow a common pipeline using the filter-refinement paradigm, see Figure 1. First, given an input image, keypoints are extracted at different image scales that aim at identifying interesting regions of the image. Then, for each of these keypoints, a *high-dimensional* feature vector is computed that describes the keypoints' surroundings in a scale-, rotation-, and translation-invariant way. For efficient query processing, these features have to be indexed together with meta-information such as image ID, position, rotation, and scale of the keypoint. Indexing is often achieved with the Bag of Visual Words (BoVW) paradigm [25]: With this approach, feature vectors are *quantized* using, for example, kMeans. Each database-feature is mapped to its closest kMeans centroid. During query processing, for each query feature the closest

---

[1] See http://www.google.com/glass/start/

A.J. Machado Traina et al. (Eds.): SISAP 2014, LNCS 8821, pp. 151–163, 2014.
DOI: 10.1007/978-3-319-11988-5_14 © Springer International Publishing Switzerland 2014

$$Image \xrightarrow{Extract} Feature\ Vectors \xrightarrow{Filter} Candidates\ Images \xrightarrow{Refine} Result$$

Fig. 1. Object Recognition Pipeline

cluster center is computed; database features falling into the same cluster center as the query vote for a given candidate image, filtering irrelevant results and allowing to rank candidates according to the number of features voting for them. Finally, during a refinement step, geometric consistancy between the matched features is checked, and a re-ranking based on the refinement result is performed.

Within the last years, due to the roughness of the binary decisions involved in the BoVW-based approach, this procedure has been softened more and more, as including the closeness between a database feature and a query feature into the ranking can further improve query performance. Therefore, not only techniques for soft assignment [22] have been investigated, but also quantization techniques, initially used for BoVW-based image retrieval, are employed for solving kNN queries on image features [11]; i.e. the focus seems to shift towards kNN queries again.

In the last years, a new kind of feature vector, not real-valued but rather binary-valued (e.g. [23,3]) has emerged. In contrast to real-valued feature vectors, binary features are less redundant, faster to compute and incur less storage overhead at the cost of a lower recognition rate. Binary features are, to the best of our knowledge, seldomly queried with BoVW-based approaches [6,7], but rather by more traditional approximate kNN techniques such as LSH [10,23], but also based on quantization [27,16].

In this paper we aim at shedding more light on this feature matching step on binary features. Equivalently to the idea of Pauleve et al. [20] in the case of real-valued features, we reduce the matching step for binary features to the idea of LSH. Given this interpretation of the problem, all of the currently used approaches only differ in the computation of the hashes, which allows high comparability. During our experiments, we evaluate these hash functions under different conditions in the context of kNN queries and range queries, and investigate under which conditions each of these approaches performs best. We see the necessity of this research for the following reasons. First, as stated before, query processing in binary space should consider its specific properties, such as its thick boundaries [27]. On the other hand, however, a variety of techniques known from real-valued features is applied to binary features, such as quantization, although a thorough evaluation of these techniques in binary space has not yet been performed. Therefore in this paper, we aim at closing this research gap.

The structure of this paper is as follows. Section 2 provides an overview over related work and the historical context. In Section 3 we review currently available techniques used for querying binary features in image databases. In Section 4 we provide an in-depth analysis of the described techniques, complementing the results from [27,16]. Section 5 concludes this work.

2   Related Work

To integrate this paper into the body of research, in this section we draw an arc over related topics, including quantization in real-valued feature spaces, binarization of real-valued features, and the direct extraction of binary features. In Section 3 we provide an in-depth review of approaches for query processing in binary space that are currently employed for object recognition. A recent survey on approximate nearest neighbor queries in general has been provided in [19]. Another survey addressing high-dimensional indexing especially in the context of image retrieval has been published as well [2].

Nearest Neighbor Search on Image Features. The idea of quantization-based nearest neighbor search in image databases has developed from Bag of Visual Words (BoVW) based retrieval techniques, initially proposed in [25]. Such techniques convert sets of feature vectors to a frequency representation similar to the word vectors in text retrieval. These word vectors can be searched by employing well-known text-retrieval techniques. For finding a frequency representation however, feature vectors have to be quantized, i.e. mapped to a set of representants. The cluster ID of these representants is then used as artificial words. Quantization is usually performed with variants of kMeans clustering [25]. With larger vocabularies, however, quantization based on kMeans clustering becomes more and more costly, as assigning a value to its representant is linear in k. To reduce the high complexity of assigning feature vectors to cluster centers during query evaluation, a variety of non-optimal but fast quantization techniques such as Hierarchical kMeans [17,24], approximate kMeans [21], Product Quantization [11] and Residual Vector Quantization [5] have been proposed. After employing quantization for BoVW-based retrieval, it has also been investigated in the context of kNN search [11], which has since then received quite some attention.

Binary Features. Related to the idea of quantization is the task of extracting distance-preserving binary codes from real-valued features. *Binarization* can be seen as some sort of compression on real-valued image features. The resulting binary codes can be queried by either retrieving features with equivalent codes from the database if the codes are relatively short, or by employing LSH-based hashing techniques [10]. The generation of binary signatures corresponds to the problem of finding a function $f : \mathbb{R}^{d_1} \to \mathbb{B}^{d_2}$ where $\forall x_1, x_2 \in \mathbb{X} : d_{L_2}(x_1, x_2) \propto d_H(f(x_1), f(x_2)))$ with $d_{L_2}(x_1, x_2)$ denoting the Euclidean distance between feature vectors, $d_H(f(x_1), f(x_2))$ corresponding to the Hamming distance between the transformed points, and $\mathbb{X}$ corresponding to the set of points stored in the database. Examples of binarization techniques include the approach from [26], Random Maximum Margin Hashing [12], Scalar Quantization [29], Spherical Hashing [9] and also combinations of kMeans quantization and binarization that assign meaningful IDs to cluster centers [8]. Note that in this paper we do not address the problem of binarization, i.e. computing $f(x)$, but rather aim at querying binary features efficiently in the context of image retrieval. The binarization can be seen as a preprocessing step before indexing in order to reduce storage overhead.

In recent times, in addition to binarization, the *direct* extraction of binary features from image data has become popular (e.g. ORB [23] and FREAK [3]), as this can increase not only storage efficiency but also computational efficiency.

3 Querying Binary Features: An Overview

As explained in Section 2, binary signatures are either derived by binarization of real-valued features or extracted directly from the image. In computer vision, they are usually queried using the Hamming Distance. Depending on the matching task and required accuracy, it is possible to use exact matching approaches (e.g. [18]), approximate solutions based on LSH [10], or approximate solutions based on quantization [16,27]. Equivalently to the case of *real-valued* features, all of these approaches can be reduced to the idea of LSH [20]; we will follow this path in this publication. The *binary space* however, according to [27] is different in its behaviour, therefore a specialized evaluation is appropriate. In the following, we review the current solutions.

Locality Sensitive Hashing. As all of the solutions for matching binary features in image databases can be reduced to Locality Sensitive Hashing (LSH) [10], we start our discussion with this approach. LSH and its derivates have been successfully used for object recognition [23] based on binary features. The approach is furthermore used as a baseline for other techniques [16,27]. Note that, given that the extracted binary signatures are very short (e.g. 24 bits), they could be used as hash keys without processing, and queried directly without employing LSH. However currently, binary vectors have a length of several hundred bits (e.g. ORB: 256 bits [23], FREAK: 512 bits [3]), and therefore an intermediate LSH-based hashing step has to be employed.

Locality sensitive hashing has been initially proposed by Indyk and Motwani [10], proposing dedicated hash functions for Hamming space and theoretical approximation guarantees. LSH is based on a family of functions $\mathcal{H} = \{h : S \to U\}$ that map values from a space S (in our case the binary vector space) to binary strings from space U. To be useful for similarity search, $h(p)$ must be (r_1, r_2, p_1, p_2)-sensitive, i.e. points closer to a given reference point have a higher probability of beeing assigned the same hash as points further away. From this family $\mathcal{H}$, a family $\mathcal{G} = \{g : S \to U^k\}$ with $g_j(p) = (h_1(p), \ldots, h_k(p)), h_i \in \mathcal{H}$ is constructed, i.e. $\mathcal{G}$ corresponds to a concatenation of different hash functions of the same family. For index generation, t of these functions $g_j(p)$ are generated and for each of them a hash table is constructed from the features stored in the database. Query processing, both kNN and range queries, can be achieved by accessing, for a query vector q the hash entry $g_j(q)$ from hash table j, concatenating the resulting candidates, and evaluating the query predicate on these candidates.

Initially, the following functions h_i (projecting a binary vector onto atomic random dimensions) and g_j (projecting on a random lower-dimensional space), were proposed for search in Hamming Space [10]:

$$\mathcal{H} = \{h : h_i(b) = b_i\} \text{ and } \mathcal{G} = \{g : g(p) = (h_{i_1}, \ldots, h_{i_n}), \{i_1, \ldots, i_n\} \subset \{1, \ldots, n\}\}$$

For example, a function $h_2(b)$ would project a 3D binary vector to its second dimension, e.g. $h_2((1, 0, 1)) = 0$. A function $g(p)$ then concatenates different functions h, producing a subvector of the initial one. For a function g mapping to the first two dimensions, we would get a vector $g((1, 0, 1)) = (1, 0)$.

To be effective, LSH has to query a relatively high number t of different hash tables with different hash functions, incurring high storage overhead, as each of the tables has to store references to all of the indexed features, resulting in a memory consumption linear in t. To solve this issue, [15] proposed to use Multi-Probe LSH that visits different hash cells close to the query point in a single hash table, similar to the soft assignment policy in computer vision [22]. This procedure significantly reduces storage overhead and has been successfully used for querying binary features in the computer vision community [23]. Trzcinski et al. [27] proposed optimized hash functions for

LSH-based search in Hamming space, aiming at improving the efficiency of this class of approaches.

Exact Solutions. Computing the Hamming distance of binary vectors is generally extremely fast, as this distance calculation can be split into an XOR operation, followed by counting the bit population in the resulting bit string. If the number of features stored in the database is moderate, matching such signatures can be performed by running a linear scan over the database. Besides this trivial solution, both ϵ-range queries and kNN queries can be solved on average in sublinear time by employing the approach from [18]. Similar to LSH, the authors apply multiple hash functions on disjunctive substrings of the b-bit binary word, building an index structure for each of the m subspaces of length b/m. ϵ-Range queries can be solved by enumerating, in each subspace, the candidates in all bins with range m/ϵ. This procedure provides a superset of the actual candidates, which is then refined exhaustively. kNN queries can be solved in a similar fashion. The technique is based on the assumption that feature vectors are close to uniformly distributed. It can be interpreted as a version of exhaustive Multi-Probe LSH, with hash functions close to the original one, but with additonal constraints:

$$subspace(g_i) \cap subspace(g_j) = \emptyset \text{ and } |\bigcup_i subspace(g_i)| = b$$

Unfortunately, given a large database, the search time can increase up to tenths of a second. Given that a single image can contain hundreds of features, the matching times become tens of seconds, significantly too much in real-world applications. Other exact solutions similar to [18] exist, mainly varying in the number of substrings to use for index generation (see [28]).

Quantization-Based. Despite the success of LSH-based methods for querying binary features in Hamming space, techniques based on quantization (and therefore variants of kMeans) have been investigated recently. The approaches are either plugged into the BoVW model [6,7], or used for nearest neighbor search [16,27]. Quantization of both binary and real-valued image features can be sketched as follows: First, perform a kMeans clustering of the extracted image features contained the database. Then, assign each database vector to its closest cluster center. During query processing, assign the query vector to its closest cluster center, and return all database vectors assigned to the same cluster.

For kMeans the probability of two features assigned to the same cluster center being spatially close is high, while the probability of two features assigned to different cluster centers being spatially close is relatively low. Therefore, according to [20], quantization-based approaches can be seen as special LSH hash functions: the hash corresponds to the ID of the closest cluster center, i.e. features assigned to the same cluster center have the same hash. This results in the following mathematical definition:

$$\mathcal{H} = \{h : h_i(b) = argmin_{c \in centers_i} d_H(v(c), b))\} \text{ and } \mathcal{G} = \{g : g(p) = h_i(p)\}$$

with $centers_i$ the cluster IDs of clustering i and $v(c)$ the vector corresponding to ID c. The hash function family $\mathcal{G}$ in this case is trivial, as each hashing function of $\mathcal{G}$ consist only of a single function h_i. On the other hand $h_i(p)$ becomes more complex as it consists of more than one bit. For standard kMeans clustering, due to the assignment step, the hash computation is exponential in the hash length, therefore approximate solutions have been developed [17,24,21,11,5]. For quantizing binary vectors, solutions based on kMedoids, kMedians or kMeans can be employed. Using kMeans directly does not optimize according to $L1$ distance, but rather according to the squared Euclidean distance, making this solution theoretically inapplicable to Hamming space. Most often, even if unknowingly, the kMedians [4] algorithm has been used which optimizes cluster centers according to the $L1$ norm, directly corresponding to the Hamming distance on binary feature vectors. kMedians can be optimized for binary features, as the median computation degenerates to a majority voting in this case [14].

Querying binary features in BoVW-based systems has most likely been initially proposed by Galvez-Lopez and Tardos [6,7]. To achieve acceptable runtime performance, the authors employed the idea of hierarchical kMeans (HkM)[17]. Hieararchical kMeans (HKM) performs kMeans clustering with a relatively low k'. It then assigns each point in the database to the corresponding cluster. For each of these sets of feature vectors, it then recursively clusters the set until a given height l is reached, resulting in a tree-like structure with k'^l leaf nodes. Due to the resulting tree structure, the computational complexity of assigning feature vectors to cluster centers decreases to $O(k'L)$, however the storage complexity of the approach remains in $O(k'^L)$ and some accuracy is lost. Utilizing the idea of HkM, Muja and Lowe [16] and Trzcinski et al. [27] simultaneously proposed to use a forest of random clustering trees on binary feature vectors. Mapping this to LSH, the leaves of a tree provide the hash functions, and each tree provides the hash function for a different hash table. Each of the trees is similar to a HkM tree, however cluster centers are not assigned by kMedians, but rather by randomly selecting cluster centers from the data: if multiple hash tables (i.e. trees) are employed, the corresponding hash functions should be *independent*. If a clustering would be performed, in the best case this clustering would represent a global optimum, generating exactly the same *non-independent* clustering for each function from $\mathcal{G}$. However, as the iterative kMedians based clustering approaches only find local minima instead of global ones, kMedians clustering could be employed with different initial seeding, as it has been done in the case of real valued vectors [20]. We will evaluate both randomized and optimized clusterings in Section 4. In [16], better retrieval performance than LSH-based approaches is achieved with multi-probing. In contrast, [27] achieved slightly worse results than LSH, however without multi-probing. Hierarchical kMedians clustering has also been employed by [14], however with optimized cluster centers.

Other quantization techniques that have been employed for real-valued image features could also be used for binary features, such as residual vector quantization [5], and product quantization [11]. In the context of residual vector quantization that involves subtraction operations, the subtraction operation would have to be replaced by an XOR operation to make the residuals remain in binary space.

4   Experimental Evaluation

For our experimental evaluation we extracted ORB [23] features from the ALOI dataset[2] using OpenCV. We aimed at extracting binary features instead of binarizing real-valued ones as this can lead to performance advantages in real-world applications. Other

---

[2] http://aloi.science.uva.nl

Parameter	Value
#probes	1, **20**, 100, 1000, 10000
#tables	**1**, 2, 4, 8
#bits (LSH)	8, 10, 12, 14, **16**, 20
k (kMedians)	$2^8, 2^{10}, 2^{12}, 2^{14}, \mathbf{2^{16}}$
h (HkM)	**4**
k' (HkM)	$2^2, 2^3, \mathbf{2^4}, 2^5, 2^6$

(a) Experimental setup. Values in bold font denote default values.

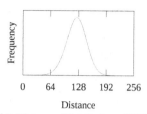

(b) Distance Distribution (ORB, 10k Feature Vectors)

binary features than ORB could be chosen as well; the decision depends on the application. From the originally $\sim$30 million features we used a random sample of size 10^7 as database content. Distances in the dataset follow a normal distribution with a maximum frequency at a distance of 124 and covering nearly all possible distances, see Figure 2(b). The different quantizations (kMedians, HkM and random HkM (RHkM)) were computed from a subset of about 7 million of the 30 million features containing, for each of the 1000 objects, 18 of the 72 images per object. We only considered the images varying in camera angle, but not the images varying e.g. in light color, as these are assumed to produce very similar features. Tailoring quantization to a specific database is reasonable as it makes quantization techniques more powerful. In contrast to [16,27], who have first proposed to use hierarchical kMeans in Hamming space, we aim at investigating the impact of system-inherent parameters on the performance of the different hashing techniques, namely the number of probes, the number of tables, the number of bits and the database size. These parameters have not been evaluated in depth or in comparison to competitors. In our experiments, we fix each of these parameters to a given value and vary a single parameter to provide a wide image of the algorithms' performance. The default parameters are provided in Table 2(a). As the parameter k is overloaded (for kMeans and for the number of neighbors), we denote the number of neighbors retrieved in a kNN query as $|NN|$.

All of the evaluated approaches are configured to generate hashes of the same length (16 bits). Equivalent to [20] we set the number of hash tables to a single one in our default setting. Although this is not realistic, it gives insight into the performance of the actual hash function. As it does not provide insights into the *independence* of different hash functions which becomes relevant when using more than one hash table (the default with LSH), we will evaluate this behaviour in Figure 3. The number of neighbors $|NN|$ was set arbitrarily to $|NN| = 10$, our experiments for $|NN| = 2$ and $|NN| = 100$ have to be omitted for space constraints. We will, however mention variations in the results for different values of $|NN|$ in the corresponding sections. In our experiments we concentrate on analyzing the *Recall* for each set of parameters, for both range- and kNN queries; a short excursus also considers the BoVW paradigm. Where necessary we also consider different performance measures such as the number of distance calculations; we favoured this measure over runtime as it provides a platform- and implementation-independent measure of the computational complexity of a given approach. However note that other costs, such as restructuring the priority queue of the quantization-based approaches, is not considered.

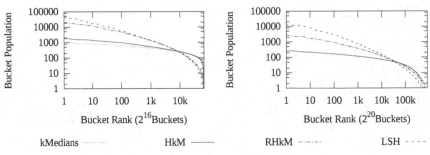

kMedians ········   HkM ——   RHkM –·–·–   LSH - - - -

Fig. 2. Populatation of buckets (log-log-space)

The optimized clusterings have been generated with a kMedoids approach. First, a random sample of 50% of the training features was selected and clustered with 5 iterations. Then, given the resulting centroids as initial cluster centers, the whole training set was clustered, stopping after an additional 10 iterations. The experimental evaluation has been conducted in the JAVA-coded framework ELKI[1] which has been specifically designed for the performance evaluation of data mining and query processing algorithms, providing variants of the kMeans algorithm, including kMedians.

4.1   Nearest Neighbor Queries

Distribution of Hash-Code Frequencies. Following [20], let us first evaluate the population of hash buckets; Figure 2 visualizes the distribution for hashes of length 16 and 20 (for 20 bits, kMedians has not been evaluated due to its high complexity). If the population is equi-distributed, this leads to a good selectivity, as regardless of the location of a query in binary space, the number of candidates in the corresponding bucket is similar. The more uneven the distribution of features over different hashes, the more irrelevant features have to be refined if a highly populated bucket is found, leading to high runtimes; if a poorly populated bucket is accessed, the recall of the query becomes unnecessarily low. Similar to the case of real-valued features [20], the original LSH functions lead to an unbalanced distribution of hashes. On the other hand, both RHkM and HkM perform better. However, there is a significant difference between the random and the optimized version; RHkM performs very close to the original LSH, while the distribution of HkM is much more uniform, quite close to kMedians that performs best. The skewed distribution of the HkM-based approaches is traded for a better computational performance: kMedians can only be seen as a theoretic solution, as it implies scanning all cluster centers during each query. The number of cluster centers scanned increases exponentially in the number of hash bits. Although these distance calculations are very fast in binary space, they can still incur a large overhead for higher bit lengths.

Number of Tables. In the following experiment we aim at investigating the effect of the number of tables on the different approaches (see Figure 3 left) in order to find out how well several hash tables complement each other. For this experiment, we kept the number of probes constant and split them between an increasing number of tables. For traditional hash functions, probes are split equally between tables. For quantization-

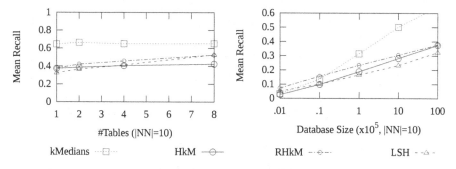

Fig. 3. Varying # Tables (left) and Database Size (right)

based approaches, the cluster centers of all tables are ranked together, resulting in a not necessarily uniform split between tables. The experimental setup aims at showing if the performance gain is actually contributed to the higher number of tables rather than the increasing number of probes.

An increasing number of tables makes mainly sense for the traditional LSH functions, and less strongly for the randomized HkM approach. The performance gain of optimized quantization-based approaches diminishes with increasing number of tables, as the different quantizations are not independent. This observation is different to the real-valued case, where Pauleve et al. [20] stated that initializing kMeans with different seeds can provide enough randomness to build independent hash tables. Note that, if the number of tables becomes sufficiently large, even the random HkM-based approach loses against the LSH-based hash functions. For $|NN| = 10$, the equilibrium was at eight tables; for $|NN| = 2$ it became lower (two tables), and for $|NN| = 100$ it became larger. Therefore, if memory is important, quantization-based approaches can be a useful solution. If recall is an issue and memory is not, LSH-based functions are the matter of choice.

Database Size. As the number of objects indexed in a database affects the nearest neighbor distance of objects, varying the database size can also affect the performance of hash functions: If the number of bits in a LSH function is too high, the NN range of queries is severely restricted, such that in small databases the NNs will be found only with small probability. However if the database size increases, the NN-range of an object shrinks. Therefore, recall is positively affected by an increasing database size, the largest increase can be observed for the kMedians based quantization approaches (see Figure 3 right). As a result, given a large database size and a specified number of bits, it can make sense to use methods based on quantization.

Number of Probes. Besides varying the number of hash tables (which trades space for recall), recall can also be traded for computational complexity by increasing the number of probes, see Figure 4 (left). Our experiments indicate that for small $|NN|$ and with an increasing number of probes, the HkM, RHkM and LSH-based approaches become very similar, leading to no significant performance gain of quantization. For larger $|NN|$, LSH catches up only for larger number of probes. The kMedians based baseline shows a better performance than the rest, reaching recall rates of greater than

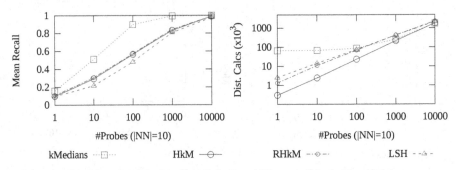

Fig. 4. Varying # Probes, Recall (left) and Distance Calculations (right)

0.9 already at a hundred probes. By comparing the number of distance calculations (see Figure 4 (right)) we aim at providing implementation-independent insights into the computational efficiency of the different approaches. The experiment shows that HkM can significantly better reduce the number of distance calculations than LSH and kMedians, as this approach leads to a much more uniform distribution of values in different buckets as LSH (cp. Figure 2), and does not have the linear complexity of computing the closest cluster center as kMedians. This, however, does not hold as strictly for RHkM, indicating a similar runtime for RHkM and LSH. For additional runtime experiments, we refer to the original papers [16,27].

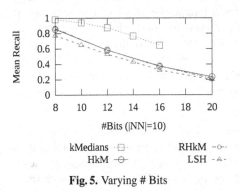

Fig. 5. Varying # Bits

Hash Length. In the following experiment, Figure 5, we aim at investigating the impact of a varying hash length on the recall of the different approaches. For the sake of easy configurability it is important that all approaches behave similarly, i.e. the choice of the best algorithm does not depend on the number of bits used, which is the case in our setting. We would like to mention that with increasing $|NN|$ the difference between quantization-based approaches and LSH becomes larger: LSH loses recall relatively to the other approaches, indicating that it can make sense for large $|NN|$ (e.g. $|NN| = 100$) to use quantization. As in image recognition it is likely that $|NN|$ is chosen relatively large to generate more candidate matches and to be more robust to noise, HkM can achieve a better performance than the LSH hashing functions.

4.2   Range Queries and BoVW

Although this paper concentrates on nearest neighbor queries, for the sake of completeness we also want to shed some light on range queries. Besides mean recall we evaluate the mean false positive rate. In Figure 6 we plot the recall given a specific radius and equivalently the false positive rate. Note that for small ranges results become more noisy

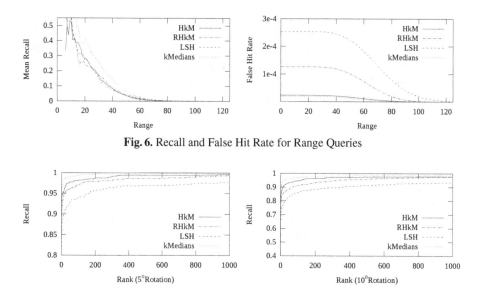

Fig. 6. Recall and False Hit Rate for Range Queries

Fig. 7. Applying the Hashing Functions to a BoVW-based Ranking

as an object beeing a result of a range query with a small range is quite improbable (see Figure 2(b)). In our setting, the mean recall of the different approaches behaves very similar. Only kMedians manages to achieve a significantly higher recall, by increasing the probability of *wanted* hash collisions.

Although the false hit rate varies significantly for the different approaches, it does not have a large effect on range queries, as unwanted results are filtered out during refinement. However, it provides useful insights into their performance when plugged into the BoVW framework: Considering that recall is similar for both HkM and LSH, would it make sense to use LSH functions for binary vectors in the BoVW-paradigm, directly treating keypoints falling into the same buckets as matches? Not necessarily, as not only recall matters, but also the false positive rate (see Figure 6 right). As the difference between BoVW and LSH is mainly that BoVW skips the pruning step, it is of major importance to avoid false positives: false positives put noise on the word vector by producing inaccurate word assignments. In Figure 7 we plugged the different hash functions into the BoVW pipeline (without geometric verification). We used the dataset from training the quantization approaches as database. As it contains all images in 20 degree angular steps, we queried with images having an angular distance of 5 and 10 degrees (Figure 7 left and right, respecively), visualizing the recall (of images, not features) with increasing rank. The different approaches have similar recall curves in Figure 6, but LSH performs significantly worse than the remaining approaches.

5   Conclusion

To conclude, given that a nearest neighbor search in Hamming space has to be performed, both quantization-based approaches and LSH provide good results, but LSH can not be beaten for larger number of tables. For a low number of tables, HkM and

RHkM are the matter of choice. For BoVW-based image retrieval, similarly to the real-valued domain, the quantization-based approaches are the matter of choice: As the probability of false hits of quantization-based approaches is significantly lower, the necessicty of refinement of these solutions decreases, making them applicable to the BoVW paradigm. In the future we would like to include different types of binary features into our evaluation, such as FREAK [3]. As these features can follow a different distribution, the performance of hashing techniques might differ.

References

1. Achtert, E., Kriegel, H.P., Schubert, E., Zimek, A.: Interactive data mining with 3d-parallel-coordinate-trees. In: Proc. SIGMOD, pp. 1009–1012 (2013)
2. Ai, L.F., Yu, J.Q., He, Y.F., Guan, T.: High-dimensional indexing technologies for large scale content-based image retrieval: a review. Journal of Zhejiang University SCIENCE C 14(7), 505–520 (2013)
3. Alahi, A., Ortiz, R., Vandergheynst, P.: Freak: Fast retina keypoint. In: Proc. CVPR, pp. 510–517. IEEE (2012)
4. Bradley, P.S., Mangasarian, O.L., Street, W.N.: Clustering via concave minimization. In: Proc. NIPS, pp. 368–374 (1996)
5. Chen, Y., Guan, T., Wang, C.: Approximate nearest neighbor search by residual vector quantization. Sensors 10(12), 11259–11273 (2010)
6. Gálvez-López, D., Tardós, J.D.: Real-time loop detection with bags of binary words. In: Proc. IROS, pp. 51–58 (2011)
7. Gálvez-López, D., Tardós, J.D.: Bags of binary words for fast place recognition in image sequences. IEEE Transactions on Robotics 28(5), 1188–1197 (2012)
8. He, K., Wen, F., Sun, J.: K-means hashing: An affinity-preserving quantization method for learning binary compact codes. In: Proc. CVPR, pp. 2938–2945 (2013)
9. Heo, J.P., Lee, Y., He, J., Chang, S.F., Yoon, S.E.: Spherical hashing. In: Proc. CVPR, pp. 2957–2964 (2012)
10. Indyk, P., Motwani, R.: Approximate nearest neighbors: Towards removing the curse of dimensionality. In: Proc. STOC, pp. 604–613 (1998)
11. Jégou, H., Douze, M., Schmid, C.: Product quantization for nearest neighbor search. IEEE PAMI 33(1), 117–128 (2011)
12. Joly, A., Buisson, O.: Random maximum margin hashing. In: Proc. CVPR, pp. 873–880 (2011)
13. Lowe, D.G.: Distinctive image features from scale-invariant keypoints. International Journal of Computer Vision 60(2), 91–110 (2004)
14. Luo, Q., Zhang, S., Huang, T., Gao, W., Tian, Q.: Scalable mobile search with binary phrase. In: Proc. ICIMCS, pp. 66–70 (2013)
15. Lv, Q., Josephson, W., Wang, Z., Charikar, M., Li, K.: Multi-probe lsh: Efficient indexing for high-dimensional similarity search. In: Proc. VLDB, pp. 950–961 (2007)
16. Muja, M., Lowe, D.G.: Fast matching of binary features. In: Proc. CRV, pp. 404–410 (2012)
17. Nistér, D., Stewénius, H.: Scalable recognition with a vocabulary tree. In: Proc. CVPR, pp. 2161–2168 (2006)
18. Norouzi, M., Punjani, A., Fleet, D.J.: Fast search in hamming space with multi-index hashing. In: Proc. CVPR, pp. 3108–3115 (2012)
19. Patella, M., Ciaccia, P.: Approximate similarity search: A multi-faceted problem. J. Discrete Algorithms 7(1), 36–48 (2009)

20. Paulevé, L., Jégou, H., Amsaleg, L.: Locality sensitive hashing: A comparison of hash function types and querying mechanisms. Pattern Recognition Letters 31(11), 1348–1358 (2010)
21. Philbin, J., Chum, O., Isard, M., Sivic, J., Zisserman, A.: Object retrieval with large vocabularies and fast spatial matching. In: Proc. CVPR, pp. 1–8 (2007)
22. Philbin, J., Chum, O., Isard, M., Sivic, J., Zisserman, A.: Lost in quantization: Improving particular object retrieval in large scale image databases. In: Proc. CVPR, pp. 1–8 (2008)
23. Rublee, E., Rabaud, V., Konolige, K., Bradski, G.: Orb: an efficient alternative to sift or surf. In: Proc. ICCV, pp. 2564–2571 (2011)
24. Schindler, G., Brown, M., Szeliski, R.: City-scale location recognition. In: Proc. CVPR, pp. 1–7 (2007)
25. Sivic, J., Zisserman, A.: Video google: A text retrieval approach to object matching in videos. In: Proc. ICCV, pp. 1470–1477 (2003)
26. Torralba, A., Fergus, R., Weiss, Y.: Small codes and large image databases for recognition. In: Proc. CVPR, pp. 1–8 (2008)
27. Trzcinski, T., Lepetit, V., Fua, P.: Thick boundaries in binary space and their influence on nearest-neighbor search. Pattern Recognition Letters 33(16), 2173–2180 (2012)
28. Zhang, X., Qin, J., Wang, W., Sun, Y., Lu, J.: Hmsearch: an efficient hamming distance query processing algorithm. In: Proc. SSDBM, pp. 1–12 (2013)
29. Zhou, W., Lu, Y., Li, H., Tian, Q.: Scalar quantization for large scale image search. ACM Multimedia, 169–178 (2012)

The Similarity-Aware Relational Intersect Database Operator

Wadha J. Al Marri[1], Qutaibah Malluhi[1], Mourad Ouzzani[2],
Mingjie Tang[3], and Walid G. Aref[3]

[1] Qatar University, Doha, Qatar
200450064@student.qu.edu.qa, qmalluhi@qu.edu.qa
[2] Qatar Computing Research Institute, Doha, Qatar
mouzzani@qf.org.qa
[3] Purdue University, West Lafayette, IN, USA
tang49@purdue.edu, aref@cs.purdue.edu

Abstract. Identifying similarities in large datasets is an essential operation in many applications such as bioinformatics, pattern recognition, and data integration. To make the underlying database system similarity-aware, the core relational operators have to be extended. Several similarity-aware relational operators have been proposed that introduce similarity processing at the database engine level, e.g., similarity joins and similarity group-by. This paper extends the semantics of the set intersection operator to operate over similar values. The paper describes the semantics of the similarity-based set intersection operator, and develops an efficient query processing algorithm for evaluating it. The proposed operator is implemented inside an open-source database system, namely PostgreSQL. Several queries from the TPC-H benchmark are extended to include similarity-based set intersetion predicates. Performance results demonstrate up to three orders of magnitude speedup in performance over equivalent queries that only employ regular operators.

1 Introduction

Diverse applications, e.g., bioinformatics [1], data compression [2], data integration [3], and statistical classification [4] mandate that their underlying database systems provide similarity-aware capabilities as a means for identifying similar objects. Several similarity-aware relational operators have been proposed that introduce similarity processing at the database engine level, e.g., similarity joins and similarity group-by's [5], [6], [7]. In this paper, we introduce similarity-aware set intersection as an extended relational database operator.

In standard SQL, relational set operations are based on exact matching. However, assume that we want to find common readings that are produced by two sensors. Assume further that the sensor readings are stored in two separate tables. The standard SQL set intersect operator is not suitable for intersecting these two tables to get the common sensor readings; sensor readings may be similar but not necessarily identical. Thus, it is desirable to perform similarity

A.J. Machado Traina et al. (Eds.): SISAP 2014, LNCS 8821, pp. 164–175, 2014.
DOI: 10.1007/978-3-319-11988-5_15 © Springer International Publishing Switzerland 2014

set intersection to find similar readings in the two tables. While the focus of this paper is on the similarity set intersection operator, we study the other similarity set operators, namely similarity-based set union and similarity-based set difference in [8]. We omit their description for space limitation.

Several relational database operators have introduced similarity into SQL. The similarity group-by operator assigns every object to a group based on a similarity condition, e.g., as in [9,10,3]. Similarity join retrieves pairs of objects that overlap based on a join attribute using a predefined threshold. Several types of similarity join have been proposed, e.g., [11,12,5,13,14]. While the similariy join reports the joining objects, similarity intersection requires union-compatible input relations and returns all similar objects from both relations. An extension to SQL to support nearest-neighbor queries has been studied extensively, e.g., see [15]. k-Nearest-neighbor can be viewed as one form of similarity as each point or tuple is connected with its k-closest (or most similar) values. SIREN [16,17] allows expressing similarity queries in SQL and executing them via a similarity retrieval engine. SIREN is a middle-tier implemented between an RDBMS and application programs that processes and answers similarity-based SQL queries issued by the application. In [18], extensions to SQL make similarity operators first-class database operators by implementing the operators inside the database engine. None of the previous work addresses similarity-based set interesection, which is the focus of our paper.

The contributions of this paper are as follows. (1) We introduce the Similarity-aware Set Intersection Operator that extends the standard SQL set intersection to produce results based on similarity rather than on equality (Section 2). (2) We develop an efficient algorithm for the proposed operator (Section 3) and implement it inside PostgreSQL, an open-source relational database management system [19]. (3) We evaluate the performance of the proposed algorithm and its scalability properties using the TPC-H benchmark [20]. We extend several queries from the TPC-H benchmark to include similarity-based set intersetion predicates. Performance results demonstrate up to three orders of magnitude enhancement in performance over equivalent queries that only employ regular relational operators (Section 4).

2   Semantics of Similarity-Based Relational Intersect

Let Q (resp. P) be a relation with k attributes denoted by $a_1, a_2, \ldots, a_k$ (resp. b) and n (resp. m) tuples $A_1, A_2, \ldots, A_n$ (resp. B), where the schemas of P and Q are compatible. To express the similarity between two tuples, one may use several possible functions to describe the distance between each pair of corresponding attribute values, e.g., edit distance, p-norm, or Jaccard distance. Let $D = \{dis_1, dis_2, \ldots, dis_r\}$ be r distance functions. For any $dis_t \in D$, let $dis_t(A_i.a_t, B_j.b_t)$ be the distance corresponding to Attribute a_t between the tuple pair (A_i, B_j) using the distance function dis_t.

We adopt the following similarity predicate: Given r thresholds $\epsilon_1, \epsilon_2, \ldots, \epsilon_r$ that are assigned to each of the attributes $a_1, a_2, \ldots, a_r$, respectively, where $r \leq$

k, we say two tuples A_i and B_j match iff: $pred(A_i, B_j) = dis_1(A_i.a_1, B_j.b_1) \leq \epsilon_1$ AND $dis_2(A_i.a_2, B_j.b_2) \leq \epsilon_2 \ldots$ AND $dis_r(A_i.a_r, B_j.b_r) \leq \epsilon_r$. If $r < k$, the set of thresholds $\epsilon_{r+1}, \ldots, \epsilon_k$ are assumed to have the value zero. An ϵ_i of value zero has to be assigned explicitly if at least one later attribute is assigned an $\epsilon > 0$. Furthermore, an ϵ_i can be assigned an infinity value.

Similarity-aware Set Intersection takes the tuples of two tables as input and returns only those tuple pairs that are similar within a threshold from both tables. More formally, given two tables, say P and Q, that have identical (or compatible) schemas, and a smilarity predicate $pred(A, B)$, the similarity-aware set intersection operation is defined as follows.

$$Q \, \widetilde{\cap} \, P = \{A \mid A \in Q, \, \exists B \in P : pred(A, B)\} \; \cup \qquad (1)$$
$$\{B \mid B \in P, \, \exists A \in Q : pred(A, B)\}$$

Example: Consider the following two tables Q and P; each having a single compatible attribute, where attribute values x and $\widetilde{x}$ are assumed to be similar. $Q = \{a, b, c, d, e, f, g, z\}$ and $P = \{\widetilde{a}, \widetilde{b}, \widetilde{c}, h, i, j, k, l, z\}$ For all calculated $pred(t_1, t_2)$ such that $t_1 \in P$ and $t_2 \in Q$, only $pred(a, \widetilde{a}), pred(b, \widetilde{b}), pred(c, \widetilde{c})$, and $pred(z, z)$ evaluate to true. Thus, $P \, \widetilde{\cap} \, Q = \{a, b, c, \widetilde{a}, \widetilde{b}, \widetilde{c}, z\}$.

Three-way similarity-aware set intersection, denoted by $\widetilde{\cap}$, is defined as follows. Let Q, P and R be three tables such that $\widetilde{\cap}(Q, P, R) = U$. Each tuple in U exists in at least one table and has two similar tuples in the two other tables such that these two tuples are also similar to each other. This can easily be extended to more than three tables. We skip the formal definition of the three-way and multi-way similarity intersect operators for brevity.

Example: In addition to the tables P and Q, given in the previous example, let $R = \{\widetilde{\widetilde{a}}, \widetilde{\widetilde{b}}, v, y\}$. Assume further that $pred(a, \widetilde{a})$, $pred(\widetilde{a}, \widetilde{\widetilde{a}})$, and $pred(b, \widetilde{b})$ hold. Thus, applying the three-way similarity set intersect operator produces: $\widetilde{\cap}(P, Q, R) = \{a, \widetilde{a}, \widetilde{\widetilde{a}}\}$. Notice that because $pred(\widetilde{b}, \widetilde{\widetilde{b}})$ does not hold, $b, \widetilde{b}, \widetilde{\widetilde{b}}$ are not part of the answer.

We extend SQL with the similarity-aware set intersect operator in the following way.

> (**SELECT** $a_1, a_2, \ldots$ **FROM** $table_1$
> **INTERSECT**
> **SELECT** $a_1, a_2, \ldots$ **FROM** $table_2$
> **INTERSECT**
>
> $\ldots$
>
> **SELECT** $a_1, a_2, \ldots$ **FROM** $table_n$
>) **WITHIN VALUES** ($\epsilon_1, \epsilon_2, \ldots$)

where the phrase **WITHIN VALUES** provides the similarity thresholds for each of the attributes participating in the similarity intersection operation. Notice that the similarity intersect operator can be expressed using standard relational operators as the query evaluation tree in Fig. 1 demonstrates.

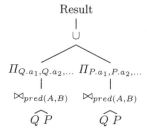

Fig. 1. Expressing Similarity Set Intersection Using Relational Operators

3   Processing the Similarity-Aware Intersect Operator

In this section, we present how the proposed similarity-aware set intersect operator is evaluated. The query processing algorithm for similarity intersect is an extension of the sort-merge join algorithm. So, the first step of the algorithm sorts both input tables unless they are already sorted. In high-level terms, similarity intersect compares tuples based on a Mark/Restore mechanism that avoids the $O(n^2)$ complexity that would result from a nested-loops implementation. To find matching tuples between two relations (named the outer and inner tables), the Mark/Restore mechanism marks the position of a tuple that may need to be restored later if some condition is satisfied as explained next.

The semantics of the similarity intersect operator is implementation independent. Therefore, the order of processing these relations will not impact the result. However, the order can impact the performance and therefore it should be part of query optimization. The current implementation simply uses left associativity to processes the relations. Since the binary and multi-way similarity set intersection operators work in the same way, we develop one algorithm for both. The result of a multi-way similarity intersect is constructed in stages, where each stage has a binary operator that produces an intermediate result that is sent to the next stage. In the first stage (first level), the intermediate result is constructed in such a way that each similar outer and inner tuples are consecutive, i.e., are next to each other in the order of emission. Similarly, results of the second stage are constructed such that the three similar tuples from the three input relations of the multi-way similarity intersect are produced in consecutive order similar to the order of the relations (i.e., the first tuple is from the first relation, the second tuple is from the second relation, and so on).

Algorithm 1 realizes the similarity-aware set intersection operator. Lines 1 and 2 initialize the outer and inner tuples. Both input relations are assumed to be sorted. Lines 4-11 advance the current inner and outer tuple(s) until a match based on the first attribute is found, i.e., when $dist(outer[0], inner[0]) \leq \epsilon_1$, where 0 refers to the index of the first attribute. Once a match is found, Line 12 marks the inner tuple position. Marking a tuple allows repositioning the inner cursor to the marked tuple later in the process.

Algorithm 1. SimIntersect(*inner, outer, nodeLevel*)

Input: outer relation, inner relation and the level of the similarity set intersection.
Output: similarity set intersection result.

```
 1: get initial outer tuple
 2: get initial inner tuple
 3: do forever {
 4: while outer[0]! ∼ inner[0] do
 5:     if outer[0] < inner[0] then
 6:         level ← nodeLevel
 7:         ADVANCEOUTER(outer,level)
 8:     else
 9:         advance inner
10:     end if
11: end while
12: mark inner position
13: do forever {
14: do{
15:     count ← COMPARE(outer,inner,nodeLevel)
16:     level ← nodeLevel
17:     if count = level then
18:         REPORTMATCHINGTUPLES(inner,outer,level)
19:     end if
20:     prevInner ← inner
21:     advance inner
22:     }
23: while inner[0] ∼ outer[0]
24: level ← nodeLevel
25: ADVANCEOUTER(outer,level)
26: if outer[0] ∼ prevInner[0] then
27:     restore inner position to mark
28: end if
29: break
30: }
31: }
```

This procedure is demonstrated in Figure 2 for the similarity intersection of tables P, Q, and R. *Level* 1 performs the similarity intersect between Q and P, and the result is intersected with R in *Level* 2. The threshold is usually determined by the application requirements. For this example, the threshold is selected to be around 10% of the attribute range of values, i.e., list={0.5,5}. Initially the outer points to tuple (0.9,10) and the inner points to tuple (0.1,5). Based on the value of the first attribute, the outer and the inner are advanced until the outer reaches (2,30) and the inner reaches (1.5,15). Then, the inner position is marked because both tuples match on the first attribute. Lines 14-23 are executed to report only the matching tuples while advancing the inner because the first attribute's value is within the outer's corresponding value and assign to *prevInner* a copy of the current inner location before advancing the inner cursor. Notice that the matching tuples are reported consecutively, i.e., tuple(s) from the outer then tuples from the inner. The reason is that in the next level, the consecutive tuples will be reported if a tuple of the next relation is similar to these consecutive similar tuples. This loop finishes when the inner reaches (5,50) as $dist(2,5) > 0.5$. Then, the outer is advanced and compared to the previous inner, and if both match on the first attribute, the inner cursor is restored to the marked position (Lines 25-28). In the example, this happens when the outer is advanced to tuple (2.5,20) and is compared to the prevInner's tuple (2.3,25). The inner is restored to the marked tuple because $dist(2.5, 2.3) \leq 0.5$. Then, the process repeats the search for other matching tuples.

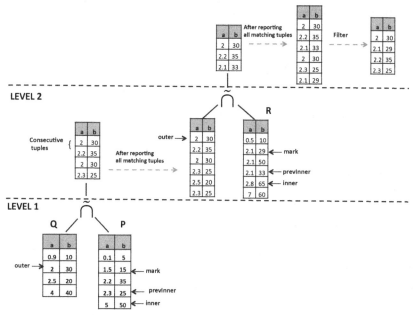

Fig. 2. Sample execution: Sim-Intersect. Threshold list={0.5,5}.

ADVANCEOUTER, COMPARE, and REPORTMATCHINGTUPLES (Algorithms 2, 3, and 4, resp.), work based on the level of the similarity intersection operator. In $Level1$, the outer is advanced once to perform any process, while in $Level2$, the outer is advanced twice, and so on. The reason is that similar tuples of the outer are consecutive to each other in the pipeline. When comparing the inner tuple to the outer, if the process is in $Level1$, the inner is only compared to the current outer whereas if the process is in $Level2$, the inner is compared to the current and the next outer tuples (i.e., the consecutive similar tuples). Referring to our example, the inner tuple (2.1,33) is similar to the outer consecutive tuples (2,30) and (2.2,35) in $Level2$. REPORTMATCHINGTUPLES produces the output by first reporting the two similar consecutive outer tuples (2,30) and (2.2,35), since they are in $Level2$, then it reports the current matching inner tuple, i.e., (2.1,33). Then, these three similar tuples are pipelined into $Level3$ for further processing, if any.

Algorithm 2 Advance Outer

1: **function** ADVANCEOUTER(*outer*,*level*)
2:     **while** *level* $\neq$ 0 **do**
3:         advance outer
4:         *level* $\leftarrow$ *level* $-$ 1
5:     **end while**
6: **end function**

Algorithm 3 Compare Tuples

1: **function** COMPARE(*inner*,*outer*,*level*)
2:     mark *outer* position
3:     *count* $\leftarrow$ 0
4:     **while** *level* $\neq$ 0 **do**
5:         **if** *outer* $\sim$ *inner* **then**
6:             *count* $\leftarrow$ *count* $+$ 1
7:             *level* $\leftarrow$ *level* $-$ 1
8:             advance *outer*
9:         **else**
10:            break
11:        **end if**
12:    **end while**
13:    restore *outer*
14:    **return** *count*
15: **end function**

Algorithm 4 Report Matching Tuples

1: **function** REPORTMATCHINGTUPLES(*inner*,*outer*,*level*)
2:     **while** *level* $\neq$ 0 **do**
3:         report *outer*
4:         advance *outer*
5:         *level* $\leftarrow$ *level* $-$ 1
6:     **end while**
7:     report *inner*
8:     restore *outer*
9: **end function**

3.1 Analysis

As mentioned in the previous section, the proposed algorithm assumes sorted inputs, and is based on a Mark/Restore mechanism that may lead to having a nested loop in the worst case. The complexity is computed as follows:

– Sorting the input relations: Assume that the outer and inner relations have n tuples, then the complexity is $O(nlogn)$.

– Processing the similarity intersect operator: Assume that the n outer tuples each iterates on average over c tuples of the inner relation, then the complexity is $O(n * c)$. The best-case scenario happens if $c = 1$, the average case is achieved when c is small with respect to the number of the inner tuples, and the worst case occurs when $c = n$. The worst-case scenario may take place when having a large similarity threshold, e.g., a big fraction of the domain range. In our algorithm, the threshold assigned to the first attribute is the one influencing the performance the most.

– Filtering the output: Filtering is usually performed by sorting the input, then grouping the duplicates. Assume that there are k output tuples, then the complexity is $O(k \log k + k)$.

Thus, the average case complexity is $O(n \log n)$ while the worst case complexity is $O(n^2)$, which is similar to sort-merge join. Typically, a threshold value is expected to be small compared to the domain size. Therefore, the complexity of the similarity intersect algorithm is closer to the average case. Thus, the performance is comparable to that of the standard set intersect, as demonstrated in the experimental section.

4   Experimental Results

We have modified PostgresSQL to support similarity intersect as an operator. We extended the Parser, Optimizer, and Executor modules of PostgreSQL for this purpose. We skip the details of how each of the PostgreSQL components is extended to support similarity intersect. The reader is referred to [8] for more details. Below, we present a summary of the performance results under various real and synthetic data sets as well as using some extensions to the TPC-H benchmark to support similarity queries.

Table 1. Equivalent regular operations

Similarity-aware Set Op.	Equivalent Query using Regular Ops.
(SELECT $a_1, a_2, \ldots, a_n$ FROM $tab1$ INTERSECT SELECT $a_1, a_2, \ldots, a_n$ FROM $tab2$) WITHIN VALUES $(\epsilon_1, \epsilon_2, \ldots, \epsilon_n)$;	SELECT $tab1.a_1, tab1.a_2, \ldots, tab1.a_n$ FROM $tab1, tab2$ WHERE $abs(tab1.a_1 - tab2.a_2) \le \epsilon_1$ and $abs(tab1.a_2 - tab2.a_2) \le \epsilon_2$ $\ldots$ and $abs(tab1.a_n - tab2.a_n) \le \epsilon_n$ UNION SELECT $tab2.a_1, tab2.a_2, \ldots, tab2.a_n$ FROM $tab1, tab2$ WHERE $abs(tab1.a_1 - tab2.a_2) \le \epsilon_1$ and $abs(tab1.a_2 - tab2.a_2) \le \epsilon_2 \ldots$ and $abs(tab1.a_n - tab2.a_n) \le \epsilon_n$

We run the experiments on an Ubuntu Linux machine with a 2.4GHz Intel Core i5 CPU and 4GB memory. Experiments are performed on real data sets [21], synthetic data, as well as using the TPC-H benchmark data [20]. We use the edit distance in our computations. We first study the effect of varying the number of attributes involved in the similarity intersect operator. Then, we compare the performance of the similarity intersect operator against (i) the standard relational intersect to demonstrate that the overhead of similarity intersect is acceptable, and (ii) the equivalent queries that use regular SQL operations to produce the same results as the corresponding similarity-aware query to demonstrate that similarity intersect yields better performance. The equivalent queries are presented in Table 1.

Impact of the Number of Attributes. We use a public dataset [21] that contains around 2.3 million readings gathered from 54 sensors deployed in the Intel Berkeley Research lab. The purpose of this experiment is to study the performance of similarity intersect as the number of involved attributes is increased. We conduct this experiment by processing the following query:

(SELECT epoch, temp, humidity, voltage FROM sensors WHERE moteid=1
INTERSECT
SELECT epoch, temp, humidity, voltage FROM sensors WHERE moteid=2)
WITHIN VALUES (10,0.1,0.1,0.1);

This query returns similar readings from mote1 and mote2. We start by querying based on one attribute, namely epoch. Then, we repeat the experiment by adding each time one more attribute. Figure 3(a) illustrates that the execution time is the highest when intersecting two datasets consisting of multiple attributes on their first attribute only and the execution time decreases as we increase the input attributes of these datasets. The reasons for this behavior are as follows. Referring to the algorithm for the similarity-aware set intersection, the number of internal comparison loops is the same for one or more attributes because the algorithm is based on the first attribute value. What differs here is the number of the returned matching tuples. When intersecting on one attribute, it is more likely to have more matching output tuples than when intersecting on two or more attributes. As the number of the output matching tuples increases, the time spent by the sort and the duplicate elimination processes increase.

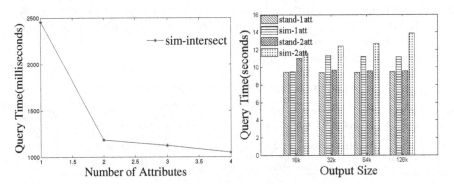

(a) Performance while increasing the num- (b) Similarity-aware set intersection vs.
ber of attributes.   standard set intersection.

Fig. 3. Effect of the number of attributes and the output size

Similarity Intersect Using Standard Relational Operators. We study the performance of the proposed similarity intersect operator against an equivalent query that performs the same functionality and that produces the same output but using only standard SQL operators. We vary the data size and the similarity threshold value while using the TPC-H data set [20]. We run the queries presented in Table 2. Through these queries, we can identify similar customer

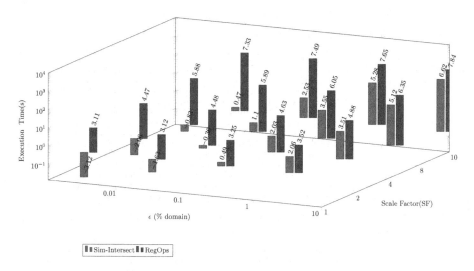

Fig. 4. Similarity-aware set intersection vs. regular operations

profiles from two countries. We may consider customer profiles to be defined
by the amount of money spent. For this case, we can run queries that use one
attribute (total price). However, some customers may spend a large amount of
money on a small quantity of items or may spend a small amount of money on
a large quantity of items. Therefore, we run a more precise query that uses two
attributes (total price and total quantity) to represent the customer profile. Notice that the assigned threshold to custkey attribute is -1. This value is used to
express the infinity value because we want to count the customers with similar
profiles regardless of whether their customer keys match or not.

We study the performance of similarity intersect when varying the similarity
threshold value from 0.01% to 10% of the attribute domain range. We vary the
threshold of the first attribute only because the algorithm is influenced highly
by its value. The threshold assigned to the second attribute is fixed to be 0.1%
of the attribute domain range. Specifically, the customer total price domain and
total quantity domain use values in the range [11020, 6289000] and [10, 4000],
respectively. We vary the input size by repeating the experiment using different
TPCH scale factors (from SF=1 to SF=8).

The results are given in Figure 4 that demonstrate a substantial query processing speedup of the similarity set intersection query over the equivalent query
that only employs regular operators. The speedup ranges between 1000 and 4
times for similarity threshold values ranging between 0.01% and 10% of the
attribute domain range, respectively.

Comparison with Standard Queries. This section evaluates the performance
of similarity intersct operator when compared to the standard SQL set intersection operator. We compare queries that have similar selectivities (i.e., queries
that produce a similar output size for a given input size). We control the output
cardinality by careful generation of synthetic input data. The details of how the
data is generated are omitted due to space limitation. The reader is referred

Table 2. Similarity-based intersect queries using TPC-H data

Operator Type	Syntax
Similarity-aware SetOp, two attributes	SELECT count(*) FROM ((SELECT p1.priceSum, p1.qtySum, p1.custkey FROM (SELECT sum(o.o_totalprice) as priceSum, sum(q.qty) as qtySum, o.o_custkey as custkey FROM orders o, customer c, (SELECT l_orderkey as o_key, sum(l_quantity) as qty FROM lineitem GROUP BY l_orderkey) q where o.o_orderkey=q.o_key and c.c_custkey=o.o_custkey and c.c_nationkey=1 GROUP BY o.o_custkey) p1 INTERSECT/EXCEPT SELECT p2.priceSum,p2.qtySum,p2.custkey FROM (SELECT sum(o.o_totalprice) as priceSum, sum(q.qty) as qtySum, o.o_custkey as custkey FROM orders o, customer c, (SELECT l_orderkey as o_key, sum(l_quantity) as qty FROM lineitem GROUP BY l_orderkey) q where o.o_orderkey=q.o_key and c.c_custkey=o.o_custkey and c.c_nationkey=2 GROUP BY o.o_custkey) p2) WITHIN VALUES $(\epsilon_1,\epsilon_2,-1)$) as result;
Equivalent Regular Operations to sim-intersect	SELECT count(*) FROM (SELECT p1.priceSum, p1.qtySum, p1.custkey FROM (SELECT sum(o.o_totalprice) as priceSum, sum(q.qty) as qtySum, o.o_custkey as custkey FROM orders o, customer c, (SELECT l_orderkey as o_key, sum(l_quantity) as qty FROM lineitem GROUP BY l_orderkey) q where o.o_orderkey=q.o_key and c.c_custkey=o.o_custkey and c.c_nationkey=1 GROUP BY o.o_custkey) p1, (SELECT sum(o.o_totalprice) as priceSum, sum(q.qty) as qtySum, o.o_custkey as custkey FROM orders o, customer c, (SELECT l_orderkey as o_key, sum(l_quantity) as qty FROM lineitem GROUP BY l_orderkey) q where o.o_orderkey=q.o_key and c.c_custkey=o.o_custkey and c.c_nationkey=2 GROUP BY o.o_custkey) p2 WHERE abs(p1.priceSum-p2.priceSum)$\leq \epsilon_1$ AND abs(p1.qtySum-p2.qtySum)$\leq \epsilon_2$ UNION SELECT p2.priceSum, p2.qtySum, p2.custkey FROM (SELECT sum(o.o_totalprice) as priceSum, sum(q.qty) as qtySum, o.o_custkey as custkey FROM orders o, customer c, (SELECT l_orderkey as o_key, sum(l_quantity) as qty FROM lineitem GROUP BY l_orderkey) q where o.o_orderkey=q.o_key and c.c_custkey=o.o_custkey and c.c_nationkey=1 GROUP BY o.o_custkey) p1, (SELECT sum(o.o_totalprice) as priceSum, sum(q.qty) as qtySum, o.o_custkey as custkey FROM orders o, customer c, (SELECT l_orderkey as o_key, sum(l_quantity) as qty FROM lineitem GROUP BY l_orderkey) q where o.o_orderkey=q.o_key and c.c_custkey=o.o_custkey and c.c_nationkey=2 GROUP BY o.o_custkey) p2 WHERE abs(p1.priceSum-p2.priceSum)$\leq \epsilon_1$ AND abs(p1.qtySum-p2.qtySum)$\leq \epsilon_2$) as result;

to [8] for further detail. From Figure 3(b), the similarity intersect operator adds a 20% overhead in the case of one-attribute-based similarity while it varies from 20% to 44% when increasing the output size from 16k to 128k in the case of the two-attribute-based similarity.

5   Conclusion

We introduced the semantics and extended SQL syntax of the similarity-based set intersection operator. We developed an algorithm that is based on the Mark/Restore mechanism to avoid the $O(n^2)$ complexity. We implemented this algorithm inside PostgreSQL and evaluated its performance. Our implementation of the proposed operator outperforms the queries that produce the same result using only regular operations. The speedup ranges between 1000 and 4 times for similarity threshold values ranging between 0.01% and 10% of the attribute domain range. We also demonstrated that the added functionality is achieved without a big overhead when compared to standard operators.

Acknowledgments. This work was supported by an NPRP grant 4-1534-1-247 from the Qatar National Research Fund and by the National Science Foundation Grants IIS 0916614, IIS 1117766, and IIS 0964639.

References

1. Narayanan, M., Karp, R.M.: Gapped local similarity search with provable guarantees. In: Jonassen, I., Kim, J. (eds.) WABI 2004. LNCS (LNBI), vol. 3240, pp. 74–86. Springer, Heidelberg (2004)
2. Wang, J., Li, G., Feng, J.: Fast-join: An efficient method for fuzzy token matching based string similarity join. In: ICDE (2011)
3. Schallehn, E., Sattler, K.U., Saake, G.: Efficient similarity-based operations for data integration. Data and Knowledge Engineering 48(3) (2004)
4. Mills, P.: Efficient statistical classification of satellite measurements. International Journal of Remote Sensing 32(21) (2011)
5. Silva, Y.N., Aref, W.G., Ali, M.H.: The similarity join database operator. In: ICDE (2010)
6. Silva, Y.N., Aref, W.G., Ali, M.H.: Similarity group-by. In: ICDE (2009)
7. Silva, Y.N., Aref, W.G., Larson, P., Pearson, S., Ali, M.H.: Similarity queries: their conceptual evaluation, transformations, and processing. VLDB J. 22(3) (2013)
8. Marri, W.J.A.: Similarity-aware set operators. Master's thesis, Qatar University (2009)
9. Wang, J., Li, G., Fe, J.: Fast-join: An efficient method for fuzzy token matching based string similarity join. In: ICDE (2011)
10. Schallehn, E., Sattler, K., Saake, G.: Advanced grouping and aggregation for data integration. In: CIKM (2001)
11. Yu, C., Cui, B., Wang, S., Su, J.: Efficient index-based knn join processing for high-dimensional data. Journal of Information and Software Technology 49(4) (2007)
12. Hjaltason, G., Samet, H.: Incremental distance join algorithms for spatial databases. In: SIGMOD (1998)
13. Arasu, A., Ganti, V., Kaushik, R.: Efficient exact set-similarity joins. In: VLDB (2006)
14. Böhm, C., Krebs, F.: The k-nearest neighbour join: Turbo charging the kdd process. Knowledge and Information Systems 6(6) (2004)
15. Gao, L., Wang, M., Wang, X.S., Padmanabhan, S.: Expressing and optimizing similarity-based queries in sql. In: Atzeni, P., Chu, W., Lu, H., Zhou, S., Ling, T.-W. (eds.) ER 2004. LNCS, vol. 3288, pp. 464–478. Springer, Heidelberg (2004)
16. Barioni, M.C.N., Razente, H.L., Traina Jr., C., Traina, A.J.M.: Querying complex objects by similarity in sql. In: SBBD (2005)
17. Barioni, M.C.N., Razente, H.L., Traina, A.J.M., Traina Jr., C.: Siren: A similarity retrieval engine for complex data. In: VLDB (2006)
18. Silva, Y.N., Aly, A.M., Aref, W.G., Larson, P.Å.: Simdb: a similarity-aware database system. In: SIGMOD (2010)
19. PostgreSQL Global Development Group: Postgresql (2014), http://www.postgresql.org/
20. TPCH: Tpc-h version 2.15.0 (2014), http://www.tpc.org/tpch
21. Intel Berkeley Research lab: Intel lab data (2014), http://db.csail.mit.edu/labdata/labdata.html

High Dimensional Search Using Polyhedral Query

Richard Connor, Stewart MacKenzie-Leigh, and Robert Moss

Department of Computer and Information Sciences,
University of Strathclyde, Glasgow, G1 1XH, United Kingdom
{richard.connor,s.mackenzie-leigh,robert.moss}@strath.ac.uk

Abstract. It is well known that, as the dimensionality of a metric space increases, metric search techniques become less effective and the cost of indexing mechanisms becomes greater than the saving they give. This is due to the so-called *curse of dimensionality*.

One effect of increasing dimensionality is that the ratio of unit hypersphere to unit hypercube volume decreases rapidly, making the solution to a similarity query (the query ball, or hypersphere) ever more difficult to identify by using metric invariants such as triangle inequality.

In this paper we take a different approach, by identifying points within a query polyhedron rather than a ball. We show how this can be achieved by constructing a surrogate metric space, such that a query ball in the surrogate space corresponds to a polyhedron in the original space. If the polyhedron contains the ball, the overall cost of the query is likely to be increased in high dimensions; however, we show that shrinking the polyhedron can capture a surprisingly high proportion of the points within the ball, whilst at the same time giving a more efficient, and more scalable, search.

We show results which confirm our underlying hypothesis. In some cases we can retrieve significant volumes of query results from spaces which are otherwise intractable.

1 Introduction

In this paper, we show a novel conceptualisation of an approximate indexing technique based on the geometry of high-dimensional metric spaces. This is based on the following observations:

1. the relationship between the shared volume of a hypersphere and a hypercube centred around the same point in high-dimensional space; especially that, as the side length of a containing hypercube is reduced, much of the hypersphere may still be contained
2. that points within an approximate hyper-polyhedron centred around an arbitrary query point can be defined, relying upon triangle inequality, by a set of inequalities based on distances from a fixed set of reference points[1]

---

[1] Corresponding to existing multiple-pivot indexing mechanisms.

A.J. Machado Traina et al. (Eds.): SISAP 2014, LNCS 8821, pp. 176–188, 2014.
DOI: 10.1007/978-3-319-11988-5_16 © Springer International Publishing Switzerland 2014

3. that a finite metric space can be re-indexed, using the Chebyshev distance over pre-calculated distances to the reference points, to allow the efficient extraction of points within a hyper-polyhedron centred around a query

The re-indexed (*surrogate*) space can often be queried more efficiently than the original: it may have smaller data points, a faster metric, and lower intrinsic dimensionality. However when queried at the same threshold, the result will be a much larger proper superset of the query in the original space, and the cost of filtering against the original space is likely to outweigh any efficiency gain.

However, observation (1) means that, as the surrogate query threshold is reduced, a corresponding increase in efficiency may be achieved without a corresponding significant loss in correct results. This makes the mechanism as proposed useful in the context of high-dimensional metric spaces, where known indexing techniques are completely ineffective. In this context, it may give a tractable and scalable approach to at least achieving some kind of imperfect search, and we show results from searching against GIST image characterisations [13] which we have been unable to otherwise achieve.

2   Dimensionality: Curse and Counter-Curse

In the domain of metric search, we are very familiar with the so-called *curse of dimensionality* [7]. The observable effect is that, as the dimensionality of a metric space increases, then for arbitrarily selected distances within the space standard deviation decreases and there are ever fewer very small values.

One explanation of this is that, as the dimensionality increases, the ratio of the volume of the unit hypersphere to the unit hypercube decreases rapidly. Points within the space fill a hypercube, while the solution to a threshold query fills a hypersphere (the query ball) centred around the query point.

It is instructive to observe the magnitude of this effect. The volume of a hypersphere with radius r in $2k$ dimensional space[2] is $\frac{\pi^k}{k!}r^{2k}$ which starts to decrease rapidly after 6 dimensions. The ratio of this volume to the volume of a containing hypercube is given by $\frac{\pi^k}{2^{2k}k!}$ which clearly becomes very small, very quickly, as k increases. At 6 dimensions the ratio is 0.08; at 10 it is 0.002 and it drops to 2×10^{-8} at 20. This sharp drop-off fits well with the generally known rule-of-thumb that metric indexing mechanisms become ineffective at an intrinsic dimensionality [7] of more than around 6-10.

2.1   Shrinking the Search Hypercube

Imagine that, for a threshold search, all points within the containing hypercube could be efficiently discovered. Even if this were true, it would be of little practical value in high dimensions, as in an evenly-distributed space almost all of the points contained would not be within the solution to the search.

---

[2] Even dimensions are used as the formula is slightly simpler, an equivalent formula exists for odd dimensions.

However, if for example all points within a hypercube of the *same volume* and centre as the query hypersphere could be discovered, these points will have a significant overlap with those in the query ball. As the number of dimensions increases, the overlap becomes proportionally smaller, but not rapidly.

The side length of this cube actually *decreases* as the dimensionality increases. Therefore, for example, in 10 dimensions a set of points which will coincide significantly with a query ball of radius t can be found by extracting points within a hypercube with a half-side length of just over $\frac{1}{2}t$, and around half of the space contained in this hypercube will also be in the hypersphere.

There is one further effect of which we can take advantage. As the half-side length increases up to the sphere radius, the contained volume of the sphere increases rapidly after a threshold of around the equivalent volume. It very slowly approaches full containment as the half-side length approaches the query radius, but includes almost all of the points within the sphere at a much smaller value than this. With higher dimensions, this effect is greater and starts at a lower threshold. Figure 1 shows the volume overlap in 10 dimensions, the graphs corresponding to recall and precision in an evenly distributed Euclidean space as the half-side length increases from 0 to 1. It can be seen that there is an overlap where both recall and precision are usefully far from zero: for example, if we could efficiently retrieve a hypercube with a half-side of little more than 0.6 of a query radius, we would retrieve over 80% of the true results, while of all the results retrieved, around 20% of them would be correct. In 10-dimensional space, this may well be a reasonable compromise if there is an associated reduction in query cost.

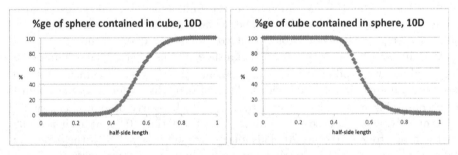

Fig. 1. With half-side ranging from 0 to 1, graphs show percentage of volumes: (a) of unit hypersphere in hypercube, and (b) of hypercube in unit hypersphere

These observations are used as follows. A surrogate metric space will be constructed such that a query ball in the surrogate space corresponds to a hyper-polyhedron within the original space. The hyper-polyhedron is expected to have similar volume-ratio properties as those determined for hypercubes. The surrogate space then allows points within an approximate hyper-polyhedron to be discovered using standard metric indexing techniques. Search at the same threshold corresponds to the minimum containing polyhedron, and will typically give a huge proportion of false positive results. However, efficiency gains from dropping the search threshold should maintain a high proportion of the true results because of the effect shown above.

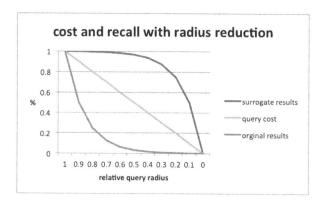

Fig. 2. Hypothesis: percentage of correct results with threshold reduction

In essence, we would hope to see a pattern as shown in Figure 2. As the query threshold is reduced, the query cost decreases approximately linearly in both original and surrogate spaces. In the original space, the amount of data returned will drop off very rapidly, due to the decrease in the hypersphere volume; however in the surrogate space, it will drop off at first very slowly, due to the effects just outlined. This should allow a high proportion of the correct results in return for a substantial reduction in query cost.

It is worth noting that, although the above discussion implicitly assumes a Euclidean space, the same patterns occur in other spaces as well, and the technique proposed works correctly over any metric space.

3 Defining Approximate Hyper-Polyhedra

There remains the issue of finding a scalable mechanism which will identify the points within the reduced hypercube. For a Euclidean space, this could be done by setting up an independent search structure for each dimension and finding the intersection of all points within the appropriate range on all dimensions; with unlimited parallel hardware this could be extremely efficient. However, our primary interests include performing search over high-dimensional, non-Euclidean metric spaces.

Instead of calculating the actual hypercube, we form an approximation of a hyper-polyhedron by use of reference points within the space. Figure 3 shows a simple example. Reference points p_1 and p_2 have been selected. For a query q, with threshold t, the property of triangle inequality means that for any u_i in the space, if $|d(q, p_1) - d(u_i, p_1)| > t$ or $|d(q, p_2) - d(u_i, p_2)| > t$, then u_i cannot be in the result set.

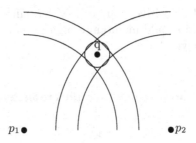

Fig. 3. An Approximate Hyper-Polyhedron in 2 dimensions

Generalising to a metric space $(\mathcal{S}, d)$, and a set of reference points $\mathcal{R}$, then an approximate hyper-polyhedron constructed for a query q with threshold t is defined as:

$$Poly(q) = \{u \leftarrow \mathcal{S} \text{ where } \forall p \in \mathcal{R}, \, |d(q,p) - d(u,p)| < t'\}$$

where $t' \leq t$, and is chosen according to the tradeoffs highlighted above.

This statement of inclusion essentially corresponds to the pivoting exclusion principle used in various multiple-pivot mechanisms, and is discussed further in Section 5. However we now show how to turn the pivot-based exclusion into a metric search in its own right. The value of doing this is that, as the value of t' is reduced, the scalability of the search increases.

4   Re-Indexing for Hyper-Polyhedral Search

If we use the denotation u^j to mean $d(u, p_j)$, then the inclusion criterion for $Poly(q)$ can be rewritten as $\max_j(|q^j - u^j|) \leq t'$, as for any $p_j \in \mathcal{R}$, $|q^j - u^j|) > t'$ means that u does not lie within the polyhedron around the query point.

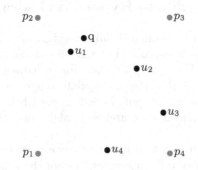

Fig. 4. Euclidean space: data u_i, reference points p_i and query q

This condition is captured by applying the Chebyshev distance (L_∞) to the ordered sets of values constructed from the distances of u and q to each reference point in turn. As Chebyshev is itself a proper metric, this means that elements of $Poly(q)$ can be found by using metric search over a metric space constructed from the original by pre-calculating these distances, and using Chebyshev as the distance metric.

Table 1. The left-hand column shows the Euclidean distances (d) from the query point in the original space (Figure 4). The right-hand column shows the corresponding Chebyshev distance (L_∞) in the surrogate space.

Original Space		Surrogate Space				
Point	$d(q,u)$	$d(p_1,u)$	$d(p_2,u)$	$d(p_3,u)$	$d(p_4,u)$	$L_\infty(q',u')$
q	0	3.68	1.52	2.67	4.28	0
u_1	0.57	3.16	1.41	3.14	4.24	0.52
u_2	1.84	3.91	3.35	1.80	2.69	1.83
u_3	3.26	3.98	4.72	2.81	1.22	3.20
u_4	3.37	2.10	4.43	4.34	1.90	2.91

Figure 4 demonstrates this by example, for data drawn in 2D Euclidean space. There are four reference points (p_i), four data points (u_i) and a single query point (q). Table 1 gives the corresponding distance values used to populate the surrogate space. It can be seen that, in all cases, the Chebyshev distance over the surrogate set gives a smaller value than the original distance, this property deriving from the triangle inequality property of the original space.

4.1   Formal Definition

Consider a metric space $(\mathcal{X}, d)$ over which a threshold search is required: that is, for some finite subset $\mathcal{S} = \{u_0, u_1, \ldots, u_n\}$ of $\mathcal{X}$, those objects within the close proximity of some $q \in \mathcal{X}$ require to be found. Note that $\mathcal{X}$ is not necessarily a Cartesian space, but d must be a proper metric.

Let $\mathcal{R}$ be an ordered set of m arbitrarily chosen points in $\mathcal{X}$, where r_j denotes the jth element of $\mathcal{R}$. $\mathcal{R}$ can be thought of as a set of *reference* points within the space.

A *surrogate* set $\mathcal{T}_\mathcal{R}$ of $\mathcal{S}$ is a set in m-dimensional Cartesian space where, for each $u_i \in \mathcal{S}$, there exists a corresponding $v_i \in \mathcal{T}_\mathcal{R}$ such that $v_i^j = d(u_i, r_j)$, where v_i^j denotes the value of the jth dimension of v_i.

The surrogate set $\mathcal{T}_\mathcal{R}$ will be used to perform queries without reference to either the actual values of $\mathcal{S}$, or the metric d. For a query q, a surrogate query $q_\mathcal{R}$ will be constructed, such that $q_\mathcal{R}^j = d(q, r_j)$.

The Chebyshev (L_∞) distance metric is defined as

$$L_\infty(x, y) = \lim_{n \to \infty} \sqrt[n]{\sum_j (|x^j - y^j|)^n}$$

which can be conveniently calculated as

$$L_\infty(x, y) = \max_j (|x^j - y^j|)$$

Being an element of the family of Lebesque metrics with $n \geq 0$, this is a proper metric. Therefore $(\mathcal{T_R}, L_\infty)$ is also a metric space.

4.2   Properties

1. $L_\infty(q_\mathcal{R}, v_i) \leq d(q, u_i)$
   That is, if $\mathcal{Q}_t(q, \mathcal{S}, d)$ denotes the set of values returned by a threshold query for metric d over $\mathcal{S}$ for the point q and the threshold t, then $\mathcal{Q}_t(q, \mathcal{S}, d) \subseteq \mathcal{Q}_t(q_\mathcal{R}, \mathcal{T_R}, L_\infty)$. The proof of this derives from the triangle inequality property of d.
2. The conditional probability of $u_i \in \mathcal{Q}_t(q, \mathcal{S}, d)$, $v_i \in \mathcal{Q}_{t'}(q_\mathcal{R}, \mathcal{T_R}, L_\infty)$ reduces at first, very slowly, from 1 as t' reduces from t downwards, while the cost of evaluating the query reduces more rapidly.

The following tradeoffs exist, according to a given search scenario:

1. Querying $L_\infty(q_\mathcal{R}, v_i)$ at the same threshold value t will always give a superset of the required results, which for a precise search will then have to be tested back in the original space before being returned as results of the threshold query. The relative sizes of the true and false results returned by a query at the same threshold depend upon the size, and individual points chosen, for the set $\mathcal{R}$; however, this aspect of query performance is likely to be in contention with choosing $\mathcal{R}$ to give the best search performance.
2. Conversely, although $L_\infty(q_\mathcal{T}, v_i) \leq d(q, u_i)$ is the only guarantee, it may be the case that, for some given ϵ, there is an acceptable probability that $L_\infty(q_\mathcal{T}, v_i) < d(q, u_i) - \epsilon$. If so, then a smaller threshold value can be used to produce an approximate result set. In many dimensions, it is extremely unlikely that a distance very close to the threshold will be reached in the surrogate space, as this can happen only with very close alignment of three points in the original space, which is increasingly less likely as the number of dimensions increases, although more likely as the size of $\mathcal{R}$ increases.

The same core mechanism can thus be used either as an accurate, or an approximate, threshold search, depending on the context of the requirements.

4.3   Choosing Reference Points

In common with other methods which use reference points, the choice of points appears to be critical to the performance of the mechanism. However, we are at an early stage of investigation in this respect.

We have tried various strategies for various spaces, and the only general deduction is that a random choice of points is relatively safe, as often the use of apparently appealing strategies only makes things worse.

For Euclidean space, it seems that the best strategy may be to choose artificial points in the "corners" of the space rather than points within the existing data. Thus in unitary space we use the origin, and then the points $(0, 1, 0, 0, \ldots)$, $(0, 1, 1, 0, \ldots)$, $(0, 1, 1, 1, \ldots)$ etc. We have not yet found equivalent series of points for other metrics, which are harder to reason about in terms of their multidimensional geometry, although we suspect they exist.

In terms of the number of reference points, we have seen some surprising results which show that many less points than might be expected can be used. This seems to depend very much upon the distribution of points within the space, and may be beyond theoretical analysis in an uneven distribution. There is however a clear law of diminishing returns: if each dimension in the surrogate space gives approximately a constant probability of excluding that point from the result set, and assuming this probability is reasonably large, then small numbers of reference points will be much more efficient than large numbers as better use of memory is made in the surrogate space. The tradeoff is that larger numbers of reference points will always give a smaller number of false positive returns, but the magnitude of this effect will depend heavily on how well the reference points can be chosen.

5   Related Work

There are already a large number of approximate methods suitable for use in higher-dimensional spaces, classified in [14], many of which use reference points.

Permutation indexing [2,4] is essentially another surrogate space mechanism. In common with our mechanism, a set of reference points is chosen and the distance to each is pre-calculated for all points in the data set. However these distances themselves are then abstracted into only their order from each point. Searching by these orders should be strongly correlated with a metric search, especially for nearest-neighbour searches. Many strategies have been suggested, with the best scaling being produced by using a relatively large number of reference points and then testing against only a much reduced view of these, allowing the resulting sparse space to be searched using inverted index techniques e.g. [1,12]. Our observation is that these techniques require more reference points, and give rather larger numbers of false positive results, than our technique, although the use of inverted indices can give impressive performance.

Re-indexing a space according to a proxy based on reference points was, to our knowledge, first suggested by Figueroa and Frediksson [10] in which a permutation space is re-indexed to give an improved metric performance.

As noted, our core semantics is identical to a set of multiple exclusions based on triangle inequality, and therefore relates closely to the pivoting exclusion principle used in various multiple-pivot mechanisms which use pre-computed distances, notably *LAESA* [15] and Extreme Pivots [16].

Our mechanism has much in common with Extreme Pivots, and in fact was derived from attempts to use this over GIST data. Perhaps because of the high dimensionality, or uneven distribution, we failed to find useful pivot groups as described in [16], and in the course of running experiments to find the best combination of pivots and pivot groups, we discovered that the best size of pivot group we could find was in fact 1. The results returned by our surrogate search are the same as those returned by a degenerate case of Extreme Pivots, using each reference point as one pivot group of a single value. However this allows the same search to be conducted in the surrogate space, rather than performing a serial scan of the data as is more generally required. The comparison of cost between our mechanism and Extreme Pivots is therefore only the increased efficiency of performing an indexed search over the space, versus a potentially greater number of false positive exclusions from using larger pivot groups. We believe our mechanism will therefore work better with larger, higher dimensional data. However, the observations based on the efficiency/recall tradeoffs via threshold reduction should apply equally to both mechanisms.

Since originally proposing this mechanism, we have discovered that exactly the same tradeoffs between efficiency and recall when reducing the search threshold within a mulitple-pivot space have been observed by Chávez and Nararro in [5]. Their explanation of the gains is based on the probability distribution function of distances within the original space, and is fully compatible with our observations on hyperspheres and hypercubes; the relationship seems worthy of further investigation. They do not propose reindexing the space using Chebyshev, which we believe gives much greater efficiency gains as the search threshold decreases.

In common with the motivation for the List of Clusters [6], the use of memory is critical in real index performance. This can be seen to be the reason for much of the performance gain we can achieve. As our indexing mechanism works over the surrogate space of the hyper-polyhedron, we can substantially reduce the query threshold without significantly reducing the number of results.

6   Results

For all experiments reported here, we have used the Euclidean "corners" strategy for Euclidean spaces, and randomly selected reference points for other metrics. We have results against three different types of data set: generated Cartesian spaces of various dimensions[3]; the SISAP *colors* data set, and a data set of GIST characterisations of images. This last set is the real target of the described mechanism: with 420 Cartesian dimensions, it is essentially intractable for metric indexing techniques.

---

[3] Not included due to space constraints, please contact the authors if interested.

In each case, we compare our technique using a balanced Vantage Point Tree (VPT) in both the original and surrogate spaces. This is just to give a point of reference against which two searches, the original and the surrogate, can be compared. It is quite likely that, for any given original or surrogate search, there are better indexing techniques available.

6.1 SISAP *colors*

The SISAP *colors* data was used with Euclidean, Cosine and SED [8, 9] distances, each of which has very different cost implications for both metric cost and scalability. 256 random points were removed from the 112,682 data points to use as queries. For each metric a query threshold sufficient to return around 1k results, i.e. mean of 4 per query, was used.

In each case, the surrogate space was searched and the results from these queries were then post-processed by comparing the original metric over the original data. These, and all other calculations, were performed on a non-optimised system, written in Java, and executed on a laptop computer, and so only the relative timings are important; each timed test was repeated until the standard error of the mean was less than 1%.

For Euclidean queries, only 5 reference points were used, this giving the best overall performance. This is many fewer than we would have expected, but as more points are used, only marginally better precision is achieved and the search cost is substantially increased. It is worth noting that each data point is therefore represented in less than one-twentieth of the memory required for query against the original 112-dimensional vectors. For Cosine and SED, the relatively high costs of the distance metrics themselves imply using larger number of reference points, to reduce the number of post-processing distances calculated.

Table 2 shows some key measurements for each metric. The figures given are: the cost of a sequential search; the cost using a standard balanced VPT, and the surrogate costs for retrieving 90%, 70% and 50% of the query results by reducing the query threshold. In all cases, achieving this through reducing the search threshold in the original space makes a negligible difference to cost. All costs are given in absolute time measured, to highlight the tradeoffs in the different metrics. The pattern of cost and recall as the surrogate threshold is decreased is shown in Figure 5.

The relative saving is quite complex, depending on a number of factors. For SED, the surrogate method is cheaper even to fetch 100% of the query results, as the number of results returned by the surrogate metric is less than the number of SED calculations performed in indexing via the VPT, and the cost of the metric makes this the dominant factor.

Depending on the context of the search, these speedups could already be quite useful. In all cases, however, the metrics over this space are already relatively tractable, with VPT indexing being substantially faster than sequential query; this is not the intended domain of our surrogate mechanism, and we turn our attention to a higher-dimensional space where this is not the case.

Table 2. Times (ms per 256 queries) for queries over the colors dataset

	Original Space		Surrogate Space		
Metric	Sequential	VPT-indexed	90% recall	70% recall	50% recall
Euclidean	1059	148	88	46	19
Cosine	9617	45	52	33	20
SED	79033	2849	1196	558	250

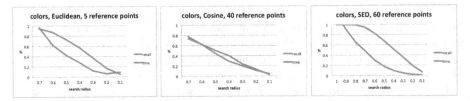

Fig. 5. Colors: three metrics, each showing cost reducing faster than recall. Time measured corresponds closely to memory use, which is optimised by choosing an appropriate number of reference points.

6.2   MirFlickr/GIST

The data used here comprises the GIST [13] characterisations of first 10k images taken from the Mir-Flickr collection [11]. A balanced VPT was built using the data, and then each value was queried against it at a threshold which returned 10k results (excluding the query itself), i.e. a mean of one per query. As would be expected with data of this complexity, the VPT gave little or no cost saving over sequential search for any metric.

Figure 6 shows the result of using polyhedral search for Euclidean and Cosine distances. The surrogate space was constructed, and a VPT used to query at different thresholds between the original, and one-tenth of the original, threshold. All values shown are relative to the cost of the original search.

Values shown are, from top to bottom in the graphs: *recall*, i.e. how many of the correct results are returned; *tree distances*, the relative number of distance calculations performed during the tree search; the actual *measured time* for the queries to complete; and the *memory use*. These last two figures include both the surrogate tree search, and the post-processing of the results using the original metric and data.

Notably for both searches even searching at the containment threshold is cheaper than the original metric search. This is because the cost is dominated by the memory cost of the original distance metric searching over the original points, each of which require 50-100 times more memory than the surrogate points. Even at the full threshold, a smaller number of results is obtained than the number of calculations performed during a tree search using the original metric and data.

For Euclidean search, the cost of retrieving 99.7% of the true results is just under half of the original, whereas 75% can be retrieved for just over one-fifth of the cost. The lack of a good strategy for choosing reference points means that more surrogate distance calculations are required for Cosine search, however this is more than compensated for by higher recall at lower relative thresholds, and the measured cost of retrieving 99.7% of the correct results is just over one-quarter of the cost of the original search, and 70% of correct results can be obtained for one-twentieth of the cost.

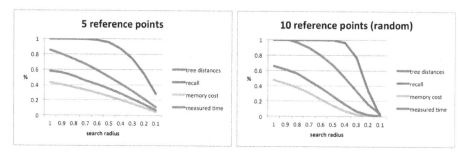

Fig. 6. Euclidean and Cosine distance over MirFlickr/GIST

7   Conclusions and Further Work

We have presented a novel strategy for approximate search in intractable, high-dimensional metric spaces. The essence of the mechanism is to re-cast the original space, via a set of reference points, into another metric space which can be usefully searched at lower thresholds. This allows, at least in some cases, a relatively predicable proportion of the correct results to be obtained for an acceptably low cost.

We are at an early stage of investigation of this technique, however we have already used it to obtain some real results that were previously unavailable to us in complex domains such as image similarity.

One area of investigation which could greatly improve the performance of the technique would be a better selection of reference points for non-Euclidean spaces, which may be possible to achieve by analysis of the geometry of these spaces as we believe we have achieved for Euclidean distance.

Finally, we have seen some interesting preliminary results from approximating nearest-neighbour (kNN) search in the surrogate space, for appropriately increased values of k. This takes advantage of an observation that there may be better correlation of the surrogate and original distances at lower threshold values, but requires further investigation.

References

1. Amato, G., Esuli, A., Falchi, F.: Pivot selection strategies for permutation-based similarity search. In: Brisaboa, et al. (eds.) [3], pp. 91–102
2. Amato, G., Savino, P.: Approximate similarity search in metric spaces using inverted files. In: Proceedings of the 3rd International Conference on Scalable Information Systems, InfoScale 2008, pp. 28:1–28:10. ICST (Institute for Computer Sciences, Social-Informatics and Telecommunications Engineering), Brussels (2008)
3. Brisaboa, N., Pedreira, O., Zezula, P. (eds.): SISAP 2013. LNCS, vol. 8199. Springer, Heidelberg (2013)
4. Chávez, E., Figueroa, K., Navarro, G.: Effective proximity retrieval by ordering permutations. IEEE Trans. Pattern Anal. Mach. Intell. 30(9), 1647–1658 (2008)
5. Chávez, E., Navarro, G.: Probabilistic proximity search: Fighting the curse of dimensionality in metric spaces. Inf. Process. Lett. 85(1), 39–46 (2003)
6. Chávez, E., Navarro, G.: A compact space decomposition for effective metric indexing. Pattern Recognition Letters 26(9), 1363–1376 (2005)
7. Chávez, E., Navarro, G., Baeza-Yates, R., Marroquín, J.L.: Searching in metric spaces. ACM Comput. Surv. 33(3), 273–321 (2001)
8. Connor, R., Moss, R.: A multivariate correlation distance for vector spaces. In: Navarro, G., Pestov, V. (eds.) SISAP 2012. LNCS, vol. 7404, pp. 209–225. Springer, Heidelberg (2012)
9. Connor, R., Simeoni, F., Iakovos, M., Moss, R.: A bounded distance metric for comparing tree structure. Inf. Syst. 36(4), 748–764 (2011)
10. Figueroa, K., Frediksson, K.: Speeding up permutation based indexing with indexing. In: Second International Workshop on Similarity Search and Applications, SISAP 2009, pp. 107–114 (August 2009)
11. Huiskes, M.J., Lew, M.S.: The mir flickr retrieval evaluation. In: MIR 2008: Proceedings of the 2008 ACM International Conference on Multimedia Information Retrieval. ACM, New York (2008)
12. Mohamed, H., Marchand-Maillet, S.: Quantized ranking for permutation-based indexing. In: Brisaboa, et al. (eds.) [3], pp. 103–114
13. Oliva, A., Torralba, A.: Modeling the shape of the scene: A holistic representation of the spatial envelope. Int. J. Comput. Vision 42(3), 145–175 (2001)
14. Patella, M., Ciaccia, P.: Approximate similarity search: A multi-faceted problem. J. of Discrete Algorithms 7(1), 36–48 (2009)
15. Ruiz, E.V.: An algorithm for finding nearest neighbours in (approximately) constant average time. Pattern Recognition Letters 4(3), 145–157 (1986)
16. Ruiz, G., Santoyo, F., Chávez, E., Figueroa, K., Tellez, E.S.: Extreme pivots for faster metric indexes. In: Brisaboa, N., Pedreira, O., Zezula, P. (eds.) SISAP 2013. LNCS, vol. 8199, pp. 115–126. Springer, Heidelberg (2013)

Generating Synthetic Data to Allow
Learning from a Single Exemplar per Class

Liudmila Ulanova, Yuan Hao, and Eamonn Keogh

Department of Computer Science & Engineering, University of California, Riverside, USA
{lulan001,yhao,eamonn}@cs.ucr.edu

Abstract. Recent years have seen an explosion in the volume of historical documents placed online. The individuality of fonts combined with the degradation suffered by century old manuscripts means that Optical Character Recognition Systems do not work well here. As human transcription is prohibitively expensive, recent efforts focused on human/computer cooperative transcription: a human annotates a small fraction of a text to provide labeled data for recognition algorithms. Such a system naturally begs the question of how much data must the human label? In this work we show that we can do well even if the human labels only a single instance from each class. We achieve this good result using two novel observations: we can leverage off a recently introduced parameter-free distance measure, improving it by taking into account the "complexity" of the glyphs being compared; we can estimate this complexity using synthetic but plausible instances made from the single training instance. We demonstrate the utility of our observations on diverse historical manuscripts.

Keywords: Classification, Semi-Supervised Learning, Historical Manuscript, Handwriting Analysis.

1 Introduction

The classification of individual glyphs is typically the first step in historical document processing. The variety of texts (hundreds of languages, tens of thousands of handwriting styles/handmade fonts), combined with the degradation often suffered by century old manuscripts, precludes the adoption of a "one-size-fits-all" off-the-shelf Optical Character Recognition (OCR) Systems.

Most Semi-Supervised Learning (SSL) techniques make explicit assumptions which are violated or only partly true in our domain of interest [3]. In particular, the *smoothness assumption* can be violated in a special way that does not seem to be well appreciated. Recall that it requires that *"(objects) which are close to each other are more likely to share a label"* [3]. However, this assumption can be violated in an unexpected way: "complex" objects tend to be closer to other objects that are "simple," at least under some distance measures such as the recently introduced CK-1 distance [2]. In Fig. 1 we show a clustering that hints at this [5, 11]. This "complexity bias" violates the notion that objects that are close to each other are more likely to share a label, at least for some classes.

A.J. Machado Traina et al. (Eds.): SISAP 2014, LNCS 8821, pp. 189–194, 2014.
DOI: 10.1007/978-3-319-11988-5_17 © Springer International Publishing Switzerland 2014

Fig. 1. (*left*) A clustering of examples from [5] suggests that the distance measure has difficulty clustering objects with different complexities, such as the simple 'i' and more complex 't'. (*right*) If we compensate for these differences in complexity we can do much better.

Given enough training data we can learn the amount of "complexity bias" for each class and compensate for it. However, this opens up a "chicken-and-egg" paradox, as we are using SSL to mitigate the *lack* of training data. As we shall show, we solve this problem by creating additional synthetic examples with a simple random distortion model. As hinted at in Fig. 2, we can easily produce plausible variations of hand-press or handwritten letters.

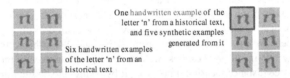

One handwritten example of the letter 'n' from a historical text, and five synthetic examples generated from it

Six handwritten examples of the letter 'n' from an historical text

Fig. 2. (*left*) Six examples of a handwritten letter. (*right*) We can take a single letter (red / highlighted example) and produce natural looking variations of it with a simple distortion model.

The rest of the paper is organized as follows. In Section 2 we discuss related work and background for our research. In Section 3 we introduce our proposed method. Section 4 presents experimental evaluations. Finally, Section 5 offers conclusions and a discussion of avenues for future research.

2   Related Work and Background

While there is a plethora of classification algorithms available, the simple Nearest Neighbor (NN) algorithm is known to be surprisingly competitive in many domains. This is because the algorithm can use *any* distance measure, including ones that can "carve out" decision boundaries that are not within the representation power of decision trees, etc. In this work we propose to leverage off a recently introduced distance measure called the CK-1 distance [2]. The CK-1 distance differs from other methods (Gabor filters, Fourier transforms, Markov random fields, wavelets, etc.) in two important ways. First, it considers shape *and* texture simultaneously. Second, it is completely parameter-free, freeing us from the need to obtain data to learn parameter settings, and greatly reducing the probability of overfitting (with no parameters to *fit*, one cannot over*fit*). The CK-1 is a compression based distance measure. The distance ranges between zero and "soft" one. If two objects are very similar to each other the distance is close to zero, whereas for very dissimilar objects the distance is close to one or slightly greater. Due to its simplicity and effectiveness we use CK-1 distance measure in this work; however, it is not a necessary condition for the utility of our ideas. CK-1 has proven its efficiency in historical documents processing domain [8], but in a bit different sense. While Hu et al. apply this distance measure to initial letters mining (intrinsically *textures*); we expand it to all glyphs (intrinsically *shapes*). However, there are two problems we must solve in order to use CK-1. The first is data scarcity. All classification algorithms benefit from more data; however, our explicit problem statement allows us to have as few as one exemplar per class. The second problem, which was hinted at in Fig. 1, is less well appreciated in the literature. At least some distance measures may overestimate the distance between "complex," but nevertheless similar, objects. For example, in our domain, letters such as $\mathscr{A}$ and $\mathscr{B}$ are complex, at least relative to the more prosaic versions, **A** and **B**. In this case the difference in complexity is related to particular typeface/handwriting. However, even *within* a single typeface, there are differences in complexity, ranging from the simple *single-stroke* letters such as **I** and **O**, to more complex *multi-stroke* letters such as **W** and **E**. The observation that differing complexities cause problems for nearest neighbor classification has been forcefully shown for time series classification [1]. Moreover, as we shall show below it is also the case for classifying glyphs with the CK-1 distance measure. Note that we are not making any claims with regard to other shape distance measures[1].

Synthetic data generation techniques are widely used to supplement datasets that do not have a sufficient number of instances for a given task [6]. If each exemplar can be described by a feature vector, then the problem of synthetic data generation can often be solved by a technique as simple as adding random Gaussian noise to copies of the original vectors, or by averaging randomly chosen vectors from the same class (i.e., SMOTE and its variants [4]). The problem becomes more complicated if we are dealing with objects that cannot be easily represented by feature vectors. In a recent work Yang et al. proposed a method of data densification in image domains [14]. Their insight is that they can forgo creating synthetic *exemplars*, and simply create synthetic *points* in the distance space. Such points make the estimation of the data manifold more accurate, and can thus improve retrieval accuracy.

While this work is closest in spirit to ours, we *do* need to create actual synthetic images in order to learn the potential biases of our distance measure, and correct them. To produce synthetic exemplars we apply transformations similar to those proposed by Ha et al. [7]. This model captures majority of variations in writing and produces plausible results shown in Fig. 2.

Wang et al. [13] introduced the Adjusted k-Nearest Neighbors Rule, considering *"influence region for each training example."* They constructed this region as a sphere centered on the example *"that is as large as possible without enclosing a training example of a different class"*. After this they rescale the distances to each training sample as distance divided by the radius of the influence region. This approach is similar to ours, because it takes into account the "density" of training items in the

---

[1] Although preliminary work suggests that other distance measures also have difficulties in datasets with classes of varying complexity.

distance space. However, this approach requires parameters to be adjusted and optimized for each particular problem, something we are anxious to avoid.

We will provide necessary definitions before describing our algorithm.

Definition 1: A *labeled example* e_i of a class E_i is a *human* annotated glyph. Similarly, an *unlabeled example* is any example of the same glyphs not labeled by human.

After selecting e_i we can generate synthetic data based on e_i by a *distortion model*:
Definition 2: A *distortion model* M is the method to modify labeled glyphs to generate synthetic data $\{S_{i,j}\} = M(e_i)$, where j denotes index of synthetic exemplars in a
particular class i ($1 \le j \le$ const).

Note: our distortion model M is only one of many methods that can generate synthetic data. Further discussion of synthetic data generation techniques is beyond the
scope of this paper; we refer the reader to [4] and [7], and therein.

To classify an unknown item we have to consider distance (similarity measure) between this item and labeled items from the training set. Our approach exploits the
correction of distance calculated by some known algorithm.

Definition 3: The *corrected distance* between the query image q and *any* object $S_{i,j}$ in
the training dataset of the ith class is a distance under some distance measure (i.e., CK-
1) divided by some correction factor μ_j.

3   Proposed Method

As the dendrogram shown in Fig. 1 (*left*) suggests, classification using the CK-1 distance measure sometimes does not correspond to the ground truth. We have observed
that in cases where both shapes are "complex" (i.e., consisting of several "strokes"
such as 'f' or 'x') the distance between them is greater than in cases where the shapes
are "simple" (i.e., containing a single stroke such as 'l' or 'o'). Fig. 3 presents a visual
intuition of this phenomenon. In terms of absolute distances the unknown object lies
slightly closer to the nearest 'o'. Intuitively however, we may feel it is likely to belong to the 'ft' class because this class is sparse and the mean distance between two
instances of this cluster is relatively larger than in 'o' class.

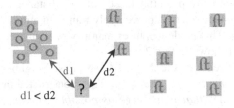

Fig. 3. While the unknown "?" object is slightly closer to the nearest o than to the nearest ft, we
intuitively feel it is more likely to belong to the latter class (exemplars are from [10])

The classic nearest neighbor algorithm does not take into account the density of
each cluster. In order to mitigate this shortcoming, we must correct for density, that
can be characterized as the mean of intraclass distances. Our approach is inspired by

the inverse-square law which is widely applied in physics. Let us consider an analogy to Newton's law of universal gravitation which follows an inverse-square law.

$$F = G\frac{mM}{r^2} \tag{1}$$

In (1) F is the force between two objects which masses are m and M, r is the distance between centers of mass of these objects and G is the gravitational constant. Analogously, in case of the classification problem we can consider the distance between the unknown object and the nearest neighbors of each class as r (denoted as r_i for each class i) and the mean of distance inside each class as M (for each class i denoted M_i). Since we simply need to compare the resulting values of F given by distance measuring between the unknown object and the nearest neighbor in each class, we are not interested in the values of G and m because they are the same in all cases. Thus, we need to compare these ratios: M_i/r_i^2 and M_j/r_j^2. Each ratio shows the "force" of attraction of the unknown object by each existing class. Therefore, the unknown object should be considered belonging to the class with the greatest "force." Recall that the nearest neighbor classifier assigns unknown objects the class label of a known object with the least distance value. Therefore, we can simply look for the least value between: $r_i/\sqrt{M_i}$ and $r_j/\sqrt{M_j}$ Thus, we can consider *division by square root* of the mean as the appropriate correction factor for the distance.

We initially imagined that creating synthetic data would be a major challenge. However, we found that simply applying tiny amounts of the affine transformations *homothety*, *rotation* and *shear mapping*, produced new images that are both visually very convincing (cf. Fig. 2) and closely modeled the true distributions of real data.

We present two different algorithms: supervised learning (SL) and semi-supervised learning (SSL). SL algorithm classifies items using only exemplars from the training set without addition of newly-classified items from the testing set to the training set. In contrast the SSL algorithm adds newly-classified instances to the training set and, therefore, performs next item classification using both training (generated) data and newly labeled instances from the testing set. For both algorithms we generate synthetic data randomly extracting one example from each class, and then distort this example applying the distortion model. After the training set is created we calculate distances between exemplars in one class with each other and find the mean of these distances to use it as our measure of class density (i.e. correction factor).

4   Experimental Evaluation

Table 1 shows the accuracy improvement of glyphs classification using distance correction over pure nearest neighbor approach. As we can see, our method demonstrated better performance than pure nearest neighbor approach.

Table 1. Classification accuracy (in percent) for the datasets: 1 – Chinese, 2 – G114 Verard Grosromain, 3 – R118 Garamount Grosromain, 4 – Liber Floridus, 5 – Petroglyphs

	Original accuracy, %					Accuracy improvement, %				
	1	**2**	**3**	**4**	**5**	**1**	**2**	**3**	**4**	**5**
SL	96.6	81.6	95.8	91.6	81.8	1.3	4.3	1.3	2.5	16.4
SSL	98.0	86.0	85.9	97.1	81.6	0.5	5.3	1.4	1.2	13.1

We have built a webpage [12] to more extensively document the experiments in this paper to ensure reproducibility. We tested our SL and SSL algorithms using a single, randomly chosen instance from each class. In every case, we averaged our results over 100 random runs. We evaluated our approach with both European, Chinese handwritten historical documents and petroglyphs from 5 datasets [5, 9, 10, 15].

5   Conclusions and Future Work

We have shown a method that allows classification of glyphs using only one exemplar of each class, by exploiting synthetic data and correcting distance calculations for the complexity of the glyph shapes. Experimental evaluation on diverse datasets demonstrated significant improvements in accuracy. We have committed to keeping a webpage with all the code and data we used online for at least five years, so others can check/reproduce and build upon our work [12].

For future work we consider expanding our techniques to other areas of images recognition as well as exploiting different distance measures for comparison of images similarity.

References

1. Batista, G., Wang, X., Keogh, E.J.: A Complexity-Invariant Distance Measure for Time Series. In: Proc. of the SDM 2011, pp. 699–710 (2011)
2. Campana, B., Keogh, E.: A Compression Based Distance Measure for Texture. In: Proc. of the SDM 2010, pp. 850–861 (2010)
3. Chapelle, O., Schölkopf, B., Zien, A.: Semi-supervised learning. MIT Press, Cambridge (2006)
4. Chawla, N., Bowyer, K., Kegelmeyer, W.: SMOTE: synthetic minority over-sampling technique. J. Artif. Intell. Res. 16, 321–357 (2002)
5. Derolez, A., Lamberti, S.: Audomari Canonici Liber Floridus, Codex Autographus Bibliothecae Universitatis Gandavensis, Ghent (1968)
6. Eno, J.: Generating Synthetic Data to Match Data Mining Patterns. IEEE Internet Computing 12(3), 78–82 (2008)
7. Ha, T., Bunke, H.: Off-line handwritten numeral recognition by perturbation method. IEEE Trans. on Pattern Analysis and Machine Intelligence 19(5), 535–539 (1997)
8. Hu, B., Rakthanmanon, T., Campana, B., Mueen, A., Keogh, E.: Image Mining of Historical Manuscripts to Establish Provenance. In: Proc. of the SDM 2012, pp. 804–815 (2012)
9. Indiana MAS Project, http://indianamas.disi.unige.it/
10. PaRADIIT Project, https://sites.google.com/site/paradiitproject/
11. Roy, P., Rayar, F., Ramel, J.Y.: An efficient coarse-to-fine indexing technique for fast text retrieval in historical documents. In: DAS 2012, pp. 150–154 (March 2012)
12. Supporting web page, https://sites.google.com/site/singleexemplar/
13. Wang, J.-G., Neskovic, P., Cooper, L.N.: An adaptive nearest neighbor algorithm for classification. In: Proc. of ICMLC 2005, pp. 3069–3074 (2005)
14. Yang, X., Bai, X., Köknar-Tezel, S., Latecki, L.J.: Densifying Distance Spaces for Shape and Image Retrieval. Journal of Mathematical Imaging and Vision, 1–17 (2012)
15. Zhang, X., Nagy, G.: The CADAL calligraphic database. In: Proc. of the HIP 2011, pp. 37–42 (2011)

Similarity for Natural Semantic Networks

Francisco Torres and Sara E. Garza

Facultad de Ingeniería Mecánica y Eléctrica, Universidad Autónoma de Nuevo León,
Cd. Universitaria, San Nicolás de los Garza, NL, Mexico
{francisco.torresgrr,sara.garzavl}@uanl.edu.mx

Abstract. A natural semantic network (NSN) represents the knowledge of a group of persons with respect to a particular topic. NSN comparison would allow to discover how close one group is to the other in terms of expertise in the topic— for example, how close apprentices are to experts or students to teachers. We propose to model natural semantic networks as weighted bipartite graphs and to extract feature vectors from these graphs for calculating similarity between pairs of networks. By comparing a set of networks from different topics, we show the approach is feasible.

Keywords: natural semantic networks, similarity, bipartite graphs, feature vectors.

1 Introduction

By means of knowledge representation, we can structure implicit information and turn it into a valuable asset. *Natural semantic networks* (NSN's) represent the knowledge of a population for a topic or domain by gathering responses from a sample group. Measuring similarity between NSN's allows to quantify a group's knowledge with respect to the experts of the domain— e.g. we could evaluate a student, job candidate, or apprentice. We propose an approach for NSN similarity calculation that is based on graph theory and document similarity; this approach, which considers both content and structure from the network, views the NSN as a bipartite graph and extracts a weighted feature vector for comparison.

The rest of this paper is organized as follows: Section 2 offers pertinent background and Section 3 briefly describes related work; our approach is explained in Section 4 and results are provided in Section 5. Section 6, finally, offers closing remarks and future work.

2 Background

This section introduces necessary vocabulary, notation, and formulas for natural semantic networks, bipartite graphs, and document similarity.

A.J. Machado Traina et al. (Eds.): SISAP 2014, LNCS 8821, pp. 195–200, 2014.
DOI: 10.1007/978-3-319-11988-5_18 © Springer International Publishing Switzerland 2014

2.1 Natural Semantic Networks

Natural semantic networks (NSN's), introduced by Figueroa et al. [6], study long-term memory by gathering a socio-cognitive perspective on a given topic. To generate a natural semantic network, a set P of *participants* (20-40) is given a set C of *target concepts* (6-10). For every $c \in C$, each participant must provide a set of individual words that come to mind when c is presented; these words are known as *definers*. The participant must also score each definer (using a scale 1-10) according to its importance within the target concept. Let us formally denote the score of participant p for definer d_i in concept c_k as $sc_k^i(p) \in \{1 \ldots 10\}$.

The total score of a definer within a given concept is known as its m-value; given d_i and c_k, this value is calculated as $m_k^i = \sum_{p \in P} sc_k^i(p)$. The ten definers with the highest m-value make up a concept's *SAM group*, where "SAM" stands for "Semantic Analysis of M-value" [7]. Let us note that a definer can be in more than one SAM group; it is thus possible to have not a single but a set of m-values for a particular definer. This also gives rise to another important metric: the f-value of a definer. The f-value is simply the number of times that the definer appears in the network. For d_i, we denote this value as f_i. A fragment of an NSN is shown in Table 1.

Table 1. Fragment of two SAM groups in a natural semantic network

F	Ecology Definer	M
1	Recycle	50
2	*Nature*	30
2	*Animals*	20
1	Plants	10

F	Environment Definer	M
2	*Nature*	100
2	*Animals*	70
1	Water	60
1	Trees	50

2.2 Bipartite Graphs

The mathematical representation for a network is a *graph*. A graph $G = (V, E)$ consists of a set V of entities known as *vertices* and a set E of connections known as *edges*. If the edges are assigned numerical weights, the graph is said to be *weighted*. A *bipartite graph*[1] is graph whose vertex set V is divided into two disjoint subsets V_1 and V_2 and whose set E only contains edges that join vertices from different subsets. A classical example of a bipartite graph is the *actor-movie* network, where the vertex subsets are conformed by actors and movies, and where each edge indicates an actor participating in a movie [8].

From a bipartite graph, it is possible to extract two *projections* or unipartite graphs (e.g. a projection where only movies are vertices and edges are common actors between them). Formally, in a projection $G_P = (V_P, E_P)$ of a bipartite graph $G_B = (V_B, E_B)$ where $V_P \subset V_B$,

$$E_P = \{\{u, v\} : (u, v \in V_P) \land (\{u, w\}, \{v, w\} \in E_B) \land (w \in V_B \setminus V_P)\}.$$

---

[1] Let us note that any graph with a single vertex set is called *unipartite* or *monopartite*.

2.3 Document Similarity with the Vector Space Model

The *vector space model* of information retrieval views a document as a *bag of words* where order is not important and extracts a *weight vector* from this bag; each vector's length is equal to the size of the document collection's vocabulary (unique words), and each weight represents the importance of a particular vocabulary word in the document (0 if the word is not present). A common metric for calculating similarity between document vectors is the *cosine similarity* [1]:

$$\operatorname{cosim}(a, b) = \frac{\boldsymbol{a} \cdot \boldsymbol{b}}{|\boldsymbol{a}| \times |\boldsymbol{b}|}, \tag{1}$$

where a and b are the documents, $\boldsymbol{a}$ and $\boldsymbol{b}$ are the vectors. A similarity of 0 indicates that the documents have no common words and a similarity of 1 indicates that the documents are identical.

3 Related Work

Network comparison is inherently related to *graph matching* [3], which can be *exact* or *inexact*. While the first addresses problems related to *graph isomorphisms* (detecting if two graphs are equal), the second attempts to provide the number of operations needed to turn one graph into another (*graph edit distance*) or a degree of resemblance between graphs (*graph similarity*). Our work and related works fall into this last category.

The works by Dehmmer and Emmert [5] and Qureshi et al. [9] both extract feature vectors for calculating graph similarity; while the former utilizes vertex degree (i.e. the number of connected edges), the latter uses statistical and symbolic features for object recognition. Meanwhile, the approach by Champin and Solnon [4] first obtains different mappings for the pair of graphs and then computes similarity with a psychologically-sustained metric. With regard to semantic data similarity, Bergmann and Gil [2] focus on semantic workflow retrieval by building graphs with different types of vertices and edges; on the other hand, Sanchez et al. [10] compare the NSN's of two distinct groups by means of an index that calculates the ratio of common edges with respect to the total amount possible (similar to the Jaccard index).

4 Measuring Similarity for Natural Semantic Networks

Our approach consists of calculating NSN pairwise similarity by compacting the networks into weighted feature vectors and obtaining cosine similarity for these vectors. Each feature is given either by a vertex or an edge of the networks, and each weight represents the importance of that feature. Because the nucleus of an NSN is given by its definers (target concepts are usually fixed along networks for the same topic), we represent the NSN as a graph where each vertex is a definer and each edge is the similarity or closeness between a pair of these. To determine which definers are related, as well as their closeness, we consider that the definer graph is a projection from a concept-definer *weighted bipartite graph*. In this other graph, there exists an edge between a concept and a definer when the latter belongs to the SAM group of the former; the weight of the edge is simply the m-value of the definer in that group.

In the definer projection, there is an edge between definers if these are found together in one or more SAM groups. To calculate edge weights, we assume that definers are closer or more similar to each other if the difference in their m-values is small. As a result, we first compute the *relative difference* between definers d_a and d_b for the SAM group of a concept c as

$$\delta_r(m_c^a, m_c^b) = \left| \frac{m_c^a - m_c^b}{m_c^{\max} - m_c^{\min}} \right|, \tag{2}$$

where $m_c^{\max}$ and $m_c^{\min}$ are, respectively, the maximum and minimum m-values of the group. Since the difference between definers is actually a *distance*, we obtain relative similarity by taking the complement of $\delta_r(m_c^a, m_c^b)$:

$$\mathrm{sim}(m_c^a, m_c^b) = 1 - \delta_r(m_c^a, m_c^b). \tag{3}$$

Also, because one same pair of definers can appear in several groups, we calculate the overall similarity between d_a and d_b as the average of their relative similarities in the set $C_{a,b} \subseteq C$ of SAM groups that contains both of them. An edge weight $w_{a,b}$ is, therefore, calculated with

$$w_{a,b} = \frac{\displaystyle\sum_{c \in C_{a,b}} \mathrm{sim}(m_c^a, m_c^b)}{|C_{a,b}|}. \tag{4}$$

Since a weight of 0 typically indicates the absence of an edge, we set $\mathrm{sim}(m_c^a, m_c^b)$ as half of the second lowest similarity in c's group when the numerator of Eq. 2 is $m_c^{\max} - m_c^{\min}$. To illustrate these calculations, an example of the bipartite and definer graphs (extracted from Table 1) is given by Figure 1.

Every edge weight of the definer graph will become a weight that corresponds to an edge feature in the NSN's feature vector. Regarding vertex features, the weight is given by the relative f-value of the definer, denoted as ϕ_a for d_a:

$$\phi_a = \frac{f_a}{f_{\max}}, \tag{5}$$

where $f_{\max}$ is the highest f-value found in the network. For Fig. 1, the vector includes, among others, a vertex feature "Recycle" (R) with weight $1/2 = 0.5$ and an edge feature "Recycle-Animals" with weight $1 - [(50 - 20)/(50 - 10)] = 0.25$.

5   Results

With the intent of showing how the proposed approach handles objects that are expected to be similar (networks from the same topic) and objects that are

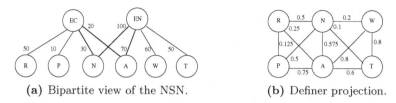

(a) Bipartite view of the NSN.   (b) Definer projection.

Fig. 1. NSN as a graph. EN and EC are concepts, while R, N, A, P, W, and T are definers

expected to be dissimilar (networks from different topics), we built a *similarity matrix* with a set of natural semantic networks from different topics; these networks were made available by a research group at the authors' university [11, 12]. The four topics covered by these networks are: *ecology* (ec1-ec6), *sentimental relationships* (lov1-lov4), *ethics* (eth1, eth2), and *scientific skills* (sk). The resulting matrix is depicted in Figure 2, where networks from the same topic were placed adjacent to each other (i.e. in blocks). We can clearly appreciate in the matrix the expected block-diagonal pattern, which indicates that similarity within the same topic (0.23 on average) is higher than similarity between different topics (0.005 on average).

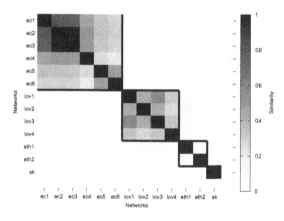

Fig. 2. Similarity matrix

6   Conclusions and Future Work

We have presented an approach for measuring similarity between natural semantic networks. The approach, which uses both content and structure, views each network as a concept-definer bipartite graph and extracts the definer projection from this graph to create a weighted feature vector; vectors are compared using cosine similarity. Future work includes comparative experiments and the use of *fuzzy graphs* for visualizing specific differences between the networks.

References

[1] Baeza-Yates, R., Ribeiro-Neto, B.: Modern information retrieval. ACM Press, New York (1999)

[2] Bergmann, R., Gil, Y.: Retrieval of semantic workflows with knowledge intensive similarity measures. In: Ram, A., Wiratunga, N. (eds.) ICCBR 2011. LNCS (LNAI), vol. 6880, pp. 17–31. Springer, Heidelberg (2011)

[3] Bunke, H.: Graph matching: Theoretical foundations, algorithms, and applications. In: Proc. Vision Interface, vol. 2000, pp. 82–88 (2000)

[4] Champin, P.-A., Solnon, C.: Measuring the similarity of labeled graphs. In: Ashley, K.D., Bridge, D.G. (eds.) ICCBR 2003. LNCS (LNAI), vol. 2689, pp. 80–95. Springer, Heidelberg (2003), http://liris.cnrs.fr/csolnon/publications/ICCBR03.pdf

[5] Dehmer, M., Emmert-Streib, F.: Comparing large graphs efficiently by margins of feature vectors. Applied Mathematics and Computation 188(2), 1699–1710 (2007)

[6] Figueroa, J., González, G., Solís, V.: Una aproximación al problema del significado: las redes semánticas. Revista Latinoamericana de Psicología 13(3), 447–458 (1981)

[7] Lopez, E.O., Theios, J.: Semantic analyzer of schemata organization (saso). Behavior Research Methods, Instruments, & Computers 24(2), 277–285 (1992)

[8] Newman, M.: Networks: An introduction. Oxford University Press (2010)

[9] Qureshi, R.J., Ramel, J., Cardot, H., Mukherji, P.: Combination of symbolic and statistical features for symbols recognition. In: International Conference on Signal Processing, Communications and Networking, ICSCN 2007, pp. 477–482. IEEE (2007)

[10] Miranda, M.P.S., de la Garza González, A., Ramirez, E.O.L.: Simulaciones computacionales sobre cuestiones ambientales en dos grupos de contraste. Liberabit 19(2), 223–233 (2013)

[11] Torres, F., López, R.: Foraging information in webpages through meaning. International Journal of Good Conscience 5(2), 308–323 (2010) ISSN 1870-557X

[12] Urdiales, M.E.: Sobre el esquema relacional de pareja en jovenes y adultos del área metropolitana de Monterrey. PhD thesis, Universidad Autónoma de Nuevo León (2009), http://uanl.vtlseurope.com/lib/item?id=chamo:224600&theme=UANL

Anomaly Detection in Streaming Time Series Based on Bounding Boxes

Heider Sanchez[*] and Benjamin Bustos

Department of Computer Science, University of Chile, Santiago, Chile
{hesanche,bebustos}@dcc.uchile.cl

Abstract. Anomaly detection in time series has been studied extensively by the scientific community utilizing a wide range of applications. One specific technique that obtains very good results is "HOT SAX", because it only requires a parameter the length of the subsequence, and it does not need a training model for detecting anomalies. However, its disadvantage is that it requires the use of a normalized Euclidean distance, which in turn requires setting a parameter ε to avoid detecting meaningless patterns (noise in the signal). Setting an appropriate ε requires an analysis of the domain of the values from the time series, which implies normalizing all subsequences before performing the detection. We propose an approach for anomaly detection based on bounding boxes, which does not require normalizing the subsequences, thus it does not need to set ε. Thereby, the proposed technique can be used directly for online detection, without any a priori knowledge and using the non-normalized Euclidean distance. Moreover, we show that our algorithm computes less CPU runtime in finding the anomaly than HOT SAX in normalized scenarios.

Keywords: Time Series, anomaly detection, indexing, streaming.

1 Introduction

Anomaly detection in time series has been studied extensively by the scientific community, who has contributed a wide variety of approaches for different types of applications [8]. In data mining, research is generally focused on searching for unusual patterns or outliers in a collection of time series using classification or clustering [21,10,9]. Most of the state-of-the-art techniques for anomaly detection in time series use a time series sample of "normal" behavior as a training model. However, data mining on subsequences from a streaming time series is a more complicated task because of particular challenges that need to be addressed. The main challenge of the subsequences is their level of overlapping: contiguous subsequences are similar to each other. This may produce a meaningless clustering result [23]. Moreover, the definition of "anomaly" is ambiguous and may be mistaken for irregularities that occur along a streaming time series, for example variations in amplitude scale and the presence of local noise.

---

[*] Work supported by a research grant from CONICYT-Chile.

A.J. Machado Traina et al. (Eds.): SISAP 2014, LNCS 8821, pp. 201–213, 2014.
DOI: 10.1007/978-3-319-11988-5_19 © Springer International Publishing Switzerland 2014

Keogh et al. [18] introduced a new anomaly concept, the "Time Series Discord", for finding the most unusual time series subsequence which does not need a training model. While similarity searching by content finds the object most similar to a query, the discord discovery process finds the time series subsequence that is least similar to all other subsequences. Moreover, Keogh proposed a generic heuristic for efficient discord discovery and a solution algorithm, the so-called HOT SAX, that is based on Symbolic Aggregate approXimation (SAX). Later, a series of related works were proposed to improve the performance of the basic heuristic [5,6,20]. All of these solutions use the normalized Euclidean distance (L2-norm), that is, each subsequence is normalized by a standard normalization procedure (Z-distribution) to obtain a symbolic sequence.

The problem with using normalized subsequences is the presence of local noise, which can result in a missed detection. This issue is easily solved by applying a parameter ε [22]. Given a subsequence $C = \{c_1, \ldots, c_m\}$, let σ and μ be the standard deviation and the mean of C, respectively. Then, if $\sigma < \varepsilon$ one sets $\forall i$, $c_i = \mu$. The problem with this approach is that one needs setting the parameter ε, which is context-dependent. On the other hand, the Euclidean distance over non-normalized subsequences (L2-raw) does not need to set ε and is more robust to local noise. Additionally, in real-time streaming, the future values of the time series are unknown. Therefore, obtaining an optimal ε is a complicated task. Moreover, there are anomalies related to amplitude changes and local oscillations (e.g., El Niño-Southern Oscillation Events [26]). These types of anomalies are at risk of not being detected in a normalized scenario. In these cases, using L2-raw is an effective option. Moreover, scalability is an important factor for many real time systems that generate large time series (e.g., seismic signals [12], electrocardiograms [19,11] and network traffic [2]).

This paper makes two contributions in streaming time series anomaly detection. First, we propose an algorithm for efficient time series discord discovery which supports both L2-norm and L2-raw distances. Second, we introduce a new automatic learning online algorithm to detect local discords in streaming time series. Our model is based on the previous works of Keogh et al. [16] and Vlachos et al. [27]. Specifically, they proposed the RTree-Index for time series using Bounding Boxes and the Dynamic Time Warping (DTW) distance. We propose modifying this structure to work directly with L2-raw and the anomaly detection algorithm. We experimentally show that our technique is faster than HOT SAX in normalized subsequences.

2   Background and Related Work

2.1   Time Series

Definition 1. *Streaming Time Series. A sequence of observations $T = \{t(i)/i = 1\cdots\infty\}$ taken at various time moments, evenly spaced and chronologically sorted.*

Definition 2. *Sliding Window. Given a time series T of length n, we use a overlapping sliding window of length $m \ll n$ to extract all possible subsequences*

C_p, $p \in \{1, \ldots, (n - m + 1)\}$, *from T. This window generates overlapping sub-sequences of contiguous position.*

Definition 3. *Normalized Subsequence. Given a subsequence $C = \{c_1, \ldots, c_m\}$, its normalized version is defined as $C' = \frac{C - \mu}{\sigma}$, where μ and σ are the mean and standard deviation of C.*

The main indexing techniques of subsequences use a reduced representation of the time series to avoid the High-Dimensionality problem. They provide a lower bounding function of the true distance between two time series. Examples are the RTree for Dynamic Time Warping (DTW) [16], iSAX for L2-norm [25] and the TS-Tree that uses the best of both techniques [4].

2.2   Discord Discovery and State-of-the-Art Solutions

Anomaly definition in a time series is ambiguous and is strongly related to the application context and data properties. In streaming time series, one usually associates an anomaly with a subsequence that produces a qualitatively significant change in the data. Furthermore, most anomaly detection approaches work on the basis of a normal behavior model of a time series. However, in many real contexts, obtaining this a priori knowledge is a difficult task. Keogh et al. [19] introduced a new definition to avoid creating workable definitions for "the most unusual subsequence", which does not require a training model.

Definition 4. *Non-self match. Given a time series T, containing a subsequence C_p of length m and a matching subsequence C_q, we say that C_q is a non-self match to C_p if $|p - q| \geq m$, where p and q are their respective starting positions in T.*

Definition 5. *Time Series Discord. Given a time series T, the subsequence C of length m is said to be the discord of T if C has the largest distance to its nearest non-self match.*

This problem can be easily solved by a brute force search using a nested loop. The outer loop takes each subsequence as a possible candidate, and the inner loop is used to search the candidate's nearest non-self match. The candidate that has the greatest such value is the discord. The computational complexity is $O(N^2)$, where N is the number of subsequences. To improve this complexity, Keogh et al. [19] proposed a generic algorithm for efficient detection. This algorithm requires two heuristics that generate two ordered lists of subsequences; one for the outer loop and the other one for the inner loop. The heuristic *Outer* is useful for quickly finding the best candidate, and the heuristic *Inner* is useful for quickly finding the best nearest non-self match. We break out of the inner loop if the distance is less than the best-so-far discord distance.

The main related methods for discord discovery are based on SAX representation, which is a discretization technique for time series introduced by Lin et al. [24] and it is used in many application domains for different purposes.

Definition 6. *Breakpoints.* "*Breakpoints are a sorted list of numbers* $\beta = \{\beta_1, \ldots, \beta_{\alpha-1}\}$ *such that the area under a* $N(0,1)$ *Gaussian curve from* β_i *to* $\beta_{i+1} = 1/\alpha$ *(* β_0 *and* β_α *area defined as* $-\infty$ *and* $+\infty$, *respectively*)." [24]

Definition 7. *SAX Representation. Given a normalized subsequence* $C = \{c_1, \ldots, c_m\}$, *first, we obtain the reduced dimension with Piecewise Aggregate Approximation (PAA [15]): the time series is divided into* D *equal sized segments, the mean value of each segment is calculated and a vector of these values becomes the reduced representation* $P = \{p_1, \ldots, p_D\}$. *Afterwards,* P *is transformed into a symbolic sequence* $W = \{w_1, \ldots, w_D\}$, *where* p_i *is mapped to symbol* w_i *of an alphabet of size* α. *It uses the predetermined breakpoints to define the symbols.*

HOT SAX: It is the first algorithm for efficient discord discovery introduced by Keogh et al. [19]. HOT SAX uses an *augmented trie* to embed all the SAX words, where leaf nodes contain a linked list index of all word occurrences that map there. The second structure is an array of SAX words of all extracted subsequences, counting the frequency of each word occurrence in the array. After building these data structures, HOT SAX uses the following heuristics: (a) *Outer loop heuristic:* It first visits the subsequences associated with the SAX words that have the smallest word count, and then it visits the rest of the subsequences in random order; (b) *Inner loop heuristic:* For each candidate in outer loop, it first searches its nearest non-self match in the leaf node of tree that has the same SAX word, and then it visits the rest of the subsequences in random order.

HOTiSAX: It extends HOT SAX for working with the iSAX index [6]. iSAX is an optimized structure for SAX binary representation [25], which provides different levels of resolution for the same SAX word changing the symbolic alphabet by binary numbers. The bits are used for building the iSAX index, which allows an efficient hierarchical access to data. The array of SAX words is refined to work effectively with the iSAX representation. Afterwards, the same heuristics of HOT SAX are used for discord discovery. Moreover, it incorporates auxiliary functions to exclude trivial matches.

3   Bounding Boxes for Discord Discovery

A minimum bounding box is a term used in geometry for enclosing a set of D-dimensional points [13]. This concept was used by Vlachos et al. [27] for bounding two-multidimensional time series by a sequence of Minimum Bounding Rectangles (MBRs). All the MBRs are indexing in a RTree in order to allow the efficient searching. Similarly, Keogh et al. [16] applied bounding boxes over the PAA of time series. The goal of this work was to generate reduced dimension vectors for the time series being indexed with a RTree and the DTW distance. Furthermore, in both works the authors provided a lower bounding function of the true distance for exact searching. In another related project, Chan et al. [7] proposed the use of a sequence of minimal bounding boxes to contain all of the training time series for detecting anomalies in trajectories.

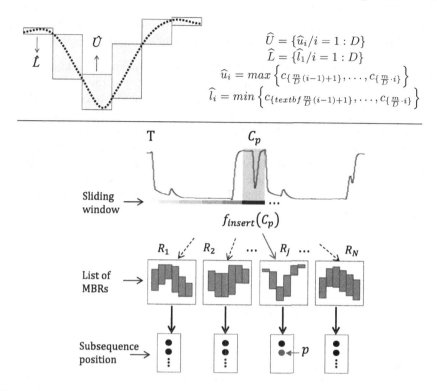

Fig. 1. Bounding of the subsequence C_p (top) and our indexing model for discord discovery (bottom)

In this paper, we use both the modeling of MBRs and the PAA representation for discord discovery in time series. To obtain efficient results on a streaming context, we use the Euclidean distance, due to its linear complexity.

3.1 List of MBRs

Figure 1 shows our indexing model of time series subsequences. We build a List of MBRs = $\{R_1, \ldots, R_N\}$, where each R_i envelops a set of similar time series subsequences. Below, we describe the insertion process phases:

1. Given a time series T, we extract a subsequence $C_p = \{c_1, \ldots, c_m\}$ using a sliding window. Next, we generate the minimum bounding boxes $(\widehat{U}, \widehat{L})$ of C_p (Figure 1.top). This function requires a reduced dimension D to split the subsequence C_p in D equal length segments, then, $\widehat{u}_i$ and $\widehat{l}_i$ is the maximum and minimum value of the ith segment respectively.
2. We search a MBRs R_j that produces the least expansion to C_p. That is, $volume(R_j \cup (\widehat{U}, \widehat{L})) < volume(R_{i,\forall i \neq j} \cup (\widehat{U}, \widehat{L}))$. We then stretch R_j to envelop the minimum bounding boxes of C_p. Finally, the position p is inserted into an integer number array, which is associated with R_j.

3. If R_j is full, we apply a splitting algorithm. We use a size threshold th_{max} to control the maximum number of elements in a MBRs.

We evaluate three classic splitting algorithms: Guttman's quadratic and linear algorithms [14], and an optimized linear algorithm [3]. For the quadratic algorithm, we considered two criteria of distribution; balanced (qua_0.50) and non-balanced (qua_0.25). Figure 2 shows the number of created MBRs and the CPU runtime in indexing and searching of the most unusual subsequence. We observe that the optimized linear algorithm generates the fewest MBRs, and therefore gets the least memory space. However, the balanced quadratic algorithm achieves the best runtime because it performs a better grouping of subsequences allowing fast discrimination in searching time. Therefore, we use the quadratic algorithm in our indexing model.

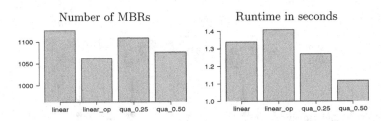

Fig. 2. Testing four splitting algorithms over a set of time series of length 16k. We show the total number of created MBRs (left) and the CPU runtime (right) for discord$_{128}$.

The time required to insert a subsequence in the list has an order of $O(N)$, where N is the total number of created MBRs. The space required for maintaining the index in memory also depends on N. For each $R_j = (H, L)$, we used two arrays of size D and another array of integer numbers for saving the subsequence positions; e.g., for a time series of length 16,000 we require 16.7KB for the index (N=1070) and 62KB for the arrays of positions.

We also build a Tree of MBRs (RTree variant) for maintaining the set of MBRs. It provides us hierarchical access to the data in order to accelerate the searching of similar objects regarding a query object. The RTree is generally used for managing large collections of data in secondary memory.

3.2   Discord Discovery Heuristics

After building the index, we reorder all subsequences for searching the best candidate using the following heuristics:

Outer Loop Heuristic: First, the algorithm visits all subsequences bounded in R_j, such that R_j contains the minimum number of subsequences. Then, the algorithm visits the rest of the subsequences in random order. This heuristic ensures that the subsequences that are most isolated will be visited at the beginning of the search as potential candidates. We then use an inner loop to search the best non-self match of each selected candidate C_q. To break the inner loop as early as possible, we need to find a subsequence that has a distance to C_q lower than the best-so-far discord distance.

Inner Loop Heuristic: First, the algorithm visits all subsequences bounded in R_j, such that $MINDIST(\bar{C}_q, R_j) < MINDIST(\bar{C}_q, R_{i,\forall i \neq j})$. Then, the algorithm visits the rest of the subsequences in random order. This heuristic allows us to first visit all the subsequences C_p most similar to C_q, increasing the probability of early termination of the loop. $MINDIST$ function is calculated by Equation 1 and illustrated in Figure 3 using $\bar{C}_q$ as PAA representation of C_q.

$$MINDIST(\bar{C}_q, R) = \sqrt{\sum_{i=1}^{D} \frac{m}{D} \begin{cases} (\bar{c}_i - h_i)^2 \ \bar{c}_i > h_i \\ (\bar{c}_i - l_i)^2 \ \bar{c}_i < l_i \\ 0 \ \text{else}, \end{cases}} \quad \begin{array}{l} \text{where:} \\ \bar{c}_i = \frac{D}{m} \sum_{j=\frac{m}{D}(i-1)+1}^{\frac{m}{D}i} c_j. \end{array} \quad (1)$$

3.3   Online Anomaly Detection

We can use discord discovery for detecting anomalous subsequences in streaming time series. A simple idea is to use normal behavior data to build a static training model. Discord discovery is then used for online detection of new inputs. This approach is very efficient, because the model retains its size, although the streaming is increased. However, it does not evolve to new states of the system. Any input that generates a new behavior is always considered anomalous.

We propose a scalable method for detecting local anomalies, where online learning is required for adapting the index model with the behavior of the input data. Our method is based on the inner loop heuristic (Algorithm 1). It requires a detection starting point in the stream, which is used to determine a "base history". Then, we apply discord discovery up to this point to obtain a threshold distance (line 1). This threshold is used to reduce the number of calls to $Dist$ in the inner loop. Our method is fed back with each data input (line 6) in order to avoid detecting recurrent anomalies. That is, if a subsequence is detected as anomalous, the next similar subsequences will not be detected as anomalous. Finally, we alert when an value input generates an anomalous subsequence.

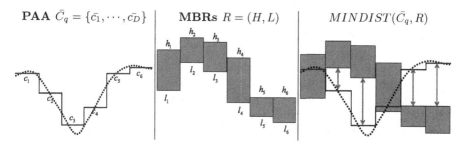

Fig. 3. An illustration of the $MINDIST$ function. The lengths of the red arrow lines, squared, scaled by m/D, summed and square rooted, are returned as the minimum distance between C_q and any sequence bounded in R [16].

Algorithm 1. Online Anomaly Detection

Require: (Streaming T, Window length m, Starting Point sp)

 1: threshold_dist $= TheMostDiscord(T[1:sp])$
 2: **while** $Input == True$ **do**
 3:     $T[i] = AcquireValue()$
 4:     $C_q = T[(i - m + 1) : i]$ ▷ *extract the subsequence*
 5:     $insert(C_q)$ ▷ *feedback*
 6:     $l_inner = $ All C_p from T ordered by **heuristic Inner**
 7:     nearest_neighbor_dist $= \infty$
 8:     **for** $C_p \in l\_inner$ **do**
 9:         **if** $|p - q| \geq m$ **then**   ▷ *non-self match?*
10:             **if** $Dist(C_p, C_q) <$ nearest_neighbor_dist **then**
11:                 nearest_neighbor_dist $= Dist(C_p, C_q)$
12:             **end if**
13:             **if** $Dist(C_p, C_q) <$ threshold_dist **then**
14:                 **Break**   ▷ *Break out of loop*
15:             **end if**
16:         **end if**
17:     **end for**
18:     **if** nearest_neighbor_dist $>$ threshold_dist **then**
19:         $Alarm(\text{"Anomaly Detected"})$
20:     **end if**
21: **end while**

The starting point can be automatically computed in real time. This is possible because the distance to the best non-self match usually reduces its value when increasing the stream length (see Figure 6.bottom). The idea is to find the stabilization point and set it as the starting point. A simple way to achieve this is by counting the distances outside the range of the deviation standard $[\mu - \sigma : \mu + \sigma]$ and applying a stop proportion.

In a long stream, it is important to update the threshold with the actual context to improve the detection of local anomalies. For this, we recommend periodically removing the past information and again, computing the new threshold with the rebuilt index. This is a scalability improvement to apply at runtime.

4   Experimental Evaluation

In this section, we evaluate the performance of our approach for discord discovery in different datasets. An Intel Core i7 3.4GHz with 8GB RAM is used for conducting all our experiments. All algorithms are implemented in C++. We define $th_{max} = 25$ as the maximum size of elements in a MBRs, which was experimentally selected from the set $\{5, 10, 15, \cdots, 120\}$. Although better efficiency is obtained when we vary the value of th_{max} according the time series length. The value of this parameter does not alter the effectiveness.

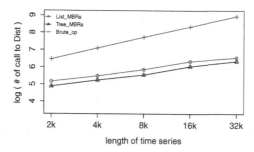

Fig. 4. The number of calls to the distance function ($Dist$) by brute force and our approach for discord$_{128}$ using L2-raw

4.1 Offline Discord Discovery

To evaluate the efficiency of our approach on static time series, we use the datasets ECG, EEG, ERP, Koski, Random Walk and Packet from "The UCR Time Series Data Mining Archive" [17]. We also use the "Time Series for Weather Data" from the National Oceanic and Atmospheric Administration in the USA [1]. For each dataset, we randomly extract time series of lengths 1k, 2k, 4k, 8k, 16k and 32k. The metrics used for comparison are the number of computed distances and the CPU runtime. The goal is to find the most unusual subsequence using the fewest distance computations and the minimum time of execution.

First, we compare our approach (List of MBRs and Tree of MBRs) with the brute force search over non-normalized subsequences using the L2-raw distance. We optimize the brute force search applying symmetry and triangle inequality properties. Figure 4 shows the performance of our approach in terms of the number of computed distances. We get the median of the results obtained for all the datasets for each time series length. From these results we note that our method clearly outperforms the brute force search, ≈ 2.5 orders of magnitude faster on all collections with time series of length 32k. This difference is correlative with the CPU runtime, where our approach was 150 times faster than the brute force.

Second, we compare our approach with the SAX techniques over normalized subsequences using the L2-norm distance. For SAX representation, the authors recommended using the following parameter setting: size of alphabet $a = 3$ and word length depending on data [19]. However, iSAX binary representation requires $a \in 2\mathbb{Z}$ (multiple of 2), we therefore set $a = 4$ [6]. Also, we set the reduced dimension based on the window length $D = \lfloor log_2(m) \rfloor$.

In Figure 5, we note our algorithms have less performance than the SAX techniques in terms of computed distances. However, our algorithms outperform HOT SAX in CPU runtime. This is explained by the low number of created

Table 1. The number of calls to MINDIST for discord$_{128}$

List of MBRs	Tree of MBRs	HOT SAX	HOTiSAX
226,259	482,256	1,950,445	314,407

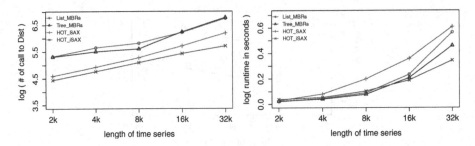

Fig. 5. The number of calls to the distance (left) and the CPU runtime (right) by SAX techniques and our algorithms for discord$_{128}$ using L2-norm and $\varepsilon = 0.05$

MBRs versus the number of created buckets by HOT SAX, resulting in the lower memory usage. Our approach uses a balanced splitting algorithm to maximize the number of elements in the MBRs. Thus, we obtain more closed groups and a better effect of the MINDIST function. Table 1 shows the number of computed MINDIST for each technique in a set of real time series.

Finally, the Tree of MBRs does not get a wide lead over the List of MBRs. The first reason is that we do not need to save all subsequences in secondary memory as it was sufficient with the referential positions in main memory. The second due to the additional cost of splitting the internal nodes. Moreover, the Tree of MBRs produces more calls to MINDIST than the List of MBRs in the search task. In practice, the discord discovery process is just applied in limited ranges of time in order not to lose local significance of the anomaly.

4.2 Online Anomaly Detection

In real-time streaming, we do not know the future values of the time series. Therefore, it is not clear how one could obtain an appropriate value for ε in normalized subsequences. For this reason, in this experiment we use L2-raw to evaluate our online anomaly detection algorithm. We cannot use SAX techniques because it is not compatible with L2-raw.

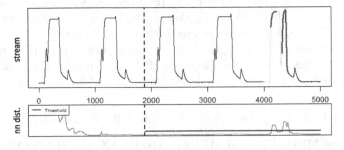

Fig. 6. Online anomaly detection on a space shuttle time series ($m = 128$)

In Figure 6, we evaluate our algorithm in a real case. The top graph shows the streaming time series and the detected anomalous subsection (red region). The bottom graphs represent the nearest non-self neighbor distance of each input subsequence where the dotted vertical line is the detection starting point that was automatically set. Also, we show the threshold distance (blue line), which is used to avoid unnecessary matches. We observe that our algorithm is successful at detecting the points that cause an anomalous subsequence at the right moment.

Figure 7 shows the number of computed distances over a set of real time series. We note a wide advantage of our indexing model vs. the brute force on all time series. In practice, our approach seems to compute far fewer distances than the quadratic brute force search algorithm. Moreover, we show the benefits of using our anomaly detection algorithm with regard to the similarity search.

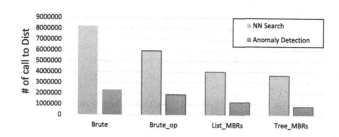

Fig. 7. The number of calls to the distance by both the nearest non-self neighbor search and the anomaly detection algorithm

5   Conclusions and Future Work

We proposed new algorithms for efficient discord discovery. We used bounding boxes for designing two indexing models List of MBRs and Tree of MBRs, which support L2-norm and L2-raw. Our approach outperforms HOT SAX in terms of the CPU runtime. In addition, we introduced a new algorithm for online anomaly detection, which does not require a training model and automatically fixes a detection starting point. We experimentally showed that this online detection algorithm is faster than the brute search force approach.

Using the discord discovery algorithm over a set of real time series [19], we obtain 79% correct detection using L2-raw. While it misses some anomalies, its advantage is that it avoids the exploration step of the data to set ε in smoothing noisy subsequences. We emphasize that L2-raw may be used as a baseline for automatic detection techniques for discord discovery, especially for quasi-periodic streaming time series.

Since we do not apply any discretization technique, we can easily extend our approach to multivariate time series, which is our next goal. Finally, we plan to optimize our online algorithm, focusing on scalability for larger amounts of data.

Acknowledgments. We would like to thank Prof. Eamonn Keogh for kindly providing us with many of the datasets used at the experimental evaluation.

References

1. Web Page for Time Series for Weather Data of National Oceanic and Atmospheric Administration in USA, http://www.esrl.noaa.gov/psd/boulder/
2. Ahmed, T., Coates, M., Lakhina, A.: Multivariate online anomaly detection using kernel recursive least squares. In: IEEE INFOCOM, pp. 625–633 (2007)
3. Ang, C.-H., Tan, T.: New linear node splitting algorithm for r-trees. In: Scholl, M.O., Voisard, A. (eds.) SSD 1997. LNCS, vol. 1262, pp. 337–349. Springer, Heidelberg (1997)
4. Assent, I., Krieger, R., Afschari, F., Seidl, T.: The TS-tree: Efficient time series search and retrieval. In: Proc. 11th Intl. Conf. on Extending Database Technology: Advances in Database Technology, pp. 252–263. ACM (2008)
5. Bu, Y., Wing Leung, O.T., Chee Fu, A.W., Keogh, E.J., Pei, J., Meshkin, S.: WAT: Finding top-k discords in time series database. In: SIAM Intl. Conf. on Data Mining, pp. 449–454 (2007)
6. Buu, H.T.Q., Anh, D.T.: Time series discord discovery based on iSAX symbolic representation. In: 2011 Third Intl. Conf. on Knowledge and Systems Engineering (KSE), pp. 11–18 (2011)
7. Chan, P.K., Mahoney, M.V.: Modeling multiple time series for anomaly detection. In: IEEE Intl. Conf. on Data Mining, pp. 90–97 (2005)
8. Chandola, V., Banerjee, A., Kumar, V.: Anomaly detection: A survey. ACM Comput. Surv. 41, 1–58 (2009)
9. Chaovalit, P., Gangopadhyay, A., Karabatis, G., Chen, Z.: Discrete wavelet transform-based time series analysis and mining. ACM Comput. Surv. 43, 1–37 (2011)
10. Chis, M., Banerjee, S., Hassanien, A.: Clustering time series data: An evolutionary approach. In: Abraham, A., Hassanien, A.-E., de Carvalho, A.P.D.L.F., Snášel, V. (eds.) Foundations of Computational, Intelligence Volume 6. SCI, vol. 206, pp. 193–207. Springer, Heidelberg (2009)
11. Chuah, M.C., Fu, F.: ECG anomaly detection via time series analysis. In: Thulasiraman, P., He, X., Xu, T.L., Denko, M.K., Thulasiram, R.K., Yang, L.T. (eds.) ISPA Workshops 2007. LNCS, vol. 4743, pp. 123–135. Springer, Heidelberg (2007)
12. Gabarda, S., Cristóbal, G.: Detection of events in seismic time series by time-frequency methods. IET Signal Processing 4(4), 413–420 (2010)
13. Gottschalk, S., Lin, M.C., Manocha, D.: OBBTree: A hierarchical structure for rapid interference detection. In: Proc. 23rd Annual Conference on Computer Graphics and Interactive Techniques, pp. 171–180. ACM (1996)
14. Guttman, A.: R-trees: a dynamic index structure for spatial searching. In: Intl. Conf. on Management of Data, pp. 47–57 (1984)
15. Keogh, E., Chakrabarti, K., Pazzani, M., Mehrotra, S.: Dimensionality reduction for fast similarity search in large time series databases. Knowledge and Information Systems 3(3), 263–286 (2001)
16. Keogh, E., Ratanamahatana, C.A.: Exact indexing of dynamic time warping. Knowl. Inf. Syst. 7(3), 358–386 (2005)
17. Keogh, E., Xi, X., Wei, L., Ratanamahatana, C.: The UCR Time Series Classification/Clustering Homepage (2011)

18. Keogh, E.J., Lin, J., Fu, A.W.: HOT SAX: Efficiently finding the most unusual time series subsequence. In: IEEE Intl. Conf. on Data Mining, pp. 226–233 (2005)
19. Keogh, E.J., Lin, J., Hee Lee, S., Herle, H.V.: Finding the most unusual time series subsequence: algorithms and applications. Knowledge and Information Systems 11, 1–27 (2007)
20. Khanh, N.D.K., Anh, D.T.: Time series discord discovery using WAT algorithm and iSAX representation. In: Proc. Third Symposium on Information and Communication Technology, pp. 207–213. ACM (2012)
21. Liao, T.W.: Clustering of time series data: a survey. Pattern Recognition 38(11), 1857–1874 (2005)
22. Lin, J., Keogh, E., Lonardi, S., Chiu, B.: A symbolic representation of time series, with implications for streaming algorithms. In: Proc. 8th ACM SIGMOD Workshop on Research Issues in Data Mining and Knowledge Discovery, pp. 2–11 (2003)
23. Lin, J., Keogh, E., Truppel, W.: Clustering of streaming time series is meaningless. In: Proc. 8th ACM SIGMOD Workshop on Research Issues in Data Mining and Knowledge Discovery, pp. 56–65. ACM (2003)
24. Lin, J., Keogh, E.J., Wei, L., Lonardi, S.: Experiencing SAX: a novel symbolic representation of time series. Data Mining and Knowledge Discovery 15, 107–144 (2007)
25. Shieh, J., Keogh, E.: iSAX: indexing and mining terabyte sized time series. In: Proc. 14th ACM SIGKDD Intl. Conf. on Knowledge Discovery and Data Mining, pp. 623–631. ACM (2008)
26. Trenberth, K.E., Hoar, T.J.: The 1990-1995 El Niño-Southern oscillation event: Longest on record. Geophysical Research Letters 23(1), 57–60 (1996)
27. Vlachos, M., Hadjieleftheriou, M., Gunopulos, D., Keogh, E.: Indexing multi-dimensional time-series with support for multiple distance measures. In: Proc. Ninth ACM SIGKDD Intl. Conf. on Knowledge Discovery and Data Mining, pp. 216–225. ACM (2003)

SVG-to-RDF Image *Semantization*

Khouloud Salameh[1], Joe Tekli[2], and Richard Chbeir[1]

[1] LIUPPA Laboratory, University of Pau and Adour Countries (UPPA)
64600 Anglet, France
{khouloud.salameh,richard.chbeir}@univ-pau.fr
[2] School of Engineering, Lebanese American University (LAU)
36 Byblos, Lebanon
joe.tekli@lau.edu.lb

Abstract. The goal of this work is to provide an original (semi-automatic) annotation framework titled *SVG-to-RDF* which converts a collection of raw Scalable vector graphic (SVG) images into a searchable semantic-based RDF graph structure that encodes relevant features and contents. Using a dedicated knowledge base, *SVG-to-RDF* offers the user possible semantic annotations for each geometric object in the image, based on a combination of shape, color, and position similarity measures. Our method presents several advantages, namely i) achieving complete *semantization* of image content, ii) allowing semantic-based data search and processing using standard RDF technologies, iii) while being compliant with Web standards (i.e., SVG and RDF) in displaying images and annotation results in any standard Web browser, as well as iv) coping with different application domains. Our solution is of linear complexity in the size of the image and knowledge base structures used. Using our prototype *SVG2RDF*, several experiments have been conducted on a set of panoramic dental x-ray images to underline our approach's effectiveness, and its applicability to different application domains.

Keywords: Vector images, SVG, RDF, semantic graph, semantic processing, image annotation and retrieval, visual features, image feature similarity.

1   Introduction

The need to index and retrieve multimedia data is becoming ever-more important, especially on the Web where image search and retrieval techniques do not seem to keep pace. Most existing Web image search engines (such as Google and AltaVista) and photo sharing sites (e.g., Flickr and Picasa) adopt the keyword (text-based) querying paradigm, usually returning a large quantity of search results, ranked by their relevance to a text-based query [27]. This can be extremely tedious and time consuming, since the returned results usually contain multiple topics mixed together, where the automated engines are guessing image visual contents using (in)direct textual clues [27]. An alternative approach is content-based image retrieval (CBIR), where images are indexed based on their visual content, using low-level color, texture, and shape descriptors, and are consequently processed via dedicated search

A.J. Machado Traina et al. (Eds.): SISAP 2014, LNCS 8821, pp. 214–228, 2014.
DOI: 10.1007/978-3-319-11988-5_20 © Springer International Publishing Switzerland 2014

engines (e.g., QBIC [5], Photobook [19], and Google *search-by-image[1]*). CBIR has been usually less successful than text-search engines since low-level features are usually unable to effectively capture the semantic meaning of the image [13]. This is known as the so-called *semantic gap* [14]: discrepancy between low-level image features and user semantics.

The main goal of our study is to convert, with as little human intervention as possible, a collection of raw images into a searchable semantic-based structure that encodes semantically relevant image content. We specifically target the semi-automatic annotation of vector images, mainly SVG (Scalable Vector Graphic) images [28]. In summary, SVG is an XML-based language for describing two-dimensional graphics and encoding three types of visual objects: vector graphic shapes, images and text. SVG images have interesting properties (resolution-independent and extremely small-size image coding) and are becoming increasingly popular in a wide range of applications covering: medical image annotation [8, 10], geographic map annotation [11, 18], manipulating graph charts as well as basic shape annotation to simplify data accessibility for the blind [1, 2].

Here, we introduce a framework titled *SVG-to-RDF* which allows to convert a collection of SVG images into an RDF (Resource Description Framework) [7] graph structure. The RDF data model is similar to classic conceptual modeling approaches such as entity–relationship or class diagrams, allowing to define statements about resources in the form of subject-predicate-object expressions. These expressions are known as triples in RDF terminology. The *subject* denotes the resource being described, the *predicate* denotes traits or aspects of the resource, expressing a relation between the subject and the object, and the *object* designates another resource or data values.

Our system automatically transforms an SVG image into an RDF graph describing the geometric objects in the image and their relations (in the form of RDF triples), and then offers the user possible semantic annotations for each geometric object encoded in the RDF graph, based on shape, color, and position similarity comparison with existing objects stored in a dedicated (RDF-based) knowledge base. The annotated RDF image graph could be in several cases integrated in the knowledge base helping extend its semantic expressiveness and hence provide more accurate annotation offers for future comparisons. Our original method presents several advantages over existing approaches, namely i) the complete *semantization* of image contents, ii) allowing sophisticated semantic-based data search and processing using standard RDF technologies (e.g., SPARQL [20]), iii) while being compliant with Web standards (i.e., SVG and RDF) in displaying images and annotation results in any standard Web browser, as well as iv) coping with different application domains by its generality and adaptability. To validate our approach, a prototype tool called *SVG2RDF* has been developed and tested on a collection of panoramic dental x-ray images. Experimental results were satisfactory and promising.

The rest of the paper is organized as follows. Section 2 presents an overview of our approach. Section 3 describes the components of our *SVG-to-RDF* image

---

[1] https://google.com/imghp

semantization framework. In Section 4, we present the experimental results obtained when evaluating our approach. Section 5 concludes the paper and discusses future directions.

2   *SVG-to-RDF* Image Semantization

An overview of our *SVG-to-RDF* annotation framework is shown in Fig. 1. It consists of five main components: i) *SVG feature extraction*, ii) *RDF graph representation*, iii) *Similarity computation*, iv) *User verification and feedback*, and v) *RDF knowledge base*. Our approach is general in that: 1) it can process both raster and vector images (since raster image contours can be automatically extracted and used to generate an SVG image), and also 2) it can be associated to different application domains.

Once an input SVG image is available, the first phase of the process consists in automatically extracting the visual features and semantic properties of the image. It is worthy to note that unlike traditional raster (visual) feature extraction methods that would require important processing time, feature extraction from SVG images (identifying geometric shapes, their colors, and related textual descriptions) can be undertaken very efficiently and quickly using XML-based parsing from the SVG source code.

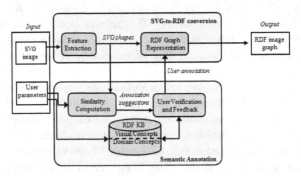

Fig. 1. Simplified activity diagram describing our SVG-to-RDF framework

After their extraction, SVG features are represented in the form of *subject-predicate-object* triples into an RDF graph and consequently compared and mapped with those already stored in the dedicated RDF knowledge base, using corresponding comparisons. The RDF knowledge base presents domain-specific reference knowledge concerning the images being annotated (cf. Section 2.1). Finally, the generated annotations could be revised/modified by the user before validating the final image graph representation. When new application domain concepts and mappings are detected within the obtained graphs, they are injected into the RDF knowledge base to incrementally update it and increase continuously its semantic expressiveness.

In the following, we present in more details *SVG-to-RDF*'s main components.

2.1   RDF Knowledge Base

The RDF knowledge base provides domain experts, who are in charge of verifying/validating image annotations, with a set of predefined concepts and relations which are then extended by creating new instances of those concepts, based on the images being annotated. It can be generated *manually* by domain experts: including application domain concepts as well as their descriptions and mapping with the visual concepts (e.g., we adopt a reference dental knowledge base from [9] in our current study, cf. Section 3), or *automatically*[2] involving some machine learning techniques, using samples (a human expert manually annotates sample images with the intended semantic concepts, which are then provided as training data for a learning algorithm that induces rules to be used for assigning concepts to other images, thus incrementally building the knowledge base) [31].

Our RDF knowledge base is represented as an RDF graph (N, E) which **nodes** N are *subjects, objects,* or *subject/object properties* representing: i) SVG visual/geometric concepts (e.g., ellipse, circle, path), ii) application domain concepts (e.g., molar tooth, planet), and iii) corresponding property values (e.g., stroke, 50); and **edges** E are *Predicates* representing: i) relations between concepts (e.g., Circle *SubClassOf* Geometric Object, circle *IsA* planet, Teeth *HasInfluentialFacts* Symptom, etc.), and ii) property and value relations (e.g., ellipse *HasRadius* 50). Fig. 2. shows an extract of a sample knowledge base used in our study.

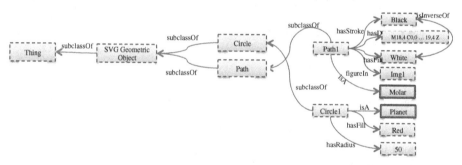

Fig. 2. Extract of the RDF knowledge base used in our study[3]

2.2   Feature Extraction

SVG allows encoding a variety of geometric objects using a set of predefined basic *shape* elements (*rectangle, circle, ellipse, line, polyline, polygon* and *path*), defining for each shape a set of descriptive attributes known as geometric object *properties*[4].

---

[2] This will be studied in a dedicated work.

[3] Subjects and Predicates URI, e.g., "http://svg2rdf.org#", are omitted here in order to simplify the graph.

[4] SVG includes a *text* element which we disregard in this paper: for clarity of presentation, and since text elements can be straightforwardly handled using traditional natural language processing techniques.

Since SVG is an XML-based coding, our SVG feature extraction component retrieve XML attributes using traditional XQuery and XPath expressions. This is a major advantage over traditional low-level feature extraction which may require extensive processing in comparison with fast XQuery/XPath processing (Fig. 3.).

2.3   RDF Graph Representation

Once the feature extraction is produced for a given SVG image, the image graph representation and annotation (using related components) are automatically executed. To do so, we adopt RDF (Resource Description Framework) [7] as a W3C data model designed to standardize the definition, the use and the reasoning of metadata, and we use it to describe and handle image representation. Initially, the graph representation contains only SVG visual elements of a given image mapped to RDF triples such that: RDF subjects represent SVG geometric elements (e.g., *circle*, *rectangle*, *path*, etc.), RDF predicates represent SVG element relations (e.g., *hasD*, *hasFill*, *hasStroke*, etc.), RDF objects represent properties values (e.g., *cx= "50"*, *stroke = "black"*, etc.).

Query: *Select the visual attributes of path elements having a fill color = "white"*			
SVG Raw image	**SVG source code**	**XPath query**	**Result:** Extracted Visual Features

Img1.svg	`<?xml version-"1,0"?>`
	`<svg width="20cm" height="20cm">`
	`<path d=" M18,4 C0,0 -26 -15 17,18`
	`C 9,33,3,18,8,31 S-1-25,7`
	`20.5 S6,19, 4,22 S15 - 31,15`
	`- 43 S26 - 4,19,4 Z"`
	`stroke="black" fill="white"/>`
	`</svg>`
		//path[@fill="white"]/@*	d =" M18,4 C0,0 -26 -15 17,18 C 9,33,3,18,8,31 S-1- 25,7 20.5 S6,19, 4,22 S15 - 31,15 - 43 S26 - 4,19,4 Z" stroke= "black" fill ="white"

Fig. 3. Example of SVG feature extraction using a typical XPath query statement

Fig. 4.a shows the extracted features of an SVG image, which is automatically transformed into an RDF graph as shown in Fig. 4.b Triple *Path1-figureIn-Img1* is added to indicate that the path is included in the image.

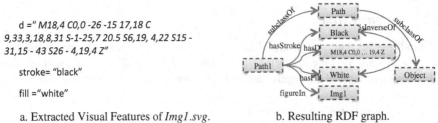

d =" M18,4 C0,0 -26 -15 17,18 C 9,33,3,18,8,31 S-1-25,7 20.5 S6,19, 4,22 S15 - 31,15 - 43 S26 - 4,19,4 Z"

stroke= "black"

fill ="white"

a. Extracted Visual Features of *Img1.svg*.   b. Resulting RDF graph.

Fig. 4. Sample SVG feature extraction and resulting RDF triple representation (before annotation)

While the mapping of SVG tags to RDF triples is straightforward at this stage, yet it is the building block required to add semantic annotations: allowing additional triples to be added when applying the semi-automatic annotation component, in order to provide user-specific semantic meaning. For instance, the triple *Path1-IsA-Molar* can be added to provide a semantic meaning for the geometric elements in the image (e.g., *Path1* represents a *molar* tooth). The annotation process is described in the subsequent sections through the following two components: *similarity computation* and *user verification and feedback*.

2.4   Similarity Computation

Although SVG coding presents the syntactic/structural properties of vector images (in the form of basic geometric objects and properties), it does not provide any semantic meaning (e.g., the SVG coding in Fig. 3. does not reflect any semantic). Once the image graph is produced for a given SVG image, the similarity process compares each of the image graph's geometric objects with those stored in the RDF knowledge base using three main similarity criteria: i) shape similarity, ii) color similarity, and iii) position similarity[5]. Given two SVG geometric objects O_1 and O_2:

$$Sim(O_1, O_2) = w_{Shape} \times Sim_{Shape}(O_1, O_2) + w_{Color} \times Sim_{Color}(O_1, O_2) + w_{Pos} \times Sim_{Pos}(O_1, O_2) \tag{1}$$

where $w_{Shape} + w_{Color} + w_{Pos} = 1$ and $(w_{Shape}, w_{Color}, w_{Pos}) \geq 0$, such that $(Sim_{Shape}, Sim_{Color}, Sim_{Pos}) \in [0, 1]$. We utilize the *weighted sum* function to combine the different similarities, allowing the user to fine-tune the weight of each criterion. Then, based on the aggregate similarity result greater than a given user or application-based predefined threshold $Thresh_{Sim}$, the system provides annotation offers for each geometric object in the image, corresponding to the most similar RDF node objects found in the knowledge base.

In the following, we briefly describe the similarity measures used by default in our system, note that the user can define his own similarity functions suitable to his domain application.

2.4.1 Shape Similarity
SVG shape similarity can be performed to compare geometric objects of the same type (comparing two *circles*, or two *rectangles*, etc.), or to compare objects of different shape types (e.g., comparing a *circle* with a *rectangle*). For this purpose, Li et al. [11] introduce a set of mathematical formulas specially tailored for the task, which we adopt in our study.

On one hand, when comparing two objects of the same type, we start by identifying the invariants of the object type, i.e., points which remain invariant even if the geometric shape undergoes a transformation (e.g., translation, rotation, or both). This results in a general mathematical equation defined based on invariant points (as

---

[5] A text similarity factor can be straightforwardly added when considering SVG *text* elements, using traditional text comparison techniques such as *string edit distance* and *N-gram* [6].

its coefficients), which can then be used to compare same-shape elements. For example, comparing two ellipses is accomplished using the quadratic conic curve similarity formula [11]:

$$\text{Dist}_{\text{Ellipse}}(O_1, O_2) = w_{\text{Major}} |a_1-a_2| + w_{\text{Minor}} |b_1-b_2| + w_{\text{Ecc}} |\varepsilon_1-\varepsilon_2| \qquad (2)$$

where $(w_{Major}, w_{Minor}, w_{Ecc}) \geq 0$ and $w_{Major} + w_{Minor} + w_{Ecc} = 1$; a_1 and a_2 are the semi-major axis of O_1 and O_2 respectively; b_1 and b_2 are their semi-minor axis; and ε_1 and ε_2 are their eccentricities. Similar formulas are provided in [11] for comparing lines and rectangles.

On the other hand, when comparing two geometric objects having different types, such as comparing a *circle* with a *path*, the proximity of their contours is computed [11]. A contour is treated as a set of points, and hence contour proximity is measured in terms of the distances between the points: two contours are more similar, if the distance between their points is smaller. Hence, considering $A = \{p_1, p_2, ..., p_n, ...\}$ and $B = \{q_1, q_2, ..., q_m, ...\}$ the set of points describing the contours of objects O_1 and O_2 respectively, the distance between O_1 and O_2 can be evaluated as:

$$\text{Dist}_{\text{DiffShapes}}(O_1, O_2) = max(h(A, B), h(B, A))$$

$$\text{Where} \quad h(A, B) = \max_{p \in A} \min_{q \in B} |p - q| \quad \text{and} \quad h(B, A) = \max_{q \in B} \min_{p \in A} |q - p| \qquad (3)$$

Note that, a particular case can be defined when comparing two objects of different types, while of them can be transformed into the other (comparing a *circle* with an *ellipse*). This can be done with a less expensive computation using the same quadratic conic curve similarity given in equation (2).

In our study, we adopt the formal definition of shape similarity as the inverse of a distance function[32], and thus deduce similarity scores from distances accordingly:

$$\text{Sim}_{\text{Shape}}(O_1, O_2) = \frac{1}{1 + \text{Dist}_{\text{Shape}}(O_1, O_2)} \in [0,1] \qquad (4)$$

2.4.2 Color Similarity

In addition to shape similarity, color is one of the most widely used features in image retrieval. On one hand, colors have been traditionally defined on a selected color space [15], such as RGB, LAB, HSV, etc., each one serving a different set of applications, where a given color is coded as a set of integer values. On the other hand, color ontologies have been recently introduced, e.g., [16, 24], in order to bridge the gap between low-level (numeric) color features and high-level (semantic) color descriptions, where colors are defined using *color names* (e.g., *red, blue, light blue,* etc.), and organized in an ontological graph structure based on their visual and semantic relatedness.

Since SVG allows coding colors in both: i) numerical format in the RGB feature space, and ii) using color names with 147 reference colors [28], we adopt both color representations in our approach to calculate the similarity between two colors by combining their semantic meaning and their visual properties as follows:

$$Sim_{FillColor/StrokeColor}(C_1, C_2) = w_{HSV} \times Sim_{HSV}(C_1, C_2) +$$
$$w_{Ont} \times Sim_{Ont}(C_1, C_2)) \quad\quad (5)$$

While SVG codes colors in numerical format in the RGB color space, yet we chose to convert RGB into the HSV color space, since HSV encoding is considered to be closer to human perception [30] and thus can be more semantically descriptive. Hence, to compare two colors (based on numerical format), we first convert their vectors from RGB to HSV using [30], and then calculate their scalar product. As for comparing color names, it can be achieved using any of several existing methods to determine the semantic similarity between concepts in a semantic network, e.g. [12, 21, 22]. These can be classified as i) edge-based: estimating similarity as the shortest path between the concepts being compared, and ii) node-based methods: estimating similarity as a function of the maximum the amount of information content concepts share in common [22]. In our approach, we combine (using weighed sum aggregation) two central edge and node-based approaches developed by *WuPalmer* [29] and *Lin* [12] (omitted here for lack of space).

Given two objects O_1 and O_2, we formally compute their color similarity as follows:

$$Sim_{Color}(O_1, O_2) = w_{FillColor} \times Sim_{FillColor}(FC_1, FC_2) +$$
$$w_{StrokeColor} \times Sim_{StrokeColor}(SC_1, SC_2) \quad\quad (6)$$

where $(w_{FillColor}, w_{StrokeColor}) \geq 0$ and $w_{FillColor} + w_{StrokeColor} = 1$ such that $(Sim_{FillColor}, Sim_{StrokeColor}) \in [0, 1]$; FC_1 and FC_2 designate the fill colors of objects O_1 and O_2 respectively; and SC_1 and $SC2$ designate their stroke colors.

2.4.3. Position Similarity

In order to compare position similarity between two geometric objects O_1 and O_2, we generate their minimum bounding rectangles (MBR_1 and MBR_2) and then compute the *Euclidian* distance between the top-left vertices of their MBRs (P_1 and P_2), where the top-left vertex serves as a reference position point for SVG rectangle objects (cf. Fig. 5.), as indicated in Equation (7).

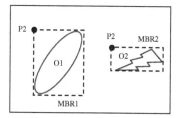

$$Sim_{Pos}(O1, O2) = \frac{1}{1 + Dist_{Euclidian}(P_1, P_2)} \in [0,1] \quad (7)$$

where $P_1(x_1, y_1)$ and $P_2(x_2, y_2)$ are the coordinates of the top-left MBR vertices.

Fig. 5. Sample MBRs & reference points

2.5   User Verification and Feedback

The similarity-based annotation suggestions, which are automatically identified for each geometric object in the RDF image graph, are presented to the user according to a (user or application-based) predefined similarity threshold $Thresh_{Sim}$. Hence, RDF object nodes which similarities are lower than $Thresh_{Sim}$ are filtered out, retaining the most similar nodes which are then ranked and presented to the user according to their similarity scores w.r.t. the geometric object being annotated. The user can then verify and/or update the annotations according to the system annotation offers. Upon accepting the annotation offers, the latter are appended to the corresponding RDF image graph describing the image, thus producing a complete semantic representation of the image. Consequently, when identifying new application domain concepts and mappings, the RDF image graph is integrated in the RDF knowledge base, by appending the image graph nodes as instance nodes under their corresponding categories in the knowledge base (e.g., nodes representing circle objects are appended as instances under the category *geometric object*, nodes representing molar are appended under the category *tooth*, etc.).

3   Experimental Evaluation

We have developed a prototype system[6], to test and evaluate our *SVG2RDF* image *semantization* framework, implemented using Java, and making use of the JENA API[7] in order to create, parse, and search RDF models (using SPARQL). While our approach is generic, yet we chose to test it in a real-world application scenario: clinical dental therapy. Our tests were designed to process a collection of dental panoramic x-ray images. After several meetings with multiple dentists specialized in dental surgery and orthodontia, we identified some of the critical information that is of interest to specialists when examining a dental panoramic image, namely: i) the shape of the tooth (e.g., the tooth looks poorly developed, decaying, etc.), ii) the tooth color (*white* for synthetic teeth, *dark gray* for decayed teeth, and *black* for lack of teeth), and iii) the position of the teeth (teeth are juxtaposed, evenly spaced, etc.). At this stage, the significance of similarity factors' weights is emphasized. Consequently, we considered that the three similarity criteria are at the same level of importance, hence we used equal weights (i.e., $w_{Shape} = w_{Color} = w_{Pos}$) However, in other applications, those criteria could have different impact in the process, so the user can change the values of the weights according to her preferences. To provide domain specific annotations, we adopted a reference dental knowledge base from [9], consisting of dental domain concepts (tooth, symptoms, etc.) and we extended it to include SVG geometric object constructs and properties (cf. Fig. 4.). Note that, initially our knowledge base does not contain any visual or semantic description of any SVG image; it only contains basic dental domain concepts and SVG basic geometric objects and properties. To simplify the process of creating SVG annotations on top of panoramic images, we used the minimum bounding rectangle (MBR) as a simple and suitable solution where MBRs designate annotated teeth (Fig. 6.).

---

[6] Available online at: http://sigappfr.acm.org/Projects/SVG-To-RDF/
[7] https://jena.apache.org/

3.1   Experimental Results

The main criteria used to evaluate the effectiveness of automatic annotation approaches is the amount of manual work required to perform the annotation task. This depends on: i) the quality of automatic annotations, and ii) the time to provide automatic annotations. It is worthy to mention that, to the best of our knowledge, there is no other existing approach that proposes to convert SVG to RDF and hence we don't provide here any comparative study.

3.1.1 Evaluating Annotation Quality
We conducted preliminary experiments on a collection of 10 panoramic dental images, each containing 16 teeth, i.e., 16 geometric shapes, consisting of the upper jaw teeth in panoramic dental X-ray images. Three dentists participated in the evaluation process.

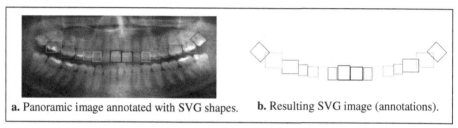

a. Panoramic image annotated with SVG shapes.   **b.** Resulting SVG image (annotations).

Fig. 6. Generating an RDF image graph annotating a raw dental panoramic image

After producing the RDF image graph, the dentist (user) annotates the first image without any help, i.e., without any automatic annotation offers since the knowledge base only contains raw concepts in the beginning (general dental concepts from the reference dental knowledge base, and basic SVG constructs). Starting from the second image, automatic annotation offers can be produced, and presented to the dentist for evaluation. The system registers the dentist's answer as: *relevant* or *irrelevant* annotation, and computes *PR*, *R*, and *F-value* accordingly. The process was repeated for each dentist, using three different similarity thresholds $Thresh_{Sim}$: 0.5, 0.7 and 0.9 ($\in$ [0, 1]), hence resulting in a total of 10×16×3×3=1440 annotation tasks. Average results are presented in Fig. 7.d Note that in all our annotation tasks, all similarities factors were considered with equal weights (i.e., w_{Shape} = w_{Color} = w_{Pos} = 0.3334, w_{Major} = w_{Minor} = w_{Ecc} = 0.3334, w_{Lengh} = W_{Slope} = 0.5, etc.). In fact, in this study, we do not address the issue of assigning similarity weights, which we report to a dedicated subsequent study.

Graphs in Fig. 7.a and Fig. 7.b show that *precision* and *recall* levels can vary from 0 (minimum) to 1 (maximum) when the system returns 0 or 16 relevant annotation offers respectively (corresponding to each of the 16 geometric objects in the image). *Precision/recall* levels increase almost regularly as the number of images being annotated increases. This is because every time the user annotates an image, the new annotations are stored in the RDF knowledge base and thus become available as potential annotation offers. Then, when it comes to annotating the following image, the number of potentially accurate annotation offers increases, thus increasing *precision* accordingly. However, we notice in Fig. 7.a and Fig. 7.b that average *precision* levels tend to increase when the $Thresh_{Sim}$ increases, whereas average *recall* levels tend to decrease. This is because increasing the similarity threshold: i) reduces the number of irrelevant annotation offers (reflected by increasing precision), yet also ii) filters out certain potentially relevant annotations which might be less similar to the object being annotated (decreasing *recall*).

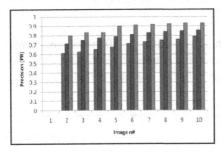

a. Precision (PR) levels.

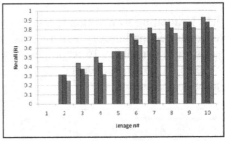

b. Recall (R) levels.

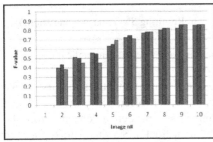

c. F-value levels

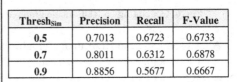

$Thresh_{Sim} = 0.5$ ■ $Thresh_{Sim} = 0.7$ ■ $Thresh_{Sim} = 0.9$ ■

$Thresh_{Sim}$	Precision	Recall	F-Value
0.5	0.7013	0.6723	0.6733
0.7	0.8011	0.6312	0.6878
0.9	0.8856	0.5677	0.6667

d. Average *PR, R,* and *F-value* results.

Fig. 7. Precision, recall, and f-value results

In cases where higher/lower *precision/recall* levels are obtained, the *f-value* measure (the harmonic mean of *precision* and *recall*) is central in evaluating the overall loss and gain in average *precision/recall*, in order to evaluate result quality. Fig. 7.c shows that *f-value* levels increase regularly w.r.t. the number of images annotated, varying from 0.4 ($\pm \varepsilon$) to 0.75 ($\pm \varepsilon$) with all three similarity thresholds. Also, results in Fig. 7.c show that *f-value* levels remain almost the same when varying the similarity threshold (e.g., *f-value* = 0.4 ($\pm \varepsilon$) for image n#1, and 0.75 ($\pm \varepsilon$) for image n#10, for all three $Thresh_{Sim}$ values). This illustrates the results mentioned above: when the threshold increases, *precision* increases yet *recall* decreases, hence inducing certain equilibrium with *f-value* levels. To sum up, *precision, recall,* and *F-value* results are clearly positive and promising, despite the similarity threshold's effect on *recall* (which tends to slowly decrease when increasing the similarity threshold), which we plan to investigate in an upcoming experimental study.

3.1.2 Evaluating Annotation Performance

In addition to testing the effectiveness of our approach in identifying meaningful annotations, we evaluate its time performance. The complexity of our method comes down to $O(|I| \times |KB|)$ where $|I|$ represents the size of the image being annotated (in number of geometric objects), and $|KB|$ the size of the reference knowledge base (in number of object nodes). Theoretical complexity analysis is omitted due to lack of space. We start by verifying our approach's linear time dependency on the size of the image and that of the RDF knowledge base. Timing experiments were carried out on a PC with an Intel Xeon 2.66 GHz processor with 2 GB RAM. Fig. 8.a shows that the time needed to produce automatic annotation offers for all geometric objects in an image grows in a linear fashion with the knowledge base size. In addition, Fig. 8.b also shows that the size of the knowledge base increases in an almost perfect linear fashion with the number of images being annotated and appended to the knowledge base. This explains the increasing slopes of the chart lines in Fig. 8.b since every geometric object is annotated and then integrated as a new geometric description in the knowledge base.

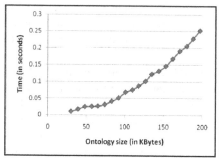 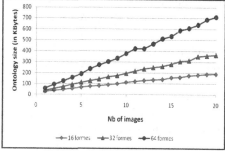

a. Images made of 16 geometric objects each. b. knowledge base graph size variation w.r.t. the number of sequentially annotated images

Fig. 8. Time performance and knowledge base size variations w.r.t. image size (in number of geometric objects per image)

4   Related Work

Many general-purpose raster image retrieval and processing systems have been proposed in the literature, and can be roughly categorized as: text-based [4, 27] and content-based [13, 14]. More recent methods have investigated XML-based solutions [25, 26] organizing images into an XML (MPEG-7) document tree hierarchy, and then applying image search and retrieval operations on the obtained XML multimedia tree. Most of these approaches suffer from the *semantic gap* [14, 23], given that the meaning of an image is rarely evident using traditional keyword and/or content-based descriptions. The interested reader can refer to surveys in [4, 13].

Few approaches have specifically targeted SVG image processing, e.g., [3, 8, 10, 17, 18]. The work in [3] suggests the organization of features extracted from SVG images in the form of an aggregation tree, where each tree node represents an SVG geometric object or an aggregated set of objects and is described by an MBR (Minimum Bounding Rectangle) and a shape description, taking into consideration the topological relationships between the objects (e.g., *disjoint, meet, overlap*, etc.). The aggregation tree is constructed using object-aggregation rules defined based on topological relations, e.g., two *disjoint* objects p and q are grouped under a higher level object n consisting of a new MBR encompassing the ones of p and q. The study in [3] presents an on-going work, aiming to index SVG images toward easier information retrieval. Another approach in [10] introduces a hierarchical SVG image abstraction layer for medical imaging, organizing low level features and high level semantic information in an image abstraction layer where content pieces are represented in XML and SVG. The authors then describe a web-based tool that visualizes, manipulates, and searches the abstraction layer using XQuery. Similar works investigating the processing and retrieval of SVG images using XML data search and manipulation techniques have been proposed in [8, 18]. In [11], the authors introduce a library of shape similarity measures designed to compare SVG geometric objects. This approach has been adopted in our framework and some of the measures are presented in Section 3. An approach which is relatively comparable to ours in presented [17] introducing a tool allowing users to manually associate semantic annotations to a sketch based query specification. Here, images are drawn and transformed into SVG coding, whereas user annotations are transformed into an RDF fragment appended to the SVG image code. Nonetheless, this approach solely focuses on manual user annotation and does not address semi-automatic annotation. Also, the resulting RDF code is appended to the SVG image source code which limits RDF semantic processing capabilities. In addition, the authors process images separately, in contrast with our approach which introduces the concept of unified reference RDF ontology to gather the collective semantics of an image repository, allowing annotation suggestions and improving image semantic processing.

5   Conclusion

This paper introduces a framework: *SVG-to-RDF* for transforming a collection of SVG images into RDF graphs. The system automatically transforms each input SVG image into a basic RDF graph, and then offers the user semantic annotation offers for each geometric object in the image, based on shape, color, and position similarity comparisons with existing objects already stored in a reference RDF knowledge base. When new concepts and mappings are detected, the annotated RDF image graph is then integrated in the knowledge base extending its semantic expressiveness. Experiments show that our approach is of average linear complexity w.r.t. image and knowledge base size, and provides promising annotation results.

We are currently investigating the extension of our approach to allow whole image search, as opposed to searching and annotating individual geometric objects within an image. In this context, dedicated reference ontologies and user-defined semantics would have to be considered to better assess image and geometric object relatedness. We also plan to study the effect of different similarity criteria (shape, color, and position) on annotation quality, proposing (if possible) weighting schemes that could help the user tune her input parameters to obtain optimal results.

References

[1]   Abu Doush, I., et al.: Multimodal Presentation of Two-Dimensional Charts: An Investigation Using Open Office XML and Microsoft Excel. ACM TACCESS 3(2), 1–50 (2012)

[2]   Awada, Y., et al.: Towards Digital Image Accessibility for Blind Users via Vibrating Touch Screen: A Feasibility Test Protocol. In: Inter. Conf. on Signal Image Tech. & Internet Systems, SITIS (2012)

[3]   Bai, S., et al.: Revised Aggregation-tree Used in Metadata Extraction from SVG Images. In: DMIN 2006, pp. 325–328 (2006)

[4]   Datta, R., Joshi, D., Li, J., Wang, J.Z.: Image Retrieval: Ideas, Influences and Trends of the New Age. ACM Computer Surveys 40(2), 1–60 (2008)

[5]   Faloutsos, C., et al.: Efficient and Effective Querying by Image Content. JIISJ 3(3:4), 231–262 (1994)

[6]   Hall, P., Dowling, G.: Approximate String Matching. Computing Survey 12(4), 381–402 (1980)

[7]   Hayes, P.: RDF Semantics. W3C Recommendation (2004), http://www.w3.org/TR/rdf-mt/ (cited May 26, 2014)

[8]   Jiang, K., et al.: Information Retrieval through SVG-based Vector Images Using an Original Method. In: Proc. of IEEE Inter. Conference on e-Business Engineering (ICEBE 2007), pp. 183–188 (2007)

[9]   Kiani, M., et al.: Ontology-Based Negotiation of Dental Therapy Options. In: Joshi, Boley, Akerkar (eds.) Advances in Semantic Computing, vol. 2, pp. 52–78 (2010)

[10]   Kim, E., et al.: A Hierarchical SVG Image Abstraction Layer for Medical Imaging. In: Society of Photo-Optical Instrumentation Engineers (SPIE) Conference, vol. 7628, p. 7 (2010)

[11]   Li, D., et al.: Shape similarity computation for SVG. Int. J. Comp. Science and Eng. 6(1/2) (2011)

[12]   Lin, D.: An Information-Theoretic Definition of Similarity. In: Inter. ICML Conf., pp. 296–304 (1998)

[13]   Liu, Y., Zhang, D., Lu, G., Ma, W.-Y.: A Survey of Content-Based Image Retrieval with High-Level Semantics. Pattern Recognition 40(1), 262–282 (2006)

[14]   Long, F., et al.: Fundamentals of Content-based Image Retrieval. MM IR Management (2003)

[15]   Manjunath, B.S.: Color and Texture Descriptors. IEEE CSVT Trans. 6, 703–715 (2001)

[16]   Mezaris, V., et al.: An Ontology Approach to Object-based Image Retrieval. In: Proceedings of the International Conference on Image Processing (ICIP), vol. 2, pp. 511–514.

[17]   Noah, S., Sabtu, S.: Binding Semantic to a Sketch Based Query Specification Tool. The International Arab Journal of Information Technology 6(2), 116 (2009)

[18] Peng, Z.R., Zhang, C.: The roles of geography markup language (GML), scalable vector graphics (SVG), and Web feature service(WFS) specifications in the development of Internet geographic information systems (GIS). Journal of Geographic Systems, 95–116 (2004)

[19] Pentland, A., Picard, R.W., Scaroff, S.: Photobook: Content-based Manipulation for Image Databases. International Journal of Computer Vision 18(3), 233–254 (1996)

[20] Prudhommeaux, E., Seaborne, A.: SPARQL Query Language for RDF. W3C Recommendation (2008), http://www.w3.org/TR/rdf-sparql-query/ (cited May 26, 2014)

[21] Resnik, P.: Using Information Content to Evaluate Semantic Similarity in a Taxonomy. In: Proceedings of the International Joint Conference on Artificial Intelligence (IJCAI), vol. 1, pp. 448–453 (1995)

[22] Richardson, R., Smeaton, A.: Using WordNet in a Knowledge-based approach to information retrieval. In: Proceedings of the BCS-IRSG Colloquium on Information Retrieval, pp. 1–16 (1995)

[23] Smeulders, A., et al.: Content-based Image Retrieval at the End of the Early Years. IEEE Trans. of Pattern Analysis and Machine Intelligence 22(12), 1349–1380 (2000)

[24] Stanchev, P.L., Green Jr., D., Dimitrov, B.: High Level Color Similarity Retrieval. Inter. Journal on Information Theory and Applications 10(3), 363–369 (2003)

[25] Torjmen, M., et al.: XML Multimedia Retrieval: From Relevant Textual Information to Relevant Multimedia Fragments. INEX: Initiative for the Evaluation of XML Retrieval (2009)

[26] Tsikrika, T., Serdyukov, P., Rode, H., Westerveld, T., Aly, R., Hiemstra, D., de Vries, A.P.: Structured document retrieval, multimedia retrieval, and entity ranking using pF/Tijah. In: Fuhr, N., Kamps, J., Lalmas, M., Trotman, A. (eds.) INEX 2007. LNCS, vol. 4862, pp. 306–320. Springer, Heidelberg (2008)

[27] Wang, S., et al.: IGroup: Presenting Web Image Search Results in Semantic Clusters. In: CHI 2007 (2007)

[28] W3C, Scalable Vector Graphics (SVG), http://www.w3.org/Graphics/SVG/ (cited May 26, 2014)

[29] Wu, Z., Palmer, M.: Verb Semantics and Lexical Selection. In: Proceedings of the 32nd Annual Meeting of the Associations of Computational Linguistics, pp. 133–138 (1994)

[30] Foley, J., van Dam, A., Feiner, S., Hughes, J.: Computer Graphics: Principles and Practice. Addison Wesley, Reading (1990)

[31] Alpaydin: Introduction to Machine Learning. MIT Press, Cambridge (2004)

[32] Ehrig, M., Sure, Y.: Ontology Mapping - an Integrated Approach. In: Bussler, C.J., Davies, J., Fensel, D., Studer, R. (eds.) ESWS 2004. LNCS, vol. 3053, pp. 76–91. Springer, Heidelberg (2004)

Employing Similarity Methods for Stellar Spectra Classification in Astroinformatics

Martin Kruliš, David Bednárek,
Jakub Yaghob, and Filip Zavoral

Parallel Architectures/Algorithms/Applications Research Group
Faculty of Mathematics and Physics, Charles University in Prague
Malostranské nám. 25, Prague, Czech Republic
{krulis,bednarek,yaghob,zavoral}@ksi.mff.cuni.cz

Abstract. In the past few years, we have observed a trend of increasing cooperation between computer science and other empirical sciences such as physics, biology, or medical fields. This e-science synergy opens new challenges for the computer science and triggers important advances in other areas of research. In our particular case, we are facing an astroinformatics challenge of analysing stellar spectra in order to establish automated classification methods for recognizing different types of Be stars. We have chosen similarity search methods, which are effectively utilized in other domains like multimedia content-based retrieval for instance. This paper presents our analysis of the problematics and proposed a solution based on Signature Quadratic Form Distance and feature signatures. We have also conducted intensive empirical evaluation which allowed us to determine appropriate configuration for our similarity model.

Keywords: similarity, SQFD, stellar spectra, feature signatures, astroinformatics, classification.

1   Introduction

The rapid development of computer systems, storage capacities, and communication technologies allowed empirical sciences – notably physics, biology, medicine, and engineering – to generate large and complex datasets. Traditional scientific methods based on individual examination of facts are not applicable anymore and the empirical sciences lean towards computer science to automate established methods, so they can be used for huge amounts of data.

Astroinformatics is a good example of this phenomenon. It is based on a synergy of modern computer science methods and advanced astronomical models systematically applied on empirical data gathered by astrophysicists. Machine learning, automated classification, clustering, and data mining yielded new discoveries and better understanding of the nature of astronomical objects [8].

A.J. Machado Traina et al. (Eds.): SISAP 2014, LNCS 8821, pp. 229–240, 2014.
DOI: 10.1007/978-3-319-11988-5_21 © Springer International Publishing Switzerland 2014

1.1 Astroinformatic Challenges

Studying the spectra of celestial objects was the key to many (if not the majority) of astronomical discoveries of the last two centuries and it still remains the most valuable instrument in stellar astronomy.

Spectra reveal significant clues about the chemical composition, temperature, and velocity of the observed object; however, the interpretation of the observed facts is difficult because different processes may result in similar observations. For instance, the shape of a spectral line corresponds to the distribution of velocities of the emitting/absorbing particles; nevertheless, the velocity may correspond to the thermal motion of gas particles as well as to the circular motion of the gas around the star. Thus, for many observed objects, the classification of their spectrum is rather a topic of discussion among astronomers than a mechanical task.

During the last decade, advances in technology allowed to produce hundreds of thousands of spectral observations every year. This data avalanche can never be processed manually; unfortunately, mechanical classification criteria exist only for the most coarse categories like *spectral class* or *luminosity class*.

A machine-learning approach may hopefully act as a multiplier of the human force in the classification task, using the manually classified spectra as the ground truth. Machine-learning approaches were already successfully used in classification of astronomical spectra; however, most of the attempts addressed the classification of nonstellar objects like galaxies or quasars [1,3].

In our research, we focus on the category of *Be stars* which are usually characterized by the following phenomena:

- spectral class B,
- the presence of strong emission lines,
- and excess luminosity in infrared band.

These basic criteria are easily determinable from the spectrum of the object (or even from five-color photometry). Since at least some of the Be stars are believed to be stars caught in their development stages, they are extraordinarily important for the understanding of the formation of stars and planetary systems.

The class of Be stars is further divided into subclasses [10] which (presumably) correspond to different geometries of the object. So far, the classification of stars into these subclasses is determined only by the consensus of astronomers and there are no exact classification criteria that could be converted to an algorithm. Furthermore, some subclasses of Be stars may naturally extend beyond the Be category, i.e., some objects may be physically similar to a subclass of Be although they do not meet the general Be-star criteria.

Our goal is to find a classification method which would automatically determine the subclass of a Be star based on its spectrum, using the objects already manually classified as the reference for the classification. In addition, the method may also be able to find similar objects among stars not classified as Be stars.

1.2 Similarity Search Methods

Our approach is to employ similarity search methods. These methods have been successfully used in a wide variety of applications including computer vision, pattern recognition, data mining, content-based image retrieval, or bioinformatics. These methods are often used for information retrieval (similarity search), where complex database objects are being looked up based on their similarity with a query object. However, a similarity model can be employed for other tasks such as object classification or database clustering.

Similarity model consists of two parts – object descriptors and a distance (dissimilarity) function. The object descriptors represent the object features that are essential for similarity comparisons. The distance function measures the dissimilarity of two object descriptors.

Since we were planning to experiment with various types of feature-based descriptors, we require an adaptive distance function, which is capable of comparing descriptors of different sizes. We have selected Signature Quadratic Form Distance (SQFD) [4,5], since it provides a good compromise between precision and efficiency. It has been very successful in multimedia content-based retrieval [6] and in various other domains. Furthermore, it can be indexed not only by metric methods but also by more effective Ptolemaic indexing [12]. Finally, the SQFD is quite suitable for GPUs [11], so it can be deployed even for larger scale experiments.

1.3 Outline

Our paper is organized as follows. Section 2 reviews previous and related work in this topic. The features examined in stellar spectra are explained in Section 3. Our similarity model is proposed in Section 4 and Section 5 summarizes its empirical evaluation. Section 6 concludes the paper.

2 Related Work

The idea of the stellar spectra clustering is not completely new. Bazaghan [2] proposed the *self organizing maps* as an unsupervised artificial neural network algorithm for classification of the stellar spectra. Jiang et al. [7] used *principal component analysis* methods to reduce dimensionality of the data, where only the first two eigenvectors are selected. Furthermore, they proposed a hierarchical clustering method for the data mining approach.

Bromová et al. [10] attempted to employ wavelets as descriptors of the stellar spectra. The spectra were sampled by discrete wavelet transformation and various transformations of the coefficients into Euclidean space were used, thus the descriptors were simple vectors. The k-means algorithm was applied on the descriptors to find similar spectra, especially to identify the spectra of Be stars. Their implementation achieved 76% precision on a sample set of 656 spectra with manually annotated ground truth.

One of the greatest problems of these approaches is that the spectra require many coefficients to be represented accurately and the number of coefficients is equal to the dimension of the feature space. Distance functions in high-dimensional metric spaces suffer from a problem called the curse of dimensionality, so the precision of the clustering is dropping as the number of the dimensions increase.

An alternative to this approach could be to use different types of descriptors. Traditional approaches like the Fourier transformation [9] are not applicable here, since the frequency analysis does not reflect the similarity properties described by the physicists. The physicists define similarity based on the shape of the histogram curve. Another option would be matching 2D curves like the Bezier curve [13] to approximate the histogram and then compare the coefficients or the defining points of the curves. Some of these alternatives are being investigated by other research teams; however, no results have been published yet to our best knowledge.

3   Stellar Spectra

Stellar spectrum is a recording of radiation intensity in the frequency domain, usually over a range of visible or near-infrared wavelengths. The most prominent features of a stellar spectrum are its general shape called *continuum*, *absorption lines*, and *emission lines*.

The continuum of a stellar spectrum usually approximately matches that of ideal black-body radiation whose temperature is dependent on the spectral class of the star.

The absorption lines are produced by relatively cold gases surrounding the star where photons of matching energies transform to excitation of gas atoms or molecules. The presence and intensity of absorption lines (which determine the spectral class of the star) depend on the chemical composition of the gas and also on its temperature and density.

The emission lines are produced by excited hot gas atoms transitioning back to lower-energy states. To be visible over the background of the star, this process requires significant amount of gas orbiting close to the star; therefore, prominent emission lines are present only in some stellar classes, including the Be stars.

Unfortunately, the signal of the observed object is affected by the interstellar medium and, in case of ground-based measurements, by the atmosphere of the Earth. The effect of the atmosphere is particularly difficult to isolate due to its dependency on geographic, meteorologic, and ionospheric conditions. The photon flux of fainter stars is often lower than the sky background and the subtraction of concurrently measured sky flux does not remove the noise inherent to the background.

The main source of noise in modern spectrographs is the fact that even a stable flux of radiation corresponds to photons randomly distributed in time – this *photon-counting* noise is thus inherent to the physical process and can be alleviated only at the cost of prolonged exposition time or using a larger telescope.

Due to the difficulty of separating the star signal from the atmospheric effects, astronomers often focus on particular lines that are prominent in stars but negligible in the atmosphere. For many types of stars, including the Be stars, the selected line is the H-alpha line at 656.281 nm (6562.81 Å), corresponding to ionized atomic hydrogen.

The experiments described in this paper are based on spectra obtained by a single-object spectrograph attached to the Ondřejov 2 m telescope. The selected spectra are high-resolution spectra covering a selected wavelength range approx. 40 nm wide centered at the H-alpha line.

The spectra were already manually *normalized*: The original signal was divided by a low-degree polynomial which was selected to best fit the signal curve. This normalization ensures that the continuum part of the normalized spectrum approximately equals to one; thus emission lines show as features above the $y = 1$ line and absorption lines under it, as shown in Figure 1. Nevertheless, the sky flux was not subtracted, which means that the normalized continuum is the sum of the star and sky continua.

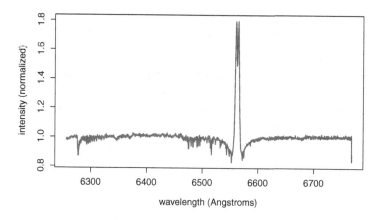

Fig. 1. Example of a normalized spectrum of Be star

Figure 1 shows features typical for Be stars: The most prominent feature is the H-alpha emission line, attributed to a disk of hot gas rotating around the star. The very center of the emission line is cut off by H-alpha absorption in colder gas surrounding the star at greater distance or in interstellar medium. The emission line itself rises from the bottom of a wide H-alpha absorption line caused by the rapidly rotating atmosphere of the star. The differences in the widths of the three superimposed lines are caused by the different thermal and rotational velocities of the associated gas masses. Different Be stars also show different ratios of the intensities of the H-alpha lines – their classification is mostly connected to these ratios.

The dense group of small absorption lines to the left of the H-alpha lines correspond to the absorption on water molecules in Earth atmosphere. Such sky effects, together with the photon-counting noise, cause that the theoretically perfectly symmetric H-alpha curves are distorted so that even finding their extremes is difficult.

The sampling at the x-axis is not equidistant as it is determined by the physical properties of the spectrograph and may vary due to different conditions of the measurement. Thus, in our case, a spectrum is a set of ordered pairs (w, i), where w is the wavelength, traditionally represented in Ångströms, and i is the intensity of the radiation measured by the corresponding sensor pixel. The intensity is an unit-less quantity due to the normalization process.

4   Similarity Search and Spectra Classification

Based on our previous experience, we have chosen similarity search methods used in multimedia databases to compare the stellar spectra. When suitable similarity model is established, we would like to apply clustering to the large spectra datasets. The ultimate objective is to determine a classification method, which will divide known stars into categories automatically.

4.1   Feature Signatures and Signature Quadratic Form Distance

The database objects are represented with descriptors, which aggregate essential object properties. We have selected feature signatures to represent stelar spectra. Formally, a feature signature S^o of an object o is defined as $S^o = \{(c_i^o, w_i^o) | i = 1 \ldots n\}$. In other words, it is a set of ordered pairs, where each pair consist of a point[1] c_i^o in the feature space $\mathbb{R}^d$ and its weight $w_i^o \in \mathbb{R}^+$. Let us emphasize that the number of features may differ for each signature, since simpler objects can be covered by fewer features than complex objects. The dimension d of the feature space $\mathbb{R}^d$ is fixed for the similarity model. In our case, we have been experimenting with low-dimensional spaces ($d = 2$ and $d = 3$).

The signature extraction is the process of extracting object features that are important for the similarity model and aggregating them into signatures. The extraction process fully determines our similarity model since we have fixed our selection of the distance function. Several approaches were implemented and tested (Section 4.2 and Section 4.3).

We have employed the Signature Quadratic Form Distance (SQFD) [5] as a measure for the signatures. The function is formally defined as

$$d_{SQFD_{f_s}}(S^q, S^o) = \sqrt{(w_q| - w_o) \cdot A_{f_s} \cdot (w_q| - w_o)^T},$$

where S^q, S^o are the compared signatures. The vector $(w_q| - w_o)$ is created by concatenation of weight vectors w_q and $-w_o$, where $-w_o$ denotes negated values of w_o. The A_{f_s} is the enumeration matrix of the similarity function f_s that compares each pair of features from the signatures. In our case, we have

---

[1] The feature point is sometimes called centroid as it represents center of a cluster of feature points.

used the Gaussian similarity function with L_2 (Euclidean distance) as a ground distance (i.e., $f_s(c_i, c_j) = e^{-\alpha \cdot L_2^2(c_i, c_j)}$). The α parameter is a tuning parameter between precision and indexability.

4.2 Direct Extraction

The first idea is to use the measured values of the spectra directly as signatures. Despite the fact the spectrum represents discrete values of a function that assigns intensity values to wavelengths, it is also a set of 2-dimensional points. There are two aspects of this approach that need to be defined: Selecting proper transformation for both dimensions and defining the best weights for the spectra points.

Both axes (the wavelength and the measured intensity) have linear properties, thus we have restricted the extraction process to linear transformations. Furthermore, the translation process has virtually no effect on the data, so the model uses only multiplicative constants. We have empirically determined that constants 1 and 4 used for wavelength and intensity dimensions respectively procude the best results.

The corresponding weights of the spectra points should reflect the importance of each point. We have tried three weight functions:

- *Uniform weight distribution* – each point has weight $1/n$, where n is the number of points. This weight function was selected only as a reference.
- *Intensity weight function* – each weight is computed as a function of the intensity. The spectrum is normalized, so that the continuum has intensity values around 1. Therefore, the interesting parts of the spectrum exhibit themselves as emissions (values >1) and absorptions (values <1). We can argue that greater emissions and absorptions are more important than values closer to the continuum. If we express this dependency linearly, the weight function will be $f(i) = |1 - i|$.
- *Normal distribution* (i.e., $f(w) = e^{-(w-\mu)^2/2\sigma^2}/\sigma\sqrt{2\pi}$) of the wavelengths w. The mean value μ was set to H-alpha (6562.8Å), since this particular part of the spectra is the most important in our case. The variance σ^2 was a variable parameter and we have empirically tested several values.

The SQFD requires that weights are normalized so their sum is equal to unity. Since both the intensity weight function and the wavelength normal distribution may produce weights that do not comply with this requirement, a linear normalization $w_i' = w_i / \sum_j w_j$ is applied after the weights are generated.

4.3 Extracting Local Extremes

A more elaborate idea is based on the approach of domain experts who manually classify the spectra. The most important are the peaks of the spectrum curve – i.e., the local extremes of the spectrum function. Since we do not have the function directly, we can select a subset of spectra points as approximate extremes. Our method works as follows.

Each point p of the measured spectrum is examined individually to determine, whether it is an approximate local extreme or not. A point is marked as extreme, if all neighbour points p_i^n (which have wavelengths in a given radius r around the wavelength of p) have their intensities either greater or lesser than the examined point. An example of this method is depicted on Figure 2.

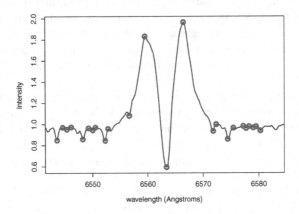

Fig. 2. Example of extracted approximate extremes from a spectrum

The signatures produced in this case use 3-dimensional feature space. The first two feature dimensions are the same as in the direct extraction approach (Section 4.2). The third dimension holds only values 1 and -1, which are assigned to distinguish local maxima and local minima respectively.

The linear transformation of the feature space is performed in the same way as in the case of the direct extraction. The optimal multiplicative constants are $1, 1, 4$ for the wavelength, intensity, and extreme orientation dimensions respectively. We have selected the normal distribution for the point weighs, since it led to the best results in the direct extraction method.

Noise Reduction. The extraction of local extremes suffers from the noise present in the data. It is difficult to determine, whether an extreme is significant for the pattern recognition, or whether it is just an outlier caused by an error in measurement. We could extend the range r that is used to identify local extremes, but such approach prunes some legitimate extremes as well as outliers.

Another approach would be to eliminate extremes that are adjacent (on the wavelength axis) and their distance (measured on the wavelength-intensity plane) is small. Unfortunately, this method does not work properly either. Figure 3 shows two spectra that demonstrate the problem. Both spectra have three local extremes close by, but whilst these points are essential in the left spectrum, they need to be eliminated in the right spectrum.

Based on the previous observations, we have selected a method that eliminates local extremes which are close by, but the threshold for this decision is dynamic.

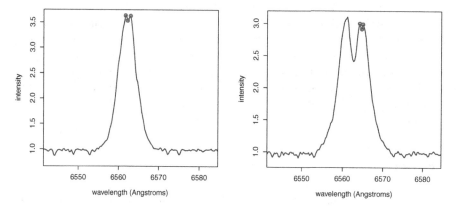

Fig. 3. Example of desired and undesired extreme extraction

The threshold is calculated based on the wavelength distance from the H-alpha, because the shape of the spectrum has more important values closer to that particular wavelength. The algorithm repeats the following steps until there are less than three extremes left or until none of the remaining extremes are close enough to be eliminated.

1. For each adjacent pair (on the wavelength axis) compute Euclidean distance $\delta_i = \sqrt{(\lambda_i - \lambda_{i+1})^2 + (I_i - I_{i+1})^2}$ in wavelength-intensity plane and a filtering threshold $t_i = \tau \cdot |H\alpha - (\lambda_i + \lambda_{i+1})/2|$, where the λ_i denotes wavelength, I_i denotes intensity and τ is the filtering parameter.
2. Create a set C of all pairs $(i, i+1)$, which pass the filtering threshold $\delta_i < t_i$.
3. If no such pair exists (set C is empty), terminate the algorithm.
4. Find and remove a pair from set C with the smallest distance δ_i.

4.4   Spectra Preprocessing

As we mentioned in Section 3, the spectra are already preprocessed, so that the continuum is normalized to intensity value 1. We apply a few additional preprocessing methods to improve the extraction process.

First of all, each spectrum is cropped to selected range of wavelengths. Even though we could achieve similar results by setting zero weights to the points out of the range, the cropping process will reduce the number of points, thus increase the efficiency of both signature extraction and distance function evaluation. We have experimentally determined that the range of 30Å with its center in the H-alpha value is the best compromise between the data size and the precision.

The second step is to normalize intensity amplitude. The values are already levelled around value 1, but the maximal and minimal values are not bounded. Since we are interested mostly in the shape of the spectra, we linearly normalize the amplitude, so that either minimal value is equal to zero, or the maximal value is equal to 2 (whichever is more distant from 1).

Finally, we have observed that some of the spectra carries a heavy load of noise caused either by specific conditions that affected the measurement or by additional physical phenomena, which are not important for this type of classification. We have tried to apply a gaussian smoothing filter, that updates each intensity by computing arithmetic average of its neighbour values within given radius. However, we have discovered that applying this filter also eliminates important information from the spectra and it had slightly negative effect on the overall precision in the most of the experiments.

5   Experiments

As mentioned in Section 3, we have used Ondřejov dataset for our experimental evaluation. The spectra were divided into four subclasses by a domain expert. Figure 4 depicts typical representatives of each class. We have used this division as a ground truth to verify our similarity model.

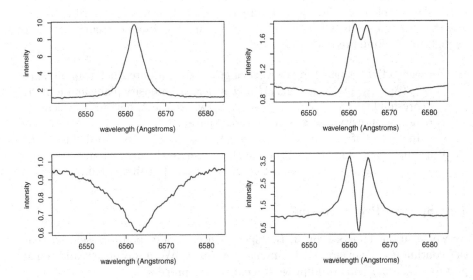

Fig. 4. Typical representatives of Be-stars subclasses

The model was verified by the following method. We iteratively select each object as a query q. SQFD distances between query and all remaining spectra, except spectra that belong to the same star as the query spectrum, are computed. These distances are used to sort the spectra in ascending order creating a list S. Then we find the first spectrum s in the list S that belongs to the same class as the query. All spectra that precedes s in the list are considered to be an error in ordering and we compute a total sum of these errors over all queries. The results are presented normalized – i.e., divided by the total amount of the spectra.

5.1   Results

The results are presented in Figure 5. Each method used optimal configuration parameters that produced the best result. These parameters were also obtained by intensive experimentation, but the details of these experiments are beyond the scope of this paper. The *simple, intens, gauss* are the results of the direct extraction method that employs uniform, intensity driven, and normal distribution of the weights respectively. The *extreme* denotes the signatures that were extracted by finding local extremes and the *extreme2* also employs the noise reduction model.

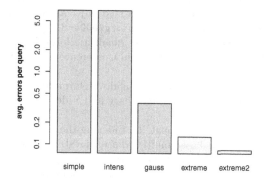

Fig. 5. Best achieved error rates of examined methods (log scale)

The results clearly indicate that the proposed method achieved the lowest error rates. Let us emphasize that the scale is logarithmic.

6   Conclusions

In this paper, we have outlined the journey to automated data mining of the stellar spectra by the means of similarity search and clustering. We hope, that this journey will take us to new discoveries in the field of astrophysics. The first step was to define and verify a similarity model, which can be used for clustering. We have established a signature extraction process that produce small signatures, which achieved good results when compared by the SQFD function.

The ultimate objective of our research is to develop a classification method which will learn from the the Ondřejov dataset, which was carefully manually annotated by domain experts. The learned classification will then be applied to a large unannotated dataset of the Sloan Digital Sky Survey (SDSS). The application to the SDSS data requires further steps like adjustment to the differences between the two spectrographs. Nevertheless, the results presented in this paper suggest that our approach is viable.

Acknowledgements. This paper was supported by Czech Science Foundation (GAČR), projects P103-14-14292P and P103-13-08195, and by Specific Research SVV-2014-260100.

References

1. Ball, N.M., Brunner, R.J.: Data mining and machine learning in astronomy. International Journal of Modern Physics D 19(07), 1049–1106 (2010), http://www.worldscientific.com/doi/abs/10.1142/S0218271810017160
2. Bazarghan, M.: Application of self-organizing map to stellar spectral classifications. Astrophysics and Space Science 337(1), 93–98 (2012)
3. Bazell, D., Miller, D.J., SubbaRao, M.: Objective subclass determination of Sloan Digital Sky Survey spectroscopically unclassified objects. The Astrophysical Journal 649(2), 678 (2006), http://stacks.iop.org/0004-637X/649/i=2/a=678
4. Beecks, C., Uysal, M., Seidl, T.: Signature quadratic form distances for content-based similarity. In: 17th ACM International Conference on Multimedia, pp. 697–700. ACM (2009)
5. Beecks, C., Uysal, M., Seidl, T.: Signature quadratic form distance. In: Proceedings of the ACM International Conference on Image and Video Retrieval, pp. 438–445. ACM (2010)
6. Beecks, C., Lokoč, J., Seidl, T., Skopal, T.: Indexing the signature quadratic form distance for efficient content-based multimedia retrieval. In: Proceedings of the 1st ACM International Conference on Multimedia Retrieval, p. 24. ACM (2011)
7. Bin, J., Chang, P.J., Ping, Y.Z., Qiang, G.: A data mining application in stellar spectra. In: International Symposium on Computer Science and Computational Technology, ISCSCT 2008, vol. 2, pp. 66–69 (2008)
8. Borne, K.D.: Scientific data mining in astronomy. In: Next Generation of Data Mining, pp. 91–114 (2009)
9. Bracewell, R.N., Bracewell, R.: The Fourier transform and its applications, vol. 31999. McGraw-Hill, New York (1986)
10. Bromová, P., Škoda, P., Zendulka, J.: Wavelet Based Feature Extraction for Clustering of Be Stars. In: Zelinka, I., Chen, G., Rössler, O.E., Snasel, V., Abraham, A. (eds.) Nostradamus 2013: Prediction, Model. & Analysis. AISC, vol. 210, pp. 467–474. Springer, Heidelberg (2013)
11. Krulis, M., Skopal, T., Lokoc, J., Beecks, C.: Combining CPU and GPU architectures for fast similarity search. Distributed and Parallel Databases 30(3-4), 179–207 (2012)
12. Lokoč, J., Hetland, M., Skopal, T., Beecks, C.: Ptolemaic indexing of the signature quadratic form distance. In: Proceedings of the Fourth International Conference on SImilarity Search and Applications, pp. 9–16. ACM (2011)
13. Yamaguchi, F., Yamaguchi, F.: Curves and surfaces in computer aided geometric design. Springer, Berlin (1988)

A Similarity-Based Method for Visual Search in Time Series Using Coulomb's Law*

Claudinei Garcia de Andrade and Marcela Xavier Ribeiro

Universidade Federal de São Carlos,
São Carlos-SP, Brazil
{claudinei.andrade,marcela}@dc.ufscar.br

Abstract. We present a method for visual search in multidimensional time series based on Coulomb's law. The proposed method integrates: a descriptor based on Coulomb's law for dimensionality reduction in time series; a system to perform similarity searching in time series; and, a module for the visualization of results. Experiments were performed using real data, indicating that the proposed method broadens the quality of through similarity queries in time series.

Keywords: Time series analysis, Index method, Similarity Search, Coulomb's law.

1   Introduction

The great early challenges to work with the analysis of temporal observations is the development of compact storage methods for series that are truly representative of the collected information, which are easy to handle and show a high level of accuracy for knowledge extraction. Thus, this paper aims to propose an integrated environment for similarity search in time series with the incorporation of a descriptor based on Coulomb's law for dimensionality reduction. In addition to it, the paper presents a system to perform similarity searching in time series and also a module for the visualization of results. Experiments with real data of varying sizes and dimensions provide validation and confirm that the system produces satisfactory results.

2   Background and Related Works

A time series can be defined as an set of observations [1], $\{Y(t), t \; \epsilon \; T\}$ in which Y is the variable of interest and T is an index set. Time series are considered complex data. There are not any way of establishing an order relation among series or their ranges. In this context, the concept of similarity is more applicable than the concept of equality.

Current similarity search methods are, in general, based on the use of descriptors in order to obtain similarity ranges. Some authors define a descriptor

* We would like to thank FAPESP, CAPES and CNPq for the financial support.

A.J. Machado Traina et al. (Eds.): SISAP 2014, LNCS 8821, pp. 241–246, 2014.
DOI: 10.1007/978-3-319-11988-5_22 © Springer International Publishing Switzerland 2014

as being formed by a pair (ϵ_D, δ_D) where ϵ_D is the component responsible for characterizing the object through the extraction of characteristics and generating a vector that will be used to analyze the data. δ_D is the function responsible for comparing the characteristics vectors, giving the amount of similarity between the object and the query [2]. There are a variety of descriptors that are effective for certain data fields but end up presenting loss of representativeness of the series data in most cases [3].

There are two basic types of similarity queries: i) (*Range queries*) which finds objects that are at a maximum r distance of the object query Q; and ii) (*k-Nearest Neighbor query* or *k-NN query*) which aim to retrieve the k objects most similar to a query object.

3   Proposed Method

The proposed system is composed of distinct modules that share data with one another and work harmoniously getting the information passed by the user to carry out the queries, applying the Coulomb descriptor to the data according to the user's interest and graphically returning objects of interest as found and listed by the descriptor.

Figure 1 illustrates the relationships among modules. Time series data serve as input to the visualization and data exploration module (VDEM) where the expert can verify the behavior and relevant characteristics of the series and select the interesting intervals for analysis. Also, they serve as input to the Coulomb descriptor module (CDM) which, by dimensionality reduction and similarity calculation, passes the ranges with some degree of

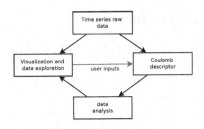

Fig. 1. Relationships among modules

similarity on to the data analysis module (DAM), according to the user's interests. From there, the data analysis module prepares information that is handed back to the VEDM, which, in turn, displays them to the user.

Coulomb's law establishes the mathematical relationship between the charges of two or more bodies and their electrical force output by calculating the existing interaction forces (attraction and repulsion) in these charges. The principles of Coulomb's law can be expressed thus: i) the intensity of the electric force is directly proportional to the product of the electric charges; and ii) the intensity of the electric force is inversely proportional to the square of the distance between the bodies. The law's formula is: $\boldsymbol{F} = K\frac{q_1 q_2}{r^2}\hat{r}$. Where: $\boldsymbol{F}$ is the force in newtons; r is the distance between the two point charges; q_1 and q_2 are the intensities of the charges; $\hat{r}$ is the unit vector; and K is Coulomb's constant.

This proposal for similarity search in series considers the observations of the time series as point charges with q constant charge values. Those are located in the coordinate plane formed by the series index and by the value of the observation. Since calculating the distance between the charges is necessary in order to ascertain the interaction between them, we take a Cartesian plane formed by the time series index (x-axis) and the value of the observations (y-axis). Thus, it is possible to calculate the distance between the charges for the calculation of forces.

Furthermore, a dummy point charge, of charge q, is inserted into the centroid composed of the sets of observations that make up the search ranges. This charge is aimed at providing an optimal representation of the range because, in addition to its being located at the geometric center of the range, it is used to calculate the interaction between itself and the other charges, thus generating the resultant force that represents the range.

As the resultant force is a vector measure, the charge's direction and magnitude influence the calculation. So, it was established that charges that are below the centroid charge are in the opposite direction to those above it, consequently, posing a negative force. Accordingly, it is possible to represent a time series through a system of electrically charged particles and to calculate the resultant force F, obtained through a vector sum of all forces that comprise the system. That way, we are able to reduce the series dimensionality, contributing to similarity search without major loss of information. In the proposed approach, the feature vector ($V = [\boldsymbol{F}, h]$) is formed both by the resultant force calculated in the range of interest and the height of the centroid.

4   Experimental Results

The experiments devised to test the proposed environment were divided into two groups: i) experiments with the Coulomb descriptor and DAM to validate the descriptor's performance in reducing data dimensionality and

	DFT	SM	Coulomb
Accuracy	20,48%	46,63%	68,95%

Fig. 2. Descriptors' accuracy comparison

with DAM in finding windows with higher similarity in the series; ii) experiments with VDEM integrated with other modules. We used randomly generated databases, a meteorological database of several Brazilian cities with minimum and maximum temperatures, along monthly precipitation indexes from the years 1961 to 2010 obtained in [8], as well as medical data obtained in [9] regarding glucose levels in patients in the course of their daily activities.

4.1   CDM and DAM Tests

Validating the Coulomb descriptor and the data analysis module aims to verify the performance of the descriptor in reducing data dimensionality and in finding windows with higher similarity in the series. We saw fit that the Coulomb descriptor was compared to the Sequential Matching *Sequential Matching* (SM) [4], [5] and *Discrete Fourier Transform* (DFT) [6] methods, since those methods are considered baselines of the work in question. The former for presenting high accuracy and the latter for having good performance in large databases. The modules were evaluated in the following aspects:

1) Computational Complexity: in this respect, we performed two experiment. The first, was a runtime test of the same *knn* query using three descriptors, with varying database sizes, graph in Figure 3. The second was a runtime test of a *knn* query with varying query window sizes, in Figure 4, we note that the Coulomb descriptor presents a shorter runtime as compared to the SM and DTF descriptors.

2) Accuracy: intended to measure the number of instances that were predicted correctly from an input query. In this test, we consulted the most similar periods (*knn*-query) to the period encompassing summer (December 21 to March 20 of the following year) and winter (from June 21 to September 21) in the city of Araraquara/SP. Queries were carried out using the descriptors mentioned and the accuracy results are presented in table of Figure 2. As shown, the accuracy displayed by the Coulomb descriptor is satisfactory for similarity queries.

3) Precision vs. Recall: proposed by [7]. From the meteorological base, we used data concerning monthly maximum temperatures of Presidente Prudente/SP, Brazil. The three aforementioned descriptors were implemented with a focus on similar seasons, periods of abnormal increase or decrease in temperatures and periods with some cyclical temperature variability. The graph in Figure 5 represents the precision and recall found. With the medical base, the tests searched periods of high and low blood glucose levels in patients before and after the administration of insulin as well as before and after meals. The precision vs. recall graph is shown in Figure 6. By analyzing the graphs, we note that the accuracy of the Coulomb descriptor remains satisfactory for good levels of recall, if compared to other methods.

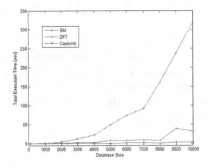

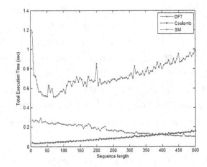

Fig. 3. Runtime per query with varying database sizes

Fig. 4. Runtime per query with varying query window sizes

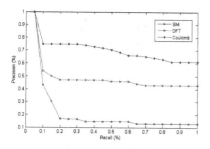
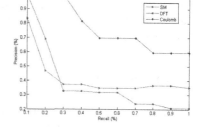

Fig. 5. Precision vs. Recall - meteorological database

Fig. 6. Precision vs. Recall - medical database

4.2 Viewing Environment Test

For carrying out tests on the visualization and data exploration module, similarity queries in meteorological time series were run for winter and summer periods. The query shown in Figure 7 was run on the time series containing the average temperature of the city of Araraquara/SP between the years 1979 and 2010. A knn-query with $n = 10$ and the winter of 1979 as period of interest (leftmost hatched period in the graph). As can be seen in the figure, the periods returned (hatched portions of the graph) by the system correspond to winter periods where there was a minimum temperature close to the selected range.

Another test carried out uses three time series regarding the maximum monthly temperatures of the cities Avaré, São Paulo and Presidente Prudente in the years 1970-2008. The similarity query with $knn = 10$ is run by selecting the winter of 1988 in Presidente Prudente as the period of interest. As shown in Figure 8, periods of greatest similarity concerning the three series are hatched.

Fig. 7. knn query = 10 applied to winters in the city of Araraquara/SP

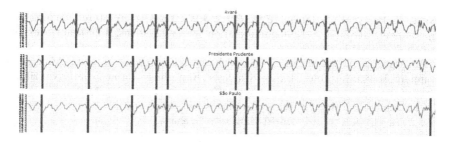

Fig. 8. knn query = 10 applied to the winter of 1988 in Presidente Prudente/SP

As shown, the visual module provides satisfactory results and allows a specialist to view similar ranges in an understandable and practical way. It makes it possible for similarity queries to be used in inferring knowledge about the series under analysis.

5 Conclusions

Upon analysis of the results obtained, we reach the conclusion that the Coulomb descriptor presents satisfactory values of accuracy and runtime for the execution of similarity queries on time series. Furthermore, a comparison of the Coulomb descriptor with traditional search methods for time series, through the analysis of accuracy vs. recall graphs, reveals significant advantages. That makes the Coulomb descriptor a potential descriptor for time series in different areas. In addition, the visualization and data exploration module allows a specialist to perform similarity queries. As a future task, the current modules will be integrated into a data-mining module in order to generate association rules using the query ranges entered by a specialist.

References

1. Wei, W.: Time series analysis: univariate and multivariate methods. Pearson Addison Wesley (2006)
2. Torres, R.D.S., Falcão, A.X.: Content-based image retrieval: Theory and applications. Revista de Informática Teórica e Aplicada 13, 161–185 (2006)
3. Zhong, S., Gang, W.: Study on algorithm of dependent pattern discovery of multiple time series data stream. In: 2011 International Conference on Computer Science and Service System (CSSS), pp. 767–769 (2011)
4. Faloutsos, C., Ranganathan, M., Manolopoulos, Y.: Fast subsequence matching in time-series databases. In: Proceedings of the 1994 ACM SIGMOD International Conference on Management of Data, SIGMOD 1994, pp. 419–429. ACM, New York (1994)
5. Keogh, E.: A fast and robust method for pattern matching in time series databases. In: Proceedings of WUSS 1997 (1997)
6. Agrawal, R., Faloutsos, C., Swami, A.N.: Efficient similarity search in sequence databases. In: Lomet, D.B. (ed.) FODO 1993. LNCS, vol. 730, pp. 69–84. Springer, Heidelberg (1993)
7. Kent, A., Berry, M.M., Luehrs, F.U., Perry, J.W.: Machine literature searching viii, operational criteria for designing information retrieval systems. American Documentation 6(2), 93–101 (1955)
8. Agrodatamine: Development of Algorithms and Methods of Data Mining to Support Researches on Climate Changes Regarding Agrometeorology (2013), http://www.gbdi.icmc.usp.br/projects/agrodatamine/index.html
9. UCI Machine Learning Repository: Diabetes Data Set (2013), http://archive.ics.uci.edu/ml/datasets/Diabetes

Classification of Epileptoid Oscillations in EEG Using Shannon's Entropy Amplitude Probability Distribution

Ronald Broberg[1] and Rory Lewis[2]

[1] Department of Mathematics, University of Colorado at Colorado Springs,
Colorado Springs, CO USA 80918, USA
[2] Department of Computer Science, University of Colorado at Colorado Springs,
Colorado Springs, CO USA 80918, USA

Abstract. This paper presents an additional tool the authors have developed to continue merging the fields of computational neuroscience with medical based neurodiagnostic clinical research, particularly those associated with machine learning in Big Electroencephalogram (EEG) Data. The authors introduce a means to identify various types of epileptic pathologic oscillations using a parameter based on the Shannon entropy of the probability distribution of the amplitudes within EEG signals. Multiple entropy and entropy-like measures have been explored to aid in epileptic seizure classification including Kolmogorov-Sinai entropy, spectral entropy, Renyi entropy, approximate entropy, and equal frequency discretization. Here we propose a more computational efficient measure which calculates a discrete probability distribution directly from the recorded amplitudes of an EEG recording over a specified window and uses an entropy-like calculation to reduce dimensionality.

1 Introduction

In previous work the authors have studied the subjective nature of what constitutes a pathological oscillation [15], and the huge dimensionality of the human brain, which has approximately 100 billion neurons each having about 1,000 connections (synapses)[16]. Moreover, neurological pathological activity may manifest itself differently from animal to animal or individual to individual [17] [7]. In a healthy human brain there is a precise interaction of neural activities, but when one develops a neurological illness (pathology) this synchronization breaks down. These abnormal synchronization processes are found in the pathological oscillations associated with several neuropsychiatric disorders including epilepsy, acute brain injury, Alzheimer's, autism post-neurosurgery Intensive Care Units (ICU) seizures, stroke, schizophrenia, dementia and basal ganglia disorders such as Parkinson's disease. In this paper we present a novel tool using Shannon's entropy function to help convert Big EEG Data into a machine learning state that will improve the efficiency of detecting seizure associated with epilepsy. Kannathal[6] grouped entropy estimators into two classes: spectral and embedded. Spectral estimators include spectral entropy[5] such as those obtained from

A.J. Machado Traina et al. (Eds.): SISAP 2014, LNCS 8821, pp. 247–252, 2014.
DOI: 10.1007/978-3-319-11988-5_23 © Springer International Publishing Switzerland 2014

Fourier Transform and Renyi entropy[6] which differs from the spectral entropies in the weighting of the lower frequencies. Embedded entropies include state space reconstructions[1], Kolmogorov-Sinai entropy[6], approximate entropy[10], and sample entropy[12]. Orhan [8] used an entropy-like method *'Equal Frequency Distribution'* where the amplitudes of the EEG signal where discretized into *'N'* bins of equal size and then applied Shannon's entropy function to the resulting discrete probability distribution. He then calculated the EFD over a range of differing *'N'* values to create a set of entropy-like values that could be used for epileptic seizure classification. Accordingly, we present a more efficient embedded entropy derived from the amplitude of EEG recordings in the classification of epileptic seizure events.

We examine a simple entropy measure in three experiments using three distinct EEG data sets. First, the amplitude entropy measure is explained. The following sections apply the method to the three data sets. In the first data set, the measure is used to classify epileptic and non-epileptic EEG segments prepared by Andrzejak.[2] The second applies the measure to two tonic-clonic, grand mal, seizure events in a pair of EEG traces made available by Quiroga.[11] In the third, the measure is applied to the 800 hours of EEG data prepared by Shoeb[14] and made available through PhysioNet.[4] In 1948 Shannon [13] defined entropy in informational theory as $H = -\sum p_i log(p_i)$ .We have used this definition to measure the entropy in the amplitude of EEG recordings after discretizing data through the straight forward conversion of the amplitude signal from floating point values to integers. The entropy measure in a given EEG segment is calculated after creating a probability distribution for a particular EEG amplitude by summing the frequency of each amplitude within the segment and dividing by the total number of amplitude measurements within the segment. In the following, Y^* is the sum of the raw frequency count for each distinct amplitude y_i within a particular given EEG segment $Y^* = \sum y_i$ where by definition, Y^* sums to number of data points within a given segment. Traditionally, the sum is normalized by dividing each amplitude frequency by the total sum of data points which results in a discrete probability distribution from which an entropy can be calculated as in (1) $p_i = \frac{y_i}{Y^*}$.

Experiment 1: Entropy Measure of the Distribution of Amplitude in Fixed Segments with Data Set 1: Andrzejak / Bonn. This canonical data set was prepared by Andrzejak et.al. and made publicly available.[3] It has been used in multiple seizure studies including Kannathal, Orhan, and Acharya. Data samples are collected at 173.6Hz and are divided into 5 labeled sets of 100 files each. The time series have an effective spectral bandwith of 0.5 Hz to 85 Hz. Each file consists of 4097 data points representing a continuous 23.6 second interval. Sets A and B are extracranial with set A comprised of recordings with eyes open and B of records with the eyes closed. Sets C, D, and E are intracranial recordings made of epileptic patients following surgical hemispheric division. Set C comes from the non-epileptic hemisphere while sets D and E are from the epileptic hemisphere. Set D consists of recordings free from seizure while set E consists of recordings with seizure. To study the entropy within each set the authors calculated the amplitude entropy of each 23.6 second segment in each 100 segment set. In the next step we aggregate the entropies in each set and test for normality using SciPy's *normaltest* which is based on D'Agostino and Pearson's test that combines skew and kurtosis to produce an omnibus test of normality. Low p-values reject hypothesis that the set is normal. Note that the aggregate of entropy of the segments within each set *in general* is not distributed normally about the mean, although the 'extra-cranial eyes open' and the 'intra-cranial seizure' both have values suggesting normal distributions. For future work the authors will study whether this could be due to artifact noise in the first case and an actual stochastic element in the latter. The statistical parameters of all 4 sets are illustrated in Table 3.

Entropy Between Each Set and EEG Classification. A training set is selected randomly from each of the sets A-E. An entropy H_j is calculated for each segment in each of the training sets. Boundary points are defined as follows: $D_{min} = min(D_{train})$, $D_{max} = max(D_{train})$, $E_{min} = min(E_{train})$ and $E_{max} = max(E_{train})$. The calculated boundary points are used to classify the test segments into the nonseizure/seizure state set as follows: no seizure(W), possible seizure (X), probable seizure(Y), seizure(Z).

$$W:=\{H_j|H_j\in[0,E_{min}]\} \quad (1)$$

$$X:=\{H_j|H_j\in(E_{min},E_{min}+\tfrac{D_{max}-E_{min}}{2}]\} \quad (2)$$

$$Y:=\{H_j|H_j\in(E_{min}+\tfrac{D_{max}-E_{min}}{2},D_{max}]\} \quad (3)$$

$$X:=\{H_j|H_j\in(E_{min},E_{min}+\tfrac{D_{max}-E_{min}}{2}]\} \quad (4)$$

The 2-fold cross-validation was repeated 100000 times. We find that this classification which includes the two indeterminate states has high precision. For the non-seizure class W, the precision is assessed as the number of non-seizures segments classified as such divided by the total number of segments assigned to the class. For the possible-, probable- and definite- seizure classes, the accuracy is assessed as the number of seizures segments assigned to the class divided by the total number of segments assigned to the class. The possible-seizure class is the least accurate by design and indicates the most mixed classification of seizure and non-seizure. As we move from class X to classes Y and Z, the confidence in the seizure classification increases. Allowing for indeterminate states X and Y, our confidence in the classification of the definite states W and Z increases. The rough set classification provides a more sensitive tool than a binary classification into seizure/non-seizure binary states with the seizure state composed of $X \cup Y \cup Z$ and the non-seizure state W.

Experiment 2: Evolution of Entropy in Time Seriesusing data Set 2: Quiroga & Caltech. Two longer EEG traces with seizure states have been made publicly available by Quiroga [11]. These files show tonic-clonic seizures of two subjects recorded with a scalp rigth central (C4) electrode (linked earlobes reference). They each contain a total of 3 minutes of data with an approximate 1 minute of pre-seizure recording followed by a seizure and some post-seizure activity. Each

Table 1. Classification with rough sets W,X,Y,Z

Set	C/E	D/E
W	98.1	97.9
X	86.3	70.1
Y	98.6	89.3
Z	99.9	97.8

Table 2. Classification with binary W+X,Y+Z

Set	C/E	D/E
W	98.1	97.9
$X \cup Y \cup Z$	97.3	89.8

Table 3. Classification with binary W+X,Y+Z

Set	C/E	D/E
W	98.1	97.9
$X \cup Y \cup Z$	97.3	89.8

signal was digitized at 409.6 Hz although after processing, the data set has an effective frequency of 102.4 Hz with an effective bandwidth of 1-50 Hz. Using windowed entropy the authors found that the longer EEG trace provides an opportunity to observe the evolution of entropy over the time series. An entropy measure of the amplitude distribution was calcuted as above for a frame of 23 seconds. This frame was moved 1 second and entropy recalculated over the length of the time series. The first deriviative of the entropy was also calcuated and is displayed in the bottom plot for each time series.

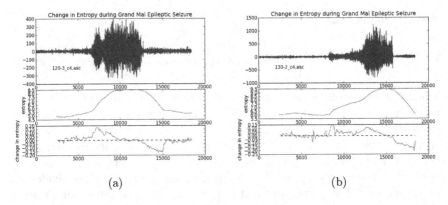

(a)   (b)

Fig. 1. Evolution of Entropy on the Time Series: Original EEG comprising both a pathological oscillation and artifact **(a)**, Entropy Evolution on Caltech data series A **(b)**, Entropy Evolution on Caltech data series B

Experiment 3: Detection of Seizures in Extended Data Series using data from Shoeb at Boston Children's Hospital. This extracranial data was collected at the Boston Children's Hospital. The database is described in Shoeb 2004[14] and made available on PhysioNet.[4][9] From the public source, 664 EDF files totaling over 44 gigabytes of compressed data were downloaded. These files contain over 800 hours of EEG data. Most files contain 23 EEG signals and they all are sampled at 256 Hz. Meta data is included with seizure times labeled. The recordings are grouped into 23 cases and are collected from 22 pediatric patients with intractable seizures following withdrawal of anti-seziure medicine during assement for surgical intervention. Using the Windowed Entropy method. we used tme evolving entropy series with non-overlapping 23 second windows are depicted in the figures below. Three exemplary plots from the CHB01 set are shown with a seizure free time series (a), a time series with a labeled seizure (b), and a non-seizure series with high noise (c). In Figure 3, an arbitrary entropy of classifier boundary of 8.2 is displayed. Additional study is being conducted to further improve this entropy analysis as a pre-processor into machine learning classifiers.

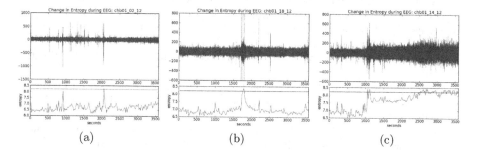

(a)   (b)   (c)

Fig. 2. Detection of Seizures in Real Data Series: Original EEG comprising both a pathological oscillation and artifact **(a)**, Entropy Evolution on PhysioNet a selected data series CHB01 without seizure **(b)**, with seizure. **(c)**, with noise and no seizure.

2 Conclusions and Future Work

These experiments show that in terms of adding a classification rule based system onto the original neuroClustering developed by the authors is a viable option so long as it will also be in a form conducive to domain adaptation. Utilizing perceptrons in the manner described in this paper to aid the neurosurgeons selecting what kind of pathological oscillations they are interested in and what they want the machine to deem as artifact, has shown to be a viable option that certainly renders the need to continue honing and refining the perceptron based method illustrated and defined in this paper and these experiments. For our future work we will test various thresholds in the perceptron algorithms against large sets of data and see where the strengths and weaknesses of timing and confidence levels pan out. Overall the results of these experiments are encouraging and are a source to drill down deeper into the methodologies presented in these experiments.

References

1. Acharya, U.R., Molinari, F., Sree, S.V., Chattopadhyay, S., Ng, K.-H., Suri, J.S.: Automated diagnosis of epileptic eeg using entropies. Biomedical Signal Processing and Control 7(4), 401–408 (2012)

2. Andrzejak, R.G., Lehnertz, K., Mormann, F., Rieke, C., David, P., Elger, C.E.: Indications of nonlinear deterministic and finite-dimensional structures in time series of brain electrical activity: Dependence on recording region and brain state. Physical Review E 64(6), 61907 (2001)

3. K. fur Epileptologie Universitat Bonn. Eeg time series (2001), http://epileptologie-bonn.de/cms/front_content.php?idcat=193&lang=3&changelang=3

4. Goldberger, A.L., Amaral, L.A., Glass, L., Hausdorff, J.M., Ivanov, P.C., Mark, R.G., Mietus, J.E., Moody, G.B., Peng, C.-K., Stanley, H.E.: Physiobank, physiotoolkit, and physionet components of a new research resource for complex physiologic signals. Circulation 101(23), e215–e220 (2000)

5. Inouye, T., Shinosaki, K., Sakamoto, H., Toi, S., Ukai, S., Iyama, A., Katsuda, Y., Hirano, M.: Quantification of eeg irregularity by use of the entropy of the power spectrum. Electroencephalography and Clinical Neurophysiology 79(3), 204–210 (1991)

6. Kannathal, N., Choo, M.L., Acharya, U.R., Sadasivan, P.: Entropies for detection of epilepsy in eeg. Computer Methods and Programs in Biomedicine 80(3), 187–194 (2005)

7. Lewis, R.A., White, A.M.: Seizure detection using sequential and coincident power spectra with deterministic finite automata. In: BIOCOMP, pp. 481–488 (2010)

8. Orhan, U., Hekim, M., Ozer, M.: Epileptic seizure detection using probability distribution based on equal frequency discretization. Journal of Medical Systems 36(4), 2219–2224 (2012)

9. PhysioNet. Chb-mit scalp eeg database (2002), http://physionet.org/pn6/chbmit/

10. Pincus, S.: Approximate entropy (apen) as a complexity measure. Chaos: An Interdisciplinary Journal of Nonlinear Science 5(1), 110–117 (1995)

11. Quiroga, R.Q., Garcia, H., Rabinowicz, A.: Frequency evolution during tonic-clonic seizures. Electromyography and Clinical Neurophysiology 42(6), 323–332 (2002)

12. Ramanand, P., Nampoori, V., Sreenivasan, R.: Complexity quantification of dense array eeg using sample entropy analysis. Journal of Integrative Neuroscience 3(03), 343–358 (2004)

13. Shannon, C.E.: A mathematical theory of communication. Bell System Technical Journal, 27 (July/October 1948)

14. Shoeb, A., Edwards, H., Connolly, J., Bourgeois, B., Ted Treves, S., Guttag, J.: Patient-specific seizure onset detection. Epilepsy & Behavior 5(4), 483–498 (2004)

15. Williams, P.A., Hellier, J.L., White, A.M., Staley, K.J., Dudek, F.E.: Development of spontaneous seizures after experimental status epilepticus: Implications for understanding epileptogenesis. Epilepsia (Series 4) 48, 157–163 (2007)

16. Williams, R.W., Herrup, K.: The control of neuron number. The Annual Review of Neuroscience 11, 423–453 (1988)

17. Zhang, X., Jiang, W., Ras, Z.W., Lewis, R.: Blind music timbre source isolation by multi-resolution comparison of spectrum signatures. In: Szczuka, M., Kryszkiewicz, M., Ramanna, S., Jensen, R., Hu, Q. (eds.) RSCTC 2010. LNCS (LNAI), vol. 6086, pp. 610–619. Springer, Heidelberg (2010)

Entity Recognition for Duplicate Filtering

J.A. Cordero Cruz[1], Sara E. Garza[1], and S.E. Schaeffer[1,2,3]

[1] FIME, UANL, Cd. Universitaria, San Nicolás de los Garza, NL, Mexico
[2] CIIDIT, UANL, PIIT Monterrey, NL, Mexico
[3] HIIT, University of Helsinki, Kumpula Campus, Helsinki, Finland
{jorge.corderocr,sara.garzavl,elisa.schaeffer}@uanl.edu.mx

Abstract. We propose a system for automatic detection of duplicate entries in a repository of semi-structured text documents. The proposed system employs text-entity recognition to extract information regarding time, location, names of persons and organizations, as well as events described within the document content. With structured representations of the content, called "metamodels", we group the entries into clusters based on the similarity of the contents. Then we apply machine-learning algorithms to the clusters to carry out duplicate detection. We present results regarding precision, recall, and F-value of the proposed system.

Keywords: entity recognition, duplicate detection, Twitter.

1 Introduction

It is now possible to easily create and share text content, which results in vast repositories of information ready to be queried and analyzed. The content of such repositories undergoes constant change: updates, deletions, and insertions of documents are frequent. Therefore, it becomes a complex task to monitor that no *duplicate* entries (that is, entries with redundant content considering the other entries present in the repository) are created as new documents arrive and the old ones are being edited.

In this work we propose a system for detecting duplicates among text documents that contain mentions of geographical locations, instances of time, and events. The operation of the proposed system can be outlined as follows: first, the words of the document are *labeled* to identify the entities contained in that document; the labeling marks locations, times, names of people and organizations, and description of events that have occurred. Then, a structured representation called a *metamodel* is created from the labeled contents, and finally, clustering algorithms are applied to the metamodels to reduce the number of comparisons necessary to identify *duplicates* among the metamodels.

As a case study to test the proposed system we chose the citizen reports received at the CIC (Center of Citizen Integration; in Spanish, *Centro de Integración Ciudadana*) in the metropolitan area of Monterrey, Mexico, available at http://www.cic.mx. The CIC reports describe, among other incidents, traffic accidents, road conditions, and need of maintenance of public infrastructure. Active citizens create reports through a web application (also available on mobile

A.J. Machado Traina et al. (Eds.): SISAP 2014, LNCS 8821, pp. 253–264, 2014.
DOI: 10.1007/978-3-319-11988-5_24 © Springer International Publishing Switzerland 2014

```
*ACCIDENTE* Leones y 18 Av carril de alta MTY #mtyfollow 19:53 via @vigila2 cc @spvmty
```

Fig. 1. An example of a CIC report via Twitter, reporting an accident at the intersection of streets called "Leones" and "18" that is blocking a lane

platforms) called Tehuan (available at `http://www.tehuan.cic.mx`) or by sending a message on Twitter (cf. `https://twitter.com`) — called a *tweet* — either mentioning the account `@Cicmty`. Upon reception, the CIC staff structures (in large part manually) the received report and inserts it into a repository.

The CIC reports include information regarding the event or condition that is being reported, the geographical location (street address or even GPS coordinates of the reporting mobile device), and the time that the event took place (possibly described verbally, but always at least as a time stamp of the report's reception). Regardless of the origin of the report, the text of the resulting document is condensed and the language used is abbreviated. Figure 1 shows an example of a Twitter-originated CIC report regarding a car accident, in Spanish. We consider two CIC reports to be *duplicates* if a human observer would interpret them as referring to a single event upon examining them both.

The remainder of this article is structured as follows: in Section 2 we discuss the identification of names within text, after which in Section 3 we discuss related literature. Section 4 details our proposed solution and then discuss the case study more closely in Section 5. Finally, in Section 6 we conclude the present work and discuss opportunities for future work.

2   Named Entity Recognition

The goal of *named entity recognition* (NER) is to extract words from text and classify them into predefined categories known as *entities*. Possible entities of interest include names of persons, names of places, and dates. There are several methods for NER and in this work we use a simple version based on hidden Markov models (HMM) [5, 10, 13], the implementation of which we now explain.

A document $\mathbf{x} = x_1 x_2 \ldots x_n$ is represented as a sequence of n words x_i. The task is to assign for each x_i a *label* y_i, resulting in a labeling sequence $\mathbf{y} = y_1 y_2 \ldots y_n$, where all labels y_i belong to a predefined set $K = \{e_1, e_2, \ldots, e_k\}$. The labels in $\mathbf{y}$ are chosen by maximizing the *joint probability* between a given text $\mathbf{x}$ and a labeling sequence $\mathbf{y}$:

$$P(\mathbf{y}, \mathbf{x}) = P(\mathbf{x} \mid \mathbf{y})P(\mathbf{y}), \tag{1}$$

where $P(\mathbf{x} \mid \mathbf{y})$ is the *conditional probability* of generating the text $\mathbf{x}$ given the labeling sequence $\mathbf{y}$ and $P(\mathbf{y})$ corresponds to an *a priori* probability distribution over the labeling sequence $\mathbf{y}$ [13, 16].

Using a second-order HMM the computation of $P(\mathbf{y}, \mathbf{x})$ is simplified to

$$P(\mathbf{y}, \mathbf{x}) = \prod_{i=1}^{n+1} P(y_i \mid y_{i-1}, y_{i-2}) \prod_{i=1}^{n} P(x_i \mid y_i). \tag{2}$$

Now, calculating the parameters $P(y_i \mid y_{i-1}, y_{i-2})$ and $P(x_i \mid y_i)$ of the HMM is easy as these are based on unigrams, bigrams, and trigrams (that is, sequences of a single label, two labels, and three labels, respectively) [4], as

$$P(s \mid u, v) = \frac{c(u, v, s)}{c(u, v)} \quad \text{and} \quad P(x \mid s) = \frac{c(s \rightsquigarrow x)}{c(s)}, \tag{3}$$

where, for a given set of labeled words, $c(u, v, w)$ is the number of occurrences of the label trigram (u, v, w), whereas $c(u, v)$ is the number of occurrences of the label bigram (u, v), and $c(u)$ the number of occurrences of the label unigram (u); the number of times that the unigram $c(s)$ corresponds to the word x is denoted by $c(s \rightsquigarrow x)$.

Once these parameters are computed, the label sequence is obtained with the *Viterbi* algorithm [6, 10], shown as Algorithm 1; the STOP label of the algorithm is introduced to allow the algorithm to operate with word sequences of different lengths [12].

Algorithm 1. Pseudocode of the Viterbi algorithm

Require: a text sequence $x_1 \ldots x_n$, parameters $P(s \mid u, v)$ and $P(x \mid s)$.
  $\forall (u, v)$ such that $(u \neq *) \vee (v \neq *)$, assign $\pi(0, *, *) = 0$
  for $k = 1 \ldots n$ **do**
    for $u \in K, v \in K$ **do**
      $\pi(k, u, v) = \max_{w \in K}(\pi(k - 1, w, u) \times P(u \mid w, u) \times P(x_k \mid v))$
      $bp(k, u, v) = \arg\max_{w \in K}(\pi(k - 1, w, u) \times P(u \mid w, u) \times P(x_k \mid v))$
    end for
  end for
  Assign $(y_{n-1}, y_n) = \arg\max_{(u,v)}(\pi(n, u, v) \times P(\text{STOP} \mid u, v))$
  for $k = (n - 2) \ldots 1$ **do**
    $y_k = bp(k + 2, y_{k+1}, y_{k+2})$
  end for
  return the labeling sequence $y_1 \ldots y_n$

---

3 Related Work

The proposed system bears similarity with existing work on Twitter analysis. Tao et al. [19] propose a method for detecting near-duplicate tweets by first determining if two tweets are considered duplicates and then assigning a level of duplicity to the pair of tweets varying from exact copy to somewhat overlapping, using five levels. Natural similarity measures to attempt on tweets include the *edit distance* (also known as the Levenshtein distance) [5, 8, 11], the proportion of shared words, as well as the proportion of shared hashtags. NER is used to obtain semantic characteristics such as the proportion of shared entities. Also, the message time stamps and the similarity of the Twitter accounts that sent the tweet are considered. The duplicate detection in itself is carried out as *logistic regression*, given the similarity data.

Another work is that of Sankaranarayanan et al. [17], extracting and analyzing news in Twitter. First the system filters the news messages from the rest of the tweets, then groups the news tweets to obtain those related to a same news story and then detects the type of news that is being reported. The filtering of news versus not news is done by a *naïve Bayesian classifier* [3, 9] that has been previously trained with a set of tweets marked as either "news" or "noise". The grouping of the news tweets is done based on the text contents as well as metadata, extracting the topic and the location (again with NER).

Agarwal et al. [1] focus on detecting local news on Twitter that report fires at factories as well as strikes. Their system operates in four stages: first, the messages that contain information relevant to the topics of interest are filtered using *regular expressions* and supervised classifiers, after which the resulting tweets are compared with those of the last 24 hours, grouping the ones that correspond to the same event, and then NER is used to extract characteristics such as the duration, the location, and the type of the event, and finally groups that have very similar characteristics are merged.

Systems that work with duplicate detection in larger text documents (and hence can base the detection on a much larger set of data per element) include the DUDE system[1] of University of Michigan for technical papers. A framework for finding duplicates among XML documents with a known schema is presented by Weis and Naumann [20], whereas a similarity calculation for text documents is presented by Schleimer et al. [18].

As the CIC reports are extremely brief, we do not expect systems designed for longer documents to be able to function well on our data set. Also, the order in which information is presented is not relevant for the CIC reports to be considered duplicates, whereas for example detecting whether one text document copies fragments of another would require several words to follow one another in a near-identical manner to detect duplicity — a CIC report is usually shorter than a normal sentence in written text. The work of Gong et al. [7] is intended for short texts, but does not incorporate the element for determining the similarity of points of time expressed in the text that is necessary for the CIC reports.

4   Proposed Solution

We propose a four-stage process for the duplicate detection. In this section, we discuss the details of each stage as well as the steps involved.

Preprocessing into Metamodels: The preprocessing aims to structure the document for posterior analysis. Information is *extracted* from the document. For example, the CIC reports are available in XML and JSON and a corresponding parser is applied to access the elements of the document and obtain the textual content. *Filters* are then applied to eliminate undesired features such as non-ASCII characters or stop words, as well as substitutions such as replacing abbreviations with full words. Using NER, *labels* are

---

[1] Available online at http://sigda.eecs.umich.edu/DUDE/.

then assigned to the remaining words. The label categories used depend on the type of documents being processed. Based on the labeled sequence of words, sets of words are created by joining those that received the same label. These are stored along with the corresponding label in a template to create the metamodel.

Metamodel Clustering: The metamodels are classified into clusters of similar metamodels (this can be done either incrementally as an online clustering or globally as a one-time static clustering). The goal is to reduce the set of metamodels with which the duplicate detection is later carried out.

Classifier Training: Within the clusters of metamodels created by the previous step, classifiers are trained to distinguish between duplicate and non-duplicate metamodels, taking into account data regarding the location, time, and type of event.

Duplicate Detection: The trained classifiers of each cluster are then presented with pairs of metamodels over all pairs of the cluster if this is a global, static analysis, and between a new metamodel and the metamodels already included in the cluster in the case of an online, incremental processing. If two metamodels are classified as duplicates, then the corresponding input documents are considered to be duplicates.

5   Case Study

In this section we discuss the application of our proposed duplicate-detection system on the CIC citizen reports (discussed already in Section 1). The reports of the CIC were downloaded in the JSON format from the developer API of the CIC (available at `http://www.developers.cic.mx/api`); an example is shown in Figure 2. The CIC reports contain the following fields: `ticket` is a unique ID assigned to each report received, `content` contains the description of the reported event, `created_at` is a time stamp of the report creation, `address_detail` is the address with some typical fields (if available), and `categories` are predefined categories of the CIC for types of reports.

In this work, we downloaded reports from the following categories (with an abbreviation indicated in parenthesis, derived from the Spanish name used by the CIC for each category): accidents (`acc`), street lights (`alu`), traffic lights (`sem`), road damage (`bac`), sewer lids (`alc`), public events (`eve`), road work or closure (`obr`), and situations of risk (`sit`). Within the categories, we employ the different criteria for defining whether two reports are to be considered duplicates. For example, traffic-accident reports are considered as duplicates when the text describing the location is similar and the time lapse between the reports does not exceed 15 minutes, whereas if the time lapse is higher (even with the location being the same), the reports are considered distinct. This is not the same for missing or damaged sewer lids, for example: the same lid may be reported several hours apart or on different days. Hence the classifiers are trained separately for each metamodel cluster.

```
"ticket": "#7YPC",
"content": "*ACCIDENTE* En gonzalitos altura de vuelta izquierda a Insurgentes MTY #mtyfollow 17:37",
"created_at": "2013-08-04T17:49:56-05:00",
"address_detail": {
    "formatted_address": "Gonzalitos 655, Sin Nombre de Colonia 31, Monterrey, NL, México",
    "zipcode": "64000",
    "county": {
        "long_name":"Monterrey", "short_name":"Monterrey",
    },
    "state": {
        "long_name": "Nuevo León", "short_name":"Nuevo León"
    },
    "neighborhood": {
        "long_name": "Centro", "short_name": "Centro"
    }
},
"categories": ["ACCIDENTE"]
```

Fig. 2. An example of a CIC report in JSON

5.1 Preprocessing

For the preprocessing phase, we employ the Python library `json`[2] to extract the following fields: 'ticket', 'content', 'created_at', 'categories', and 'address_detail' for each report. From the date, the UTC date and time was parsed into fields for year, month, day, hour, minute, and second. Then, we apply the filters to clean up the data; we mention some examples of the filters used:

– Replace all accented characters with their ASCII equivalent.
– Eliminate everything that begins with `http`.
– Eliminate special characters such as *, :, ?, ;, -.
– Eliminate (with regular expressions) all hashtags and Twitter accounts.

At this point *no* stop-word elimination has yet been applied, as the prepositions are important for correct identification of place names in Spanish with NER: expressions such as `entre Avenida P. Livas y Las Americas` (between two specific avenues), `en Paseo de los Leones` (at a specific street), `rumbo a Lazaro Cardenas` (near a specific avenue). The categories used for the NER labeling of the reports are the following: places (`LOC`), time (`TIME`), persons (`NAME`), organizations (`ORG`), and event descriptions (`DESC`); a label for irrelevant information (`IRR`) was also employed, as done in the work of Ratinov and Roth [14]. The Vitebri algorithm (Algorithm 1 on page 255) was employed to assign the labels. Given the labeling, the metamodel is composed (as described in Section 4). Figure 3 shows an example of an input, the resulting labeling, and the created metamodel.

The clean-up carried out upon creating the metamodel out of the labeled sequence involves the elimination of stop words, making all words lowercase, replacing plural nouns by their singular forms, and elimination of word repetition.

---

[2] Available at https://docs.python.org/2/library/json.html.

ACCIDENTE en Ave. Garza	[ACCIDENTE, DESC], [en, LOC], [Ave., LOC], [Garza,
Sada sin lesionados 6:30	LOC], [Sada, LOC], [sin, DESC], [lesionados, DESC],
pm MTY NL gracias	[6:30, TIME], [pm, TIME], [NL, LOC], [gracias, O]
(a) Cleaned text	(b) Labeled text

```
<metamodel>
    <desc>ACCIDENTE sin lesionados</desc>
    <loc>en Ave. Garza Sada</loc>
    <time>6:30 pm</time>
</metamodel>
```

(c) Metamodel

Fig. 3. Cleaned text sequence, the corresponding labeled sequence, and the resulting metamodel

The information in the metamodels is then accessed by querying on three text elements: `tinfo` that describes the reported event (formed by cleaning the content of the labels `DESC`, `NAME`, and `ORG` of the metamodel), `tloc` that indicates the location (simply the cleaned-up content of the label `LOC` in the metamodel), and `ttime` that states the time stamp at level of minutes as UNIX time. Time differences are measured (for purposes of evaluating their similarity) as

$$\Delta_{l_i,t_j} = 1 - \left(1 + \log_{10}(|l_i - t_j|)\right)^{-1}, \tag{4}$$

where t_i is the UNIX time of the first metamodel and t_j that of the second.

5.2 Classifier Training

Eight support vector machine classifiers [3] are trained to detect duplicate pairs of metamodels, one per CIC category. The training commences by creating h training triples

$$\mathcal{T} = [(m_1^1, m_2^1, \delta^1), (m_1^2, m_2^2, \delta^2), \ldots, (m_1^h, m_2^h, \delta^h)], \tag{5}$$

where for the ith triplet, m_1^i y m_2^i are two (distinct) metamodels and $\delta^i \in [0, 1]$ is a binary decision variable: zero indicates that they are not duplicates whereas one indicates that the two metamodels are considered duplicates of one another. Each triple is then processed:

1. The fields `desc`, `loc`, and `time` are accessed for both metamodels.
2. Two weighted vectors are created for both metamodels using *term frequency - inverse document frequency* or tf-idf [12]: vector $\mathcal{I}_j$ is based on `tinfo` and vector $\mathcal{L}_j$ is based on `tloc` for metamodel $j \in \{1, 2\}$. The vocabulary employed for the terms was created manually from a sample of 1,784 meta-models and another distinct sample set was used for training each classifier.
3. The *cosine similarity* [2], defined for two vectors $\mathbf{v}$ and $\mathbf{w}$ as

$$(\mathbf{v} \cdot \mathbf{w})/(|\mathbf{v}||\mathbf{w}|), \tag{6}$$

is computed for $\mathcal{I}_1$ versus $\mathcal{I}_2$ (we denote the result by $\rho_{\mathcal{I}}$) and also for $\mathcal{L}_1$ versus $\mathcal{L}_2$ (yielding $\rho_{\mathcal{L}}$).

4. The time difference is computed (cf. Equation (4)); we denote this simply by Δ when the two metamodels used to obtain the time stamps are implicitly clear.

Then, a characteristic matrix of dimension $h \times 3$ is created together with a $1 \times h$ column vector:

$$\mathbf{X} = \begin{bmatrix} \rho_{\mathcal{I}}^1 \ \rho_{\mathcal{L}}^1 \ \Delta^1 \\ \vdots \ \ \vdots \ \ \vdots \\ \rho_{\mathcal{I}}^h \ \rho_{\mathcal{L}}^h \ \Delta^h \end{bmatrix} \quad \text{and} \quad \boldsymbol{y} = \begin{bmatrix} \delta^1 \ \delta^2 \dots \delta^h \end{bmatrix}^{\mathrm{T}}. \tag{7}$$

Using these two, a classifier is then trained for a specific category with the scikit-learn[3] Python-library. The resulting classifier for category ℓ is denoted by $\mathcal{C}_\ell$.

5.3 Duplicate Detection

The metamodel clustering for the case study is done simply based on the category assigned by the CIC (we have also carried out experiments using k-means variants to recover the categories based on document similarity with reasonable success). We hence apply the classifier $\mathcal{C}_\ell$ for each category $\ell \in \{\text{acc}, \text{alu}, \dots \text{sit}\}$ (done either using all pairs of metamodels within that category or upon the introduction of a new metamodel to a set of existing metamodels) for the test set of documents (those used for training and dictionary-creating are excluded). The test set contained 105 metamodels corresponding to the category acc, 20 to alc, 85 to alu, 90 to bac, 45 to eve, 45 to abr, 75 to sem, and 40 to sit. The pseudocode for the process is shown in Algorithm 2 for the case of adding a *single* new metamodel into a set of existing metamodels of the same category.

5.4 Results

Our results include the evaluation of the NER-labeler (alone) and the evaluation of, properly, the duplicate detection system. With respect to the former, the cleaned-up CIC reports contained a total of 123,583 words (3,823 reports). All these words were manually labeled using the labels discussed in Section 5.1: LOC, TIME, NAME, ORG, DESC, and IRR. Then, the parameters of the NER-labeler were computed. The labeler was tested with a set of 5,099 words extracted from a new set of CIC reports; we created one labeling with the trained labeler and another manually, obtaining a 92% precision on the automated labeling with respect to the manual one. The success was notable in identifying places, times, and event descriptions, possibly attributable to the limited vocabulary employed in the CIC reports. As pointed out by Ritter et al. [15], existing NER tools tend to perform poorly on Twitter messages; we hence conclude that our labeler has a sufficient performance with the current precision.

---

[3] Available at http://scikit-learn.org.

Algorithm 2. Duplicate-detection algorithm outline

Require: incoming metamodel m, existing metamodels M, trained classifier $\mathcal{C}$

1: $\mathcal{I} \leftarrow$ tf-idf($\texttt{tinfo}(m)$)
2: $\mathcal{L} \leftarrow$ tf-idf($\texttt{tloc}(m)$)
3: $t \leftarrow \texttt{ttime}(m)$
4: $D = \emptyset$ (list of duplicates of m detected within M)
5: **for** $m' \in M$ **do**
6:     $\mathcal{I}' \leftarrow$ tf-idf($\texttt{tinfo}(m')$)
7:     $\mathcal{L}' \leftarrow$ tf-idf($\texttt{tloc}(m')$)
8:     $t' \leftarrow \texttt{ttime}(m')$
9:     $\rho_{\mathcal{I}} \leftarrow \text{sim}(\mathcal{I}, \mathcal{I}')$ with Equation (6)
10:    $\rho_{\mathcal{L}} \leftarrow \text{sim}(\mathcal{L}, \mathcal{L}')$ with Equation (6)
11:    $\Delta \leftarrow \Delta_{t,t'}$ with Equation (4)
12:    $\delta \leftarrow \mathcal{C}(\rho_{\mathcal{I}}, \rho_{\mathcal{L}}, \Delta)$ (classifier output)
13:    **if** $\delta = 1$ **then**
14:       $D \leftarrow D \cup \{m'\}$ (add to results the detected duplicate)
15:    **end if**
16: **end for**
17: **return** D

For evaluating the reliability of the duplicate-detection system, we performed modifications on CIC reports to produce duplicates, then testing whether the modified duplicates were correctly identified by the classifiers. We used a total of 201 original reports and created two artificially modified duplicates for each, also creating two artificial non-duplicated by applying drastic modification. The modifications were made manually to ensure that the resulting reports make sense and that those intended as duplicates are in fact semantically similar whereas the non-duplicates have differences that permit a human observer to conclude that they are clearly distinct.

The set of metamodels thusly obtained was divided into a training set and a test set as follows: the original 201 metamodels were grouped according to their respective CIC categories (acc, alu, sem, bac, alc, eve, obr, and sit). With the K-iterations method [9], one half of each category was assigned to a training set and the other half to a test set. The modified metamodels (two duplicates and two non-duplicates) were then inserted in the same category and set as their corresponding original.

The classifier training was repeated ten times to study possible variations in the end result; Figure 4 illustrates the precision (on average 55%), recall (on average 84%), and F-value (on average 66%) obtained for the duplicate detector. Recent related work on detecting duplicate tweets by Tao et al. [19] obtained 48% precision and 43% recall, in the light of which our system seems to perform quite satisfactorily given the similarity between the input data in their work and ours. As the authors know of no other system for duplicate detection adaptable to the CIC context, we do not present a comparison between our method and another one; for example, using the system of Tao et al. [19] for comparison would

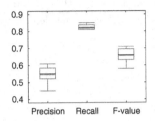

Fig. 4. Box-whiskers plots of the statistics of three performance measures over 10 repetitions of the classifier training

be unfair as the mere textual similarity that suffices for tweets is expected to perform poorly on the CIC reports that may in fact be written by two different people, simply describing the same indicent in different words.

6   Conclusions

We have presented an approach for detecting duplicates within document repositories, based on named entity recognition and supervised classification. The proposed method is tested on a case study using citizen-reported urban incidents in the metropolitan area of Monterrey, Mexico. The similarity between two reports is evaluated in terms of locations, times, events, and names present in the documents. The computational results obtained are better than expected from the performance of state-of-the-art solutions for similar data sets.

Improvements to the present system, left as future work, include parsing verbal expressions of time (phrases like "last Friday" or "at noon") and estimating geographical coordinates (latitude and longitude) based on textual address information (for example through the Google Maps API) to estimate the distance between locations when GPS coordinates are not included in the reports.

The integration of the proposed system as an automated step at CIC upon report reception is left as future work — presently the staff attempts to notice duplicates as a human effort and tend to struggle around personnel-shift changes. Failure to notice a duplicate document may result in a CIC staff member calling the fire department after a colleague already reported the same fire a few moments earlier before heading home.

As future work, also the generalization of the system towards other types of repositories such as scientific publications (to detect attempts of double submission of a single work as well as plagiarism) is of interest. This would require the design of a label set for the NER phase and a redefinition of what similarity means for this type of documents.

References

[1] Agarwal, P., Vaithiyanathan, R., Sharma, S., Shroff, G.: Catching the long-tail: Extracting local news events from Twitter. In: Proc. of the 6th International AAAI Conference on Weblogs and Social Media, Palo Alto, CA, USA, pp. 379–382. AAAI (2012)

[2] Baeza-Yates, R., Ribeiro-Neto, B.: Modern information retrieval. ACM Press, New York (1999)

[3] Bishop, C.: Pattern Recognition and Machine Learning. Information Science and Statistics. Springer, New York (2006)

[4] Church, K.: A stochastic parts program and noun phrase parser for unrestricted text. In: Proc. of the 2nd Conference on Applied Natural Language Processing, Stroudsburg, PA, USA, pp. 136–143. Association for Computational Linguistics (1988)

[5] Duda, R., Hart, P., Stork, D.: Pattern Classification, 2nd edn. Wiley-Interscience, New York (2000)

[6] Esposito, R., Radicioni, D.: CarpeDiem: Optimizing the Viterbi algorithm and applications to supervised sequential learning. The Journal of Machine Learning Research 10, 1851–1880 (2009)

[7] Gong, C., Huang, Y., Cheng, X., Bai, S.: Detecting near-duplicates in large-scale short text databases. In: Washio, T., Suzuki, E., Ting, K.M., Inokuchi, A. (eds.) PAKDD 2008. LNCS (LNAI), vol. 5012, pp. 877–883. Springer, Heidelberg (2008)

[8] Hall, P., Dowling, G.: Approximate string matching. ACM Computing Surveys 12(4), 381–402 (1980)

[9] Hastie, T., Tibshirani, R., Friedman, J.: The Elements of Statistical Learning: Data Mining, Inference, and Prediction, 2nd edn. Springer Series in Statistics. Springer, New York (2009)

[10] Jurafsky, D., Martin, J.: Speech and Language Processing: An Introduction to Natural Language Processing, Computational Linguistics, and Speech Recognition. Prentice Hall PTR, Upper Saddle River (2000)

[11] Levenshtein, V.: Binary codes capable of correcting deletions, insertions and reversals. Soviet Physics Doklady 10, 707 (1966)

[12] Manning, C., Raghavan, P., Schütze, H.: Introduction to Information Retrieval. Cambridge University Press (2008)

[13] Ponomareva, N., Rosso, P., Pla, F., Molina, A.: Conditional random fields vs. hidden Markov models in a biomedical named entity recognition task. In: Proc. of International Conference Recent Advances in Natural Language Processing, Borovets, Bulgaria, pp. 479–483. RANLP 2007 Organising Committee (2007)

[14] Ratinov, L., Roth, D.: Design challenges and misconceptions in named entity recognition. In: Proc. of the 13th Conference on Computational Natural Language Learning, Stroudsburg, PA, USA, pp. 147–155. Association for Computational Linguistics (2009)

[15] Ritter, A., Clark, S., Mausam, Etzioni, O.: Named entity recognition in tweets: An experimental study. In: Proc. of the Conference on Empirical Methods in Natural Language Processing, Stroudsburg, PA, USA, pp. 1524–1534. Association for Computational Linguistics (2011)

[16] Ross, S.: A first course in probability. Pearson Prentice Hall, Harlow (2010)

[17] Sankaranarayanan, J., Samet, H., Teitler, B., Lieberman, M., Sperling, J.: TwitterStand: News in tweets. In: Proc. of the 17th ACM SIGSPATIAL International Conference on Advances in Geographic Information Systems, pp. 42–51. ACM, New York (2009)

[18] Schleimer, S., Wilkerson, D.S., Aiken, A.: Winnowing: local algorithms for document fingerprinting. In: Proc. of the 2003 ACM SIGMOD International Conference on Management of Data, New York, NY, USA, pp. 76–85. ACM (2003)

[19] Tao, K., Abel, F., Hauff, C., Houben, G.-J., Gadiraju, U.: Groundhog Day: Near-duplicate Detection on Twitter. In: Proc. of the 22nd International Conference on World Wide Web, Republic and Canton of Geneva, Switzerland, pp. 1273–1284. International World Wide Web Conferences Steering Committee (2013)

[20] Weis, M., Naumann, F.: DogmatiX tracks down duplicates in xml. In: Proc. of the 2005 ACM SIGMOD International Conference on Management of Data, New York, NY, USA, pp. 431–442. ACM (2005)

A Bayesian Ensemble Classifier for Source Code Authorship Attribution

Matthew F. Tennyson[1] and Francisco J. Mitropoulos[2]

[1] Bradley University, Peoria, IL, USA
mtennyson@bradley.edu
[2] Nova Southeastern University, Fort-Lauderdale-Davie, FL, USA
mitrof@nova.edu

Abstract. Authorship attribution of source code is the task of deciding who wrote software, given its source code, when the author of the software is not explicitly known. There are numerous scenarios in which it is necessary to identify the author of a piece of software whose author is unknown, including software forensics investigations, plagiarism detection, and questions of software ownership. A number of methods for authorship attribution of source code have been presented in the past, including two state-of-the-art methods: SCAP and Burrows. Each of these two state-of-the-art methods was individually improved, and – as presented in this paper – an ensemble method was developed from them based on the Bayes optimal classifier. An empirical study was performed using a data set consisting of 7,231 open-source and textbook programs written in C++ and Java by thirty unique authors. The ensemble method successfully attributed 98.2% of all documents in the data set, compared to 88.9% by the Burrows baseline method and 91.0% by the SCAP baseline method.

Keywords: authorship attribution, software forensics, plagiarism detection, information retrieval, Bayesian probability theory.

1 Introduction

The term "authorship attribution" refers simply to "the task of deciding who wrote a document" [1]. Authorship attribution of source code, then, is the task of deciding who wrote a source code document. Source code authorship attribution is a tenet of software forensics, which is the process of analyzing software to identify characteristics of its authors for use in forensics activities [2]. The ultimate aim of software forensics is typically author identification, and it is usually applied to malicious code when analyzing software remnants left by an attacker in order to identify the origin of the attack or characteristics of the one who originated the attack.

Source code authorship attribution has many other applications outside of software forensics. In academia, the most obvious application is that of plagiarism detection on programming assignments. In programming courses, students often plagiarize solutions to programming problems by "borrowing" code from outside sources such as the Web, friends, or "Rent-A-Coder" services [3-5]. In industry, applications include

A.J. Machado Traina et al. (Eds.): SISAP 2014, LNCS 8821, pp. 265–276, 2014.
DOI: 10.1007/978-3-319-11988-5_25 © Springer International Publishing Switzerland 2014

activities related to configuration management and software ownership. In regards to configuration management, it could be important for author tracking or change control. In regards to software ownership, it could be important for the protection of trade secrets, patent claims, copyright infringement, or cases of software theft [6].

An authorship attribution problem usually proceeds as follows: A document is encountered of unknown authorship. The document is compared to a corpus of documents of known authorship. The author from the corpus that is most similar to the unknown document is attributed to be its author. The measure of similarity is usually based in some way on style so as to answer the question, "Which author's style best matches the style in which the unknown document was written?"

An authorship attribution experiment typically proceeds in a similar fashion: Several sample documents (of known authorship) are selected for experimental purposes. These documents are excluded from the corpus. An author is attributed to each of the samples using the experimental technique. The success is measured as a percentage of documents correctly attributed. In a closed form of the experiment, the actual author of the document is guaranteed to exist in the corpus. In an open form of the experiment, the actual author may not exist in the corpus. The open form of the experiment is obviously much harder, and no such studies of source code authorship attribution are known to have been published. Issues such as programming language, size of the samples, size of the corpus, etc., are variables that are either controlled or whose effects are measured experimentally. In most studies of source code authorship attribution, a "sample document" refers to a single source file.

Several methods of source code authorship attribution have been proposed [6-20]. In 2010, Burrows performed a comparative study of these methods [15] and determined that the Burrows method [14] is the most effective and the SCAP method [9] is the second most effective. The Burrows study is the first known controlled, comprehensive comparative study of source code authorship attribution. An extended version of this study was later published in [16]. In 2013, Tennyson performed the first known *independent* comparative study of source code authorship attribution [21] and found that the SCAP method is the most effective. In either case, the Burrows and SCAP methods of source code authorship attribution are considered state-of-the-art.

This paper presents an overview of the baseline SCAP and Burrows methods, individual improvements that can be made to each of the methods, how the methods can be combined to create an ensemble method based on the Bayes optimal classifier, and an empirical study gauging the effectiveness of this ensemble method.

2   Literature Review

This section presents an overview of the baseline SCAP and Burrows methods and describes individual improvements made to each.

2.1   The Baseline SCAP Method

The SCAP method [9-13] of source code authorship attribution utilizes byte-level n-grams to represent programs. That is, every byte contained in the source document is included in the n-gram representation of that document. All programs known to have been written by each candidate author are concatenated together, represented as n-grams, and the frequency of each n-gram is stored in a table. This table of n-gram frequencies is considered to be the profile of that author, or the Source Code Author Profile (SCAP). Only the L-most frequently occurring n-grams are retained, so that L is referred to as the profile length. The SCAP method utilizes a similarity measure referred to as the Simplified Profile Intersection (SPI), which is simply the number of n-grams that an author profile and a program have in common:

$$| P_A \cap P_P |$$

where P_A represents the author profile and P_P represents the program profile (i.e., the set of n-grams that occur in that program). To determine the author of a query program, the program is compared using the SPI similarity measure to all of the author profiles in the data set. The author of the most-similar profile is considered the author of the query program. So, in essence, it is the author whose profile is most similar to the query program that is attributed to be the author. More precisely, it is the author who often uses the n-grams that appear in the query program that is attributed.

Sample Author Profile

Programs	n-grams	Freq.

`int main()`
`{`
`  cout << "Hi";`
`}`
 	
`int main()`
`{`
`  for(int i=0; i<5; i++)`
`    cout << i;`
`}`
	) \n { \n _ _	2

	_ _ c o u t	2
	_ m a i n (	2
	a i n ( ) \n	2
	i n ( ) \n {	2
	c o u t _ <	2
	i n t _ m a	2
	n ( ) \n { \n	2
	u t _ < < _	2
	m a i n ( )	2

Sample Program Profile

Program	n-grams	Freq.
	t _ x ; \n }	1
	i n t _ x ;	1
	_ i n t _ x	1
	{ \n _ _ _ i	1
	) _ { \n _ _	1
	a i n ( ) _	1
	n t _ m a i	1
	_ { \n _ _ _	1

`int main() {`
`    int x;`
`}`
	_ m a i n (	1

	_ _ i n t _	1
	_ _ _ i n t	1
	( ) _ { \n _	1
	\n _ _ _ i n	1
	n t _ x ; \n	1
	m a i n ( )	1
	t _ m a i n	1
	i n ( ) _ {	1
	n ( ) _ { \n	1
	i n t _ m a	1

Fig. 1. A trivial SCAP example

Figure 1 shows an example, where the entire known corpus of the fictitious author consists of only two short programs. The corresponding author profile is shown as a table of n-gram frequencies, where $L=10$, so that only the 10 most frequent n-grams are retained. In the example, the SPI is 3, because there are precisely three n-grams that appear in both the author profile and the program profile. The intersecting n-grams are highlighted in the figure.

2.2   An Improved SCAP Method

In 2014, Tennyson et al. presented an improvement to the SCAP method [22]. Tennyson showed that instead of choosing a discrete value for the profile length L, it is more effective to simply to retain all n-grams in a profile except those that appear exactly once. Prior to this work, the choice for L was difficult. In the work of Frantzeskou et al. [9-13], various values for L had been used and recommended, ranging from 200 to 10000, while the value 2000 was most often used. In [15], Burrows suggested that the best value for L is effectively infinite, so that the author profiles are not truncated at all and all n-grams used by the authors are retained in their profiles. Tennyson performed a comparison of these and other approaches, and concluded that retaining all n-grams in a profile except those that appear exactly once was most effective.

2.3   The Baseline Burrows Method

The Burrows method of source code authorship attribution [14] utilizes token-based n-grams and a similarity measure to determine authorship. Tokens include certain operators, keywords, and whitespace. Programs are scanned, and the token stream is broken into n-grams using a sliding window approach. Based on empirical results, the authors chose $n=6$ for the n-gram size and Okapi BM25 [24] as the similarity measure. To determine the author of a program, that program is considered to be a query. The query is compared using a similarity measure to all of the programs in the data set. The author of the most-similar program is considered the author of the query program. So, in essence, it is the author who wrote the program that is most similar to the query program that is attributed to be its author.

The tokens used by Burrows were selected by creating six feature classes: operators, keywords, input/output, functions, white space, and literals. Basic programming features were categorized into these classes. Sets of features were formed from all possible combinations of the classes and empirical means were used to select the most significant feature classes. In the end, the feature classes selected were operators, keywords, and white space tokens. A list of these tokens, grouped by class, can be found in [15]. The Okapi BM25 metric is used primarily in search engines, and calculates the likelihood that a document is relevant to the information need expressed in the query [24]:

$$Okapi(Q, D_d) = \sum_{t \in Q} w_t \cdot \frac{(k_1 + 1)f_{d,t}}{K + f_{d,t}} \cdot \frac{(k_3 + 1)f_{q,t}}{k_3 + f_{q,t}}$$

$$w_t = \ln\left(\frac{N - f_t + 0.5}{f_t + 0.5}\right), \qquad K = k_1 \cdot \left((1 - b) + \frac{b \cdot W_d}{W_D}\right)$$

where Q is the query document, D_d is the document to which the query document is being compared, t is a term in the query that also appears in document D_d, N is the number of documents in the collection, W_d is the document length, W_D is the average document length, f_t is the collection frequency of the term, $f_{d,t}$ is the within-document frequency of the term, and $f_{q,t}$ is the within-query frequency of the term. k_1, k_3, and b are parameters. The values k_1 and b default to 1.2 and 0.75, respectively. In long queries, k_3 is often set to 1000, which is meant to be effectively infinite. In the Burrows method, the default values for k_1 and b are used, while $k_3 = 10^{10}$.

2.4   An Improved Burrows Method

In 2013, Tennyson et al. [23] improved the Burrows method in two ways: (1) by adding additional features to the feature set used for program representation and (2) by selecting different values for the Okapi parameters k_1 and b. Features that were added included additional keywords, symbols, and frequently-used identifiers. Based on empirical testing, Tennyson found that the Okapi parameter values k_1=0.2 and b=1.6 provided better results than the default values for the particular data set used.

3   The Ensemble Method

In this paper, two ensemble methods are presented with the aim of developing a new method that is even more effective than the current state-of-the-art methods independently are. The first ensemble method is based on Bayesian maximum a posteriori, and the second is based on the Bayes optimal classifier. Both of these approaches are presented in this section.

3.1   Maximum a Posteriori (MAP)

In Bayesian probability theory, the maximum a posteriori (MAP) is the most independently probable outcome given the observation data:

$$F_{MAP} = \underset{F \in H}{\operatorname{argmax}} P(F|E)$$

where F is a hypothesis (e.g., Author X wrote the document), H is the set of all possible hypotheses, E is the observable data (e.g., Author X was attributed to be the author of the document), and $P(F/E)$ is the probability of F given E (e.g., the probability that Author X wrote the document given Author X was attributed to be the author).

The first ensemble method is based on this concept. Using this approach, each of the individual methods is used to attribute each document. The posterior probabilities are calculated for each author using each method according to Bayes Theorem:

$$P(F|E) = \frac{P(E|F)P(F)}{P(E)}$$

where E is the observation and F is the hypothesis. The probabilities $P(E/F)$, $P(F)$, and $P(E)$ are determined as follows, based on the training data:

1. $P(E/F)$ is the probability that Author X was attributed to be the author given that Author X wrote the document. This is the accuracy for that individual author (i.e., the percentage of documents written by that author that were correctly attributed).

2. $P(F)$ is the probably that Author X wrote the document. This is simply the percentage of documents written by the author. Note that this assumes that the relative number of documents written by each author in the training data is truly representative of the total population (i.e., the distribution of documents in the sample is identical to the overall distribution in reality).

3. $P(E)$ is the probability that Author X was attributed to be the author. This is the percentage of documents attributed to have been written by the author.

Each of the individual methods is used to attribute an author to each document. If they agree, then it is the chosen author that is attributed. If they do not agree, then a final decision is made based on the F_{MAP}. That is, it is the author that was attributed by whichever method has the greater $P(F/E)$ for that author that is attributed to be the author. For example, if Burrows attributed Author X and SCAP attributed Author Y, and the probability that Author X wrote the document when Burrows attributes Author X is greater than the probability that Author Y wrote the document when SCAP attributes Author Y, then Author X is attributed.

3.2   Bayes Optimal Classifier

Attributing based on MAP is not optimal. The most probable classification is not always F_{MAP} because F_{MAP} only takes into account part of the observed data. It is the greatest of all the *independently*-calculated outcomes. A better approach would be to calculate the probability that Author X wrote the document given that Burrows attributed Author X and SCAP attributed Author Y versus the probability that Author Y wrote the document given those same *complete* observations.

This is the approach taken in the second ensemble classifier. The probability that the author wrote the document taking *all* observations into account is calculated:

$$P(F \mid E_B \& E_S) = \sum_{h \in H} P(F|h)P(h| E_B \& E_S)$$

where F is a classification, h is an hypothesis, H is the set of all hypotheses, E_B is Burrows' observation, and E_S is SCAP's observation. Therefore, $P(F/E_B \& E_S)$ is the probability that F is correct given both Burrows' and SCAP's observations.

It can be shown that the summation given above equates to $P(F/E_B \& E_S)$, given the fact that in this particular application the classifications and hypotheses are equivalent. For example, the hypothesis $h=$"Author X wrote the document" is equivalent to the classification $F=$"Author X" and so $P(F/h)=1$. In other words, the probability that Author X is the correct classification given that Author X wrote the document is certainty. On the other hand, when the h and F don't correspond (e.g., the probability that Author X is correct given Author Y wrote the document) is nil. When F corresponds to h, $P(F/h)=1$; otherwise $P(F/h)=0$. Therefore, the summation can be simplified:

$$P(F \mid E_B \& E_S) = P(h|E_B \& E_S)$$

The most probable classification is then obtained by determining which classification F gives the highest probability:

$$\underset{F \in Authors}{\text{argmax }} P(F \mid E_B \& E_S)$$

This is accomplished by determining $P(F/E_B \& E_S)$ for every combination of observations E_B and E_S. In this ensemble method, each of the individual methods is used to attribute an author to each document. If they agree, then it is the chosen author that is attributed. If they do not agree, then a final decision is made based on $P(F/E_B \& E_S)$. It is the author that has the greater $P(F/E_B \& E_S)$ that is attributed.

4   Methodology

In this study, the baseline Burrows method, the improved Burrows method, the baseline SCAP method, the improved SCAP method, and the two ensemble methods are compared. Note that both of the ensemble methods utilize the improved versions of the Burrows and SCAP methods.

The basic experimental design consisted of a 15-class experiment, utilizing a leave-one-out cross validation, and the results were measured in terms of accuracy. A 15-class experiment means that the author was determined from a set of fifteen candidate authors. A leave-one-out cross validation means that each program in the data set was selected, in turn, as a query program while the remaining programs were used as the training data. This approach maximizes the size of the training data with each query, while simultaneously maximizing the number of queries made. Accuracy was measured simply as a percentage of programs correctly identified.

4.1   The Data Set

The data set consisted of 7,231 programs written in C++ and Java. The programs consisted of both open-source programs and sample programs from programming and data structures textbooks. The data set was split into four segments: open-source C++ programs (SegA), open-source Java programs (SegB), textbook C++ programs (SegC), and textbook Java programs (SegD). Each segment contained programs from fifteen unique authors, while the data set consisted of 30 unique authors overall.

The open-source programs were collected using a procedure established by Burrows [15]. The textbook programs were collected using a procedure established by

Tennyson [22-23]. The reader is encouraged to consult those sources for detailed descriptions of the collection procedures, which are not repeated here.

A high-level statistical breakdown of the data set is provided in Table 1. A "sample" refers to a single document – in this case, a single source file. The min, median, mean, and max samples describe characteristics of the authors. The min, median, mean, and max LOC describe characteristics of the samples themselves.

Table 1. Statistical overview of the data set

	SegA	SegB	SegC	SegD
Total Authors	15	15	15	15
Total Samples	521	536	3134	3040
Min Samples	4	4	19	29
Median Samples	24	19	185	219
Mean Samples	35	36	209	203
Max Samples	101	172	453	509
Min LOC	6	6	2	1
Median LOC	78	92	36	37
Mean LOC	149	184	48	50
Max LOC	3265	2724	548	1079

4.2   Experimental Design and Approach

The experiment was executed independently for each of the four segments of data. Because each segment of data contained fifteen unique authors, each execution of the experiment was a 15-class experiment. The leave-one-out cross validation approach required that when a program was selected as the query program, that program was excluded from the author's profile in order to eliminate any resulting bias. Accuracy was measured as a percentage of programs correctly identified. Only the aggregate results across all four segments of data are reported in the Results section of this paper for the sake of clarity and concision.

5   Results

The experimental results are shown in Figure 2 and Table 2. The results show that the ensemble method based on the Bayes optimal classifier outperforms all the other methods. Using the optimal ensemble classifier, the performance improved to 98.2% compared to 97.3% using the improved version of the SCAP method. This improvement is statistically significant based on a chi-square statistical test for significance. The statistical improvement is measured relative to the improved version of the SCAP method because the ensemble method is based on the improved versions of the SCAP and Burrows method, and the SCAP method itself individually outperforms the Burrows method. The goal was to create an ensemble method that outperforms both SCAP and Burrows individually. So, the improved version of the SCAP method is the standard against which the ensemble method is measured.

6   Analysis and Conclusion

An ensemble method was created based on the Bayes optimal classifier that improved the accuracy to 98.2% compared to 88.9% by the Burrows baseline method and 91.0% by the SCAP baseline method based on an empirical study whose results are shown in Figure 2 and Table 2.

Table 2. Overview of experimental results

Method	Total	Correct	Percent	Chi-Square
Burrows Baseline	7231	6426	88.9%	N/A
Burrows Improved	7231	6778	93.7%	N/A
SCAP Baseline	7231	6581	91.0%	N/A
SCAP Improved	7231	7037	97.3%	1
Ensemble (MAP)	7231	7040	97.4%	0.827167
Ensemble (Optimal)	7231	7099	98.2%	6.41E-06

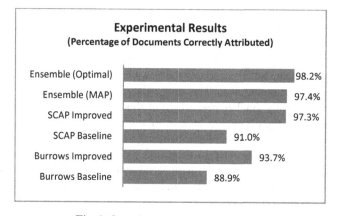

Fig. 2. Overview of experimental results

6.1   Analysis

The first ensemble classifier was based on the maximum a posteriori (MAP) attribution, which is the most independently probable attribution given independently observed data. The issue is that it is based on an independently-determined probability, which is not optimal in an ensemble classifier where there are multiple methods and multiple observations. The posterior probability is based on data when using only one respective method and it only takes into account the observation made by that method. The method did perform marginally better than the improved SCAP individually performed. However, it was not statistically significant.

The second ensemble classifier is based on the Bayes optimal classifier. This type of classifier is guaranteed to be optimal based on the training data, because it takes into account *all* observations and the posterior probabilities are calculated based on all observations and attributions from the training data.

Although it clearly performs the best, there are two issues with this ensemble method. The first issue is a general problem that applies to all Bayes optimal classifiers. They are susceptible to overfitting. Certainly, this case is no exception. There are fifteen unique authors for each segment of data in the data set. There are two individual classifiers being combined in the ensemble classifier. So there are 15×15=225 possible permutations, or combined observations. The vast majority of these observations never occur in the training data. In fact, in the SegA segment of data (picked arbitrarily as an example), only 21% (48 out of 225) of the possible observations actually occurred in the training data. Of those 48 observations, fifteen of those were matching observations (i.e., when both methods attributed the same author). That leaves only 33 observations of interest. Those 33 observations applied to only 46 documents (less than 1.4 documents per observation). This means that most of those observations applied to only a single document in the data set. So, in those cases, the calculated posterior probably was based on a single observation of a single document.

The second issue deals with the leave-one-out experimental design. In this type of experiment, the entire data set is used both for training and querying. Calculating each posterior probability 7231 times (once for each document in the data set, as each is chosen in turn to be the query document) is computationally not practical. So, the classifier was implemented such that the probabilities were calculated only once. As a result, the query document was left in the posterior probability calculations used to attribute that document. So, there exists the *potential* for bias in the results. As described previously, the potential for bias only exists in an extremely small percentage of observations. A vastly smaller percentage of those observations would result in a different attribution if the query document were omitted from the probability calculations. In fact, a manual inspection of the results did not reveal a single attribution that would have been made differently had the query document been omitted from the calculations. However, it can't be said with certainty that such an attribution does not exist. Therefore, an unavoidable – but certainly nominal – bias potentially exists in the posterior probability calculations.

Despite these issues, the results are indicative that such an ensemble classifier could be used in a practical application. In a practical setting, the issue regarding the leave-one-out validation would not be applicable. The known corpus of the authors would be given, and the probabilities would only need to be calculated once based on that data. The issue regarding overfitting, on the other hand, is difficult to mitigate. The probabilities would simply be calculated based on the best and only information available, and the classifier is guaranteed to be optimal with respect to that data.

6.2   Future Work

Student-submitted programs are often used in authorship attribution but are problematic for several reasons. They cannot generally be shared due to privacy issues. Minimally, the programs would have to be made anonymous, such that all explicitly-identifying information is stripped. Student-submitted programs are also not a good source of "perfect ground truth" because a large portion of student programs are copied and plagiarized. Furthermore, the programs should come from students with varied backgrounds and from various institutions, but typically consist only of programs from a single institution because such programs cannot be shared. Also, programming assignments often contain program segments that are provided by the instructor as part of the assignment, which would obviously skew experimental results.

However, a study could be executed expressly for the purpose of collecting student-submitted programs to be shared and used by researchers. The study would need to be widespread, covering numerous and varied institutions. The programs would need to be developed in a controlled environment to prevent copying and plagiarism. Although it would be a large undertaking, such a project would be feasible and applicable to more than just authorship attribution experiments. Any research requiring the use of student-submitted programs would be benefitted.

A universal data set suitable for standard use in authorship attribution experiments (as well as other fields also requiring similar collections of programs) is a common vision among researchers. Perhaps the use of open-source programs collected using the Burrows approach, textbook programs that are freely available and accessible for download via the Web, and student-submitted programs collected in a controlled environment as suggested would together satisfy that vision.

Other areas of future work include issues such as code reuse, multiple authors, code obfuscation, the use of IDEs, the application of conventions and guidelines, inconsistent programming style, and programmer background and education. Another major factor is the typical closed form presentation of the problem, such that authors are attributed from one of a set of finite candidate authors. A variation for further study is the much harder open form of the problem: to attribute whether the author was one of a finite set or an unknown author from outside that set.

References

1. Zhao, Y., Zobel, J.: Effective and scalable authorship attribution using function words. In: Lee, G.G., Yamada, A., Meng, H., Myaeng, S.-H. (eds.) AIRS 2005. LNCS, vol. 3689, pp. 174–189. Springer, Heidelberg (2005)
2. Spafford, E., Weeber, S.: Software forensics: Can we track code to its authors? Computers & Security (COMPSEC) 12(6), 585–595 (1993)
3. McCabe, D.: Levels of cheating and plagiarism remain high. Technical report, Center for Academic Integrity, Duke University (2005)
4. Bull, J., Collins, C., Coughlin, E., Sharp, D.: Technical Review of Plagiarism Detection Software Report.Technical report, Joint Information System Committee (2001)
5. Culwin, F., MacLeod, A., Lancaster, T.: Source Code Plagiarism in UK HE Computing Schools, Issues, Attitudes and Tools.Technical report, South Bank University (2001)
6. MacDonell, S., Gray, A., MacLennan, G., Sallis, P.: Software forensics for discriminating between program authors. In: Proceedings of the 6th International Conference on Neural Information Processing, pp. 66–71 (1999)
7. Krsul, I., Spafford, E.: Authorship analysis: Identifying the author of a program. Computers & Security (COMPSEC) 16(3), 233–257 (1997)
8. Ding, H., Samadzadeh, M.: Extraction of java program fingerprints for software authorship identification. The Journal of Systems and Software 72, 49–57 (2004)

9. Frantzeskou, G., Stamatatos, E., Gritzalis, S.: Supporting the cybercrime investigation process: Effective discrimination of source code authors based on byte-level information. In: Proceedings of the Second International Conference on E-business and Telecommunication Networks, pp. 283–290 (2005)
10. Frantzeskou, G., Stamatatos, E., Gritzalis, S., Katsikas, S.: Source code author identification based on n-gram author profiles. In: Maglogiannis, I., Karpouzis, K., Bramer, M. (eds.) Artificial Intelligence Applications and Innovations. IFIP AICT, vol. 204, pp. 508–515. Springer, Heidelberg (2006)
11. Frantzeskou, G., Stamatatos, E., Gritzalis, S., Katsikas, S.: Effective identification of source code authors using byte-level information. In: Proceedings of the Twenty-Eighth International Conference on Software Engineering, pp. 893–896 (2006)
12. Frantzeskou, G., Stamatatos, E., Gritzalis, S., Chaski, C., Howald, B.: Identifying authorship by byte-level n-grams: The source code author profile (SCAP) method. International Journal of Digital Evidence 6(1), 1–18 (2007)
13. Frantzeskou, G., MacDonell, S.G., Stamatatos, E., Gritzalis, S.: Examining the significance of high-level programming features in source code author classification. Journal of Systems and Software 81(3), 447–460 (2008)
14. Burrows, S., Tahaghoghi, S.: Source code authorship attribution using n-grams. In: Proceedings of the 12th Australasian Document Computing Symposium, pp. 32–39 (2007)
15. Burrows, S.: Source Code Authorship Attribution. Doctoral thesis, RMIT University, Melbourne, Victoria, Australia (2010)
16. Burrows, S., Uitdenbogerd, A., Turpin, A.: Comparing techniques for authorship attribution of source code. Software: Practice and Experience 44(1), 1–32 (2014)
17. Lange, R., Mancoridis, S.: Using code metric histograms and genetic algorithms to perform author identification for software forensics. In: Proceedings of the 9th Annual Conference on Genetic and Evolutionary Computation, pp. 2082–2089 (2007)
18. Kothari, J., Shevertalov, M., Stehle, E., Mancoridis, S.: A probabilistic approach to source code authorship identification. In: Proceedings of the Fourth International Conference on Information Technology, pp. 243–248 (2007)
19. Elenbogen, B., Seliya, N.: Detecting outsourced student programming assignments. Journal of Computing Sciences in Colleges 23(3), 50–57 (2008)
20. Shevertalov, M., Kothari, J., Stehle, E., Mancoridis, S.: On the use of discretized source code metrics for author identification. In: Proceedings of the 1st International Symposium on Search Based Software Engineering, pp. 69–78 (2009)
21. Tennyson, M.: A Replicated Comparative Study of Source Code Authorship Attribution. In: Proceedings of the International Workshop on Replication in Empirical Software Engineering Research, pp. 76–83 (2013)
22. Tennyson, M., Mitropoulos, F.: Choosing a Profile Length in the SCAP Method of Source Code Authorship Attribution. In: 2014 Proceedings of the IEEE Southeastcon (2014)
23. Tennyson, M., Mitropoulos, F.: Improving the Burrows Method of Source Code Authorship Attribution. In: Proceedings of the IADIS International Conference on Applied Computing, pp. 3–9 (2013)
24. Robertson, S., Walker, S.: Okapi/Keenbow at TREC-8. In: Proceedings of the 8th Text Retrieval Conference, pp. 151–162 (1999)

Multi-Core (CPU and GPU) for Permutation-Based Indexing

Hisham Mohamed, Hasmik Osipyan, and Stéphane Marchand-Maillet

Université de Genève, Geneva, Switzerland
{hisham.mohamed,stephane.marchand-maillet}@unige.ch,
hasmik.osipyan@etu.unige.ch

Abstract. Permutation-based indexing is a technique to approximate k-nearest neighbor computation in high-dimensional spaces. The technique aims to predict the proximity between elements encoding their location with respect to their surrounding. The strategy is fast and effective to answer user queries. The main constraint of this technique is the indexing time. Opening the GPUs to general purpose computation allows to perform parallel computation on a powerful platform. In this paper, we propose efficient indexing algorithms for the permutation-based indexing using multi-core architecture GPU and CPU. We study the performance and efficiency of our algorithms on large-scale datasets of millions of documents. Experimental results show a decrease of the indexing time.

Keywords: K-NN, Similarity Search, GPU, Permutation-Based Indexing, Big-Data.

1 Introduction

Searching for similar objects in a database is a fundamental problem for many applications, such as information retrieval, visualization, machine learning and data mining. Several techniques have been proposed for improving the performance of searching. One of the promising routes is the approximate searching methodologies [1].

Approximate searching provides fast response time, while accepting some imprecision in the output results. *Permutation-based indexing* [2, 3] is a recent technique for approximate searching. The idea behind it is to represent each object by a list of permutations of selected neighboring items (reference points). The similarity between any two objects is then derived by comparing the two corresponding permutation lists. Several data structures have been proposed to handle these permutation lists [4, 3, 5–7]. They manage to answer users queries in fast and effective way. On the other hand, the indexing process consumes a lot of time, which makes the algorithm not effective for large-scale data (aka Big-Data). To handle this big data, a parallel platform is needed. The increase in performance of graphic processing units (GPU) using the NVIDIA Compute Unified Device Architecture API (CUDA) allows users from different communities to perform parallel computation on a powerful platform.

Studying the performance of the permutation-based indexing on GPU was proposed in [8]. In [8], the authors showed how can the GPU be used to answer multiple queries requests for permutation-based indexing.

A.J. Machado Traina et al. (Eds.): SISAP 2014, LNCS 8821, pp. 277–288, 2014.
DOI: 10.1007/978-3-319-11988-5_26 © Springer International Publishing Switzerland 2014

In this work, we propose a multi-core implementation of the indexing algorithm on GPU and CPU. Our proposal, can be adapted easily to work for any of the available data structures for permutation-based indexing [4, 3, 5–7]. To validate our claims, we test our techniques on number of high dimensional large dataset containing several millions of objects. Hence, The main difference between this work and the work presented in [8], that we target the indexing process, while in [8] the searching process is targeted.

The rest of the paper is organized as follows. Section 2 introduces the permutation-based indexing model. Section 3 provides an overview of the GPU architecture. Section 4 introduces parallel indexing algorithms on GPU and CPU. Finally, we present our results in section 5 and conclude in section 6.

2   Permutation-Based Indexing

2.1   Indexing Model

Definition 1. *Given a set of N objects o_i, $D = \{o_1, \ldots, o_N\}$ in m-dimensional space, a set of reference objects $R = \{r_1, \ldots, r_n\} \subset D$, and a distance function which follows the metric space postulates, we define the* ordered list *of R relative to $o \in D$, $L(o, R)$, as the ordering of elements in R with respect to their increasing distance from o:*

$$L(o, R) = \{r_{i_1}, \ldots, r_{i_n}\} \text{ such that } d(o, r_{i_j}) \leq d(o, r_{i_{j+1}}) \ \forall j = 1, \ldots, n-1$$

Then, for any $r \in R$, $L(o, R)_{|r}$ indicates the position of r in $L(o, R)$. In other words, $L(o, R)_{|r} = j$ such that $r_{i_j} = r$. Further, given $\tilde{n} > 0$, $\tilde{L}(o, R)$ is the pruned ordered list of the $\tilde{n}$ first elements of $L(o, R)$.

Figures 1(b) and 1(c) give the ordered lists $L(o, R)$ ($n = 4$) and the pruned ordered lists $\tilde{L}(o, R)$ ($\tilde{n} = 2$), for D and R illustrated in Figure 1(a).

In K-NN similarity queries, we are interested in ranking objects (to extract the K first elements) and not in the actual inter-object distance values. Permutation-based indexing relaxes distance calculations by assuming that they will be approximated in terms of their ordering when comparing the ordered lists of objects. To efficiently answer users queries, authors in [4, 3, 5–7] proposed several strategies to compare the ordered lists efficiently. The main weakness of these techniques is the indexing time, especially when the number of references and objects increases.

2.2   Technical Implementation

Algorithm 1 details the indexing process. For each object in D, the distance $d(o_i, r_j)$ with all the references in the reference set R is calculated (lines 1-4). After sorting the distances in increasing order using the suitable sorting algorithm (*QuickSort* is used), the full ordered list $L(o_i, R)$ is created (Line 5). In line 6, partial lists $\tilde{L}(o_i, R)$ are generated by choosing the top $\tilde{n}$ references from $L(o_i, R)$. In line 7, $\tilde{L}(o_i, R)$ are stored in the appropriate data structure [4, 3, 5–7]. Theoretically, the sorting complexity is $O(n\log n)$ which leads to $O(N(n + n\log n))$ indexing complexity.

$L(o_1,R) = (r_3,r_4,r_1,r_2)$ $L(o_2,R) = (r_3,r_2,r_4,r_1)$ $\tilde{L}(o_1,R) = (r_3,r_4)$ $\tilde{L}(o_2,R) = (r_3,r_2)$
$L(o_3,R) = (r_3,r_4,r_1,r_2)$ $L(o_4,R) = (r_2,r_3,r_1,r_4)$ $\tilde{L}(o_3,R) = (r_3,r_4)$ $\tilde{L}(o_4,R) = (r_2,r_3)$
$L(o_5,R) = (r_2,r_1,r_3,r_4)$ $L(o_6,R) = (r_2,r_3,r_1,r_4)$ $\tilde{L}(o_5,R) = (r_2,r_1)$ $\tilde{L}(o_6,R) = (r_2,r_3)$
$L(o_7,R) = (r_1,r_2,r_4,r_3)$ $L(o_8,R) = (r_4,r_1,r_3,r_2)$ $\tilde{L}(o_7,R) = (r_1,r_2)$ $\tilde{L}(o_8,R) = (r_4,r_1)$
$L(o_9,R) = (r_1,r_4,r_2,r_3)$ $L(o_{10},R) = (r_4,r_1,r_3,r_2)$ $\tilde{L}(o_9,R) = (r_1,r_4)$ $\tilde{L}(o_{10},R) = (r_4,r_1)$
$L(q,R) = (r_4,r_3,r_1,r_2)$ $\tilde{L}(q,R) = (r_4,r_3)$

(a)   (b)   (c)

Fig. 1. a) White circles are data objects o_i; black circles are reference objects r_j; the gray circle is the query object q b) Ordered lists $L(o_i, R)$, $n = 4$. c) Pruned ordered lists $\tilde{L}(o_i, R)$, $\tilde{n} = 2$.

Algorithm 1 (Permutation-Based Indexing)

IN: D of size N, R of n and $\tilde{n} \leq n$
OUT: $\tilde{L}(o_i, R) \forall i = 1, \ldots, N - 1$
1. For $o_i \in D$
2.   For $r_j \in R$
3. $b[j].dis = d(o_i, r_j)$
4. $b[j].indx = j$
5. $L(o_i, R) = quicksort(b, n)$
6. $\tilde{L}(o_i, R) = partiallist(L(o_i, R), \tilde{n})$
7.   Store the ID i of o_i and its $\tilde{L}(o_i, R)$ for other processing.

3   GPU Architecture

The GPU architecture follows the Single Instruction Multiple Thread (SIMT) model. That allows to execute single instruction through different threads on different data. GPUs were dedicated to handle graphics primitives. With the evolution of the NVIDIA CUDA API, it becomes applicable to access the GPUs for powerful computation. A *device* is the GPU card and a *host* is the computer that hosts the GPU. The GPU contains a large number of computing cores with limited specification to perform multiple operations in parallel. A kernel (function) launches thousands of threads which are organized in two levels: *grids* and *blocks*. The grids are two or three-dimensional of blocks. Every block consists of an upper limit of threads (512 or 1024) depending on the device. GPU has *global* and *local* memory. Accessing the local memory is faster than the global memory. On the other hand, the local memory size (kilo bytes) is smaller than the global memory (Giga bytes). In order to get a good performance from the GPU, the data that need to be computed should be small enough in order to fit in the local memory, otherwise all the time is consumed in fetching the data from the global memory. Accordingly, algorithms like sorting are not effective on the GPU as the data that need to be sorted cannot be fitted once in the local memory, hence most of the time is consumed in fetching the data from the global memory.

4   Multi-Core Indexing

In sections 4.1 and 4.2, we propose parallel indexing algorithms for our permutation-based indexing strategy on multicore architecture GPU and CPU, respectively. A general baseline, as we are working with large-scale data, the data can not be allocated neither on the GPU global memory nor on the CPU random-access memory (RAM). The indexing process is done by portion. Hence, the indexing process is organized as follows. We read a portion of the data N_l. The read portion is indexed and stored in the RAM. When the RAM is full, the data are stored to the hard-disk.

4.1   Exploiting GPU Architecture

As explained in section 2, the indexing process is composed of two procedures. First, we calculate the distance between each object o_i and each reference point r_j in R. Then, we select the closest $\tilde{n}$ reference points to each object o_i by sorting the references based on their distances (building $\tilde{L}(o_i, R)$). The GPU architecture is based on single instruction multiple threads (SIMT) model. All the threads perform the same operation, but for different data, which meets our algorithm. In the next three subsections, we propose three parallel strategies that work on different level for improving the running time and increasing the throughput.

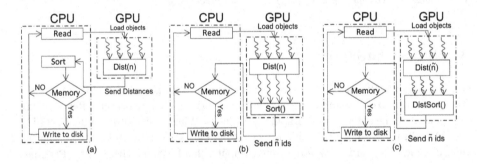

Fig. 2. Our 3 GPU algorithms a) PDSS b) PDPS c) PIOF

Parallel Distance Calculation and Sequential Sorting (PDSS). In *PDSS*, the work is shared between the CPU and the GPU. On the GPU, there is one kernel (*Dist()*), which calculates the distances between each N_l object and the reference set R. The distance information is then sent back to the CPU to be sorted and stored (Figure 2(a)). Using this technique, only the distances are calculated in parallel. Hence, the complexity of the algorithm is reduced to $O(\frac{Nn}{P} + N(nlogn)) + t_1$, where P is the number of the threads that can run in parallel at the same time and t_1 is the time needed to transfer the data between the CPU and GPU. The data size that can be processed in parallel on the GPU has to satisfy this equation:

$$s \times ((N_l \times m) + (n \times m) + (N_l \times n)) < \text{GPU Global Memory Size,} \qquad (1)$$

where $(N_l \times m)$, $(n \times m)$ and $(N_l \times n)$ represent the data objects, the reference points and the distance information respectively. s represents a 4 byte data-type (*float* for distance information and *int* for object id). The distance calculation kernel on the GPU is built using 2 dimensional block with sizes equal to 16 and 2D grid with sizes equal to N_l divided by the block size. Hence, every thread calculates the Euclidean distance between a given object and a given reference point based on the thread coordinates [9].

Parallel Distance Calculation and Sorting (PDPS). Here, the complete process is done on the GPU. We parallelize the sorting step to minimize the size of the data transferred from the GPU to the CPU.

We have two kernels. The first kernel (*Dist()*) calculates the distances similarly to the previous technique, between the N_l read objects and the reference set R in parallel. In the second kernel (*Sort()*), each thread sorts a separate ordered list using the QuickSort algorithm. The sorting process of one list on one GPU core is slower compared to using the CPU core, as the GPU cores are less powerful. On the other hand, on GPU, multiple lists are sorted in parallel, which should improve the running time (Figure 2(b)).

The complexity of the algorithm is therefore reduced to $O(\frac{N(n+(n\log n))}{P})+t_2$, where t_2 is the time needed to send the ranks to the host. t_1 in *PDSS* is greater than t_2, because the number of elements that are sent to the CPU is $N_l \times n$ while, for this new technique, it is $N_l \times \tilde{n}$.

The main constraint of this technique is that all the data needs to reside on the GPU (the partial objects, the references, the ids of references and the distances values). Since, the GPU global memory size is limited. Hence, equation (1) becomes:

$$s \times ((N_l \times m) + (n \times m) + (N_l \times n) + (N_l \times \tilde{n})) < \text{GPU Global Memory Size,} \quad (2)$$

where $(N_l \times \tilde{n})$ represents the reference id (4 bytes). Accordingly, with a high number of reference points, the number of objects processed at the same time N_l has to be decreased, which decreases the throughput and affects the speedup.

Parallel Indexing On the Fly (PIOF). As mentioned in section 3, a good performance can be achieved using the GPU, if the data that needs to be computed is small enough to fit in the registers of the GPU cores. Otherwise all the time is consumed in moving the data from and to the global memory. Hence, the sorting process on the GPU for the *PDPS* is a major bottleneck, since the distances are located in the GPU global memory.

We propose algorithm 2 to reduce the complexity of sorting to pick up the closest $\tilde{n}$ references. Instead of sorting the whole distances between an object and the list of references, to get the closest references, we sort only the first $\tilde{n}$ references of the list (line 1) by their distance to the object. Since, there might be other references in the list closer to the object than the objects in this pilot sorted partial list. The algorithm checks the distances starting from position $\tilde{n} + 1$ to n (lines 2-5). If one of the distance is smaller than the distance at position $\tilde{n}$, the partial list is updated (lines 5-9). Lines 3-6 in algorithm 1 are simply replaced by algorithm 2.

Theoretically, the indexing sequential complexity is reduced from $O(N(n+n\log n))$ to $O(N(n + (\tilde{n}\log\tilde{n}) + n))$, where $O(\tilde{n}\log\tilde{n})$ represents the partial sorting and $O(n)$ represents the updating.

Algorithm 2 (Picking $\tilde{n}$ nearest references)

IN: Distance information b and $\tilde{n}$
OUT: $\tilde{L}(o_i, R)$

1. $quicksort(b, \tilde{n})$
2.   For $j = \tilde{n} + 1 \rightarrow n$
3.   If $(b[\tilde{n}].dis > b[j].dis)$
4. $i = BinarySearch(0, \tilde{n}, b[j].dis)$
5. $tmp = \tilde{n}$
6. $while(tmp \geq i)$
7. $b[tmp].dis = b[tmp - 1].dis$
8. $b[tmp].indx = b[tmp - 1].indx$
9. $tmp - -$

To map this strategy on the GPU, we have two kernels (Figure 2(c)). In the first kernel (*Dist()*), the distance is calculated similarly to the previous techniques, but between the loaded objects and the first $\tilde{n}$ references only, to get primary distance information (line 1 in algorithm 2). Accordingly, the size of the distance information array is reduced to $(N_l \times \tilde{n})$. Then, in the second kernel (*DistSort()*, lines 2-9 in algorithm 2), each thread calculates the distance between an object and the rest of references from position $\tilde{n} + 1$ to position n. If a relative smaller distance information is found, the distance array is updated. Otherwise, the distance information is ignored. The complexity of the algorithm is reduced to $O(\frac{N(2n+(\tilde{n}\log\tilde{n}))}{P}) + t_2$. For the memory constraint, equation (2) becomes:

$$s \times ((N_l \times m) + (n \times m) + 2(N_l \times \tilde{n})) < \text{GPU Global Memory Size}, \qquad (3)$$

With this technique, we gain three privileges. The first gain is decreasing the size of the distance information array from $N_l \times n$ to $N_l \times \tilde{n}$. That allows to process more objects in parallel compared to *PDSS* and *PDPS*. The second gain is in decreasing the dependency between the threads by sharing the work between the kernels, which allows faster processing of the objects. The third gain is decreasing the latency between the cores and the global memory. In *PDPS*, for sorting the lists, the distance information is loaded from the global memory. With *PIOF*, the distance information is calculated instantly. Hence, it is located in the local memory and accessing the local memory is much faster than the global memory in the GPU architecture, which improves the performance.

In the second kernel (*DistSort()*), as the distance is calculated between an object and the rest of references (from $\tilde{n} + 1$ *to* n), we reduce the latency by taking a copy of the object under processing from the global memory to the local memory.

4.2   Multi-Core CPU

The most common library for CPU multi-core parallelization is OpenMP [10]. We use it for the indexing process. When deploying algorithm 1, a bottleneck in the access of the data file appears, because the OpenMP library does not permit parallel disk access. Accordingly, accessing the hard disk by different threads has to be organized in order to avoid random disk access. We propose two strategies.

Single-Disk Access. In this procedure, there is only one thread which accesses the hard disk. The main thread reads N_l objects at each iteration. After that, P threads start to process the objects in parallel (creating the ordered list, sorting and storing it). The main issue with this technique is that, if one of the threads finishes its work before the others, it has to wait to synchronize with the master thread, since the master thread is the only thread that can access the hard disk per-time.

The complexity of the algorithm is therefore reduced to $O(\frac{N(n+(nlogn))}{P})+t_s$, where t_s is the time needed for synchronization.

Multiple-Disk Access. Unlike in the previous procedure, all the threads can access the hard disk. Each thread reads multiple objects per-time and processes them separately without waiting for the master thread. To avoid random disk access, we setup a pipeline. Hence, when the algorithm starts, the first thread reads a portion from the file, then the second thread reads a second portion and so on. Once a thread is done with its portion, it goes directly to the disk and reads the next portion. The total output size from all the threads should be less than the maximum size of the RAM:

$$s \times ((n \times m) + P((N_p \times m) + 2(N_p \times \tilde{n}))) < RAMSize, \qquad (4)$$

where P is the number of active processes and N_P is the number of objects read by each processor ($N_p=\frac{N_l}{P}$).

The algorithm complexity is similar to that of the previous technique, but the pipeline strategy gives more flexibility for the threads, reduces the synchronization time t_s and avoids random disk access.

5   Experimental Results

We use the CoPhIR dataset [11] to evaluate the proposed algorithms. This dataset consists of 106-million MPEG-7 global visual descriptors extracted from 106-million images. The descriptors are scalable color, color structure, color layout, edge histogram and homogeneous texture. In our experiments, we extracted three data sets of sizes 1-million, 10-millions and 106-millions. The vectors in each dataset are constructed by combining the first three descriptors, which result in vectors of size 208-dimension.

We applied our indexing algorithms on the *metric permutation table* [4] as it is the recent data structure for permutation-based indexing. The experiments were performed for pruned ordered lists of size $\tilde{n} = 10$ and $\tilde{n} = 100$ for $n = 1000$ and $n = 2000$.

The algorithms were implemented in C. We used CUDA 5.1 and OpenMP for GPU and CPU parallelization respectively. The experiments were run on two different machines. The first machine C_1 holds 4 cores (3GHZ) Intel i7 processor, 4MB L3-Cache, 16GB RAM, NVIDIA NVS 5200M (96 CUDA Cores, 2 Streaming Multiprocessor, 1 GB global memory), and linked with 512GB storage capacity (SSD Disk). The second machine C_2 holds 32 Cores (2.70GHz) Intel Xeon processor, 20MB L3-Cache, 128GB RAM and linked with 512GB storage capacity (SSD Disk). On the two machines, the maximum RAM size which is used is 1.5 GB. Hence, writing to the hard disk is done when the size of the permutation lists produced exceeds 1.5 GB.

For a fair comparison, algorithm 2 is applied for all the implementations (sequential and multicore CPU and GPU) to pick up the closest $\tilde{n}$ reference objects. As the two machines do not have the same specifications, we run the sequential algorithm on both of them. On the first machine Seq_{C1}, to compare it with the GPU implementations. On the second machine Seq_{C2} to compare it with the CPU multicore implementation. For equity, we fixed the size of N_l, with respect to the dataset for all the algorithms.

5.1 Indexing Using GPU

Table 1 shows the sequential indexing time Seq_{C1}, the parallel indexing time, the speedup ($S_p = \frac{Sequential\ time}{Parallel\ time}$) and the average memory used at each iteration for the three proposed GPU algorithms in section 4.1.

In terms of indexing time, we note that the *PIOF* algorithm gives the best indexing time and speedup compared to the *PDSS* and *PDPS* algorithms. The reason is the partial sorting process that is achieved through the picking algorithm that we proposed (algorithm 2). That helps to calculate the distances instantly, which makes the distances reside in the local memory. Accordingly, the latency of sorting which is consumed in getting the distance values from the global memory is reduced.

For the *PDSS* algorithm, we note that the speedup decreases when the number of nearest references $\tilde{n}$ increases, as the sorting step is done sequentially. For the *PDPS*, with small number of nearest references $\tilde{n} = 10$, the algorithm performs better than the *PDSS* algorithm. On the other hand, when $\tilde{n} = 100$, *PDPS* becomes similar or slower than *PDSS* in performance. The main reason is that for the *PDPS*, the distances are pre-calculated and saved in the global memory. As a result, the sorting process on GPU becomes not efficient due to increasing the latency for getting the distances from the global memory. Hence, when $\tilde{n}$ increases, the parallelization which is offered by the GPU for sorting becomes not as efficient as the sequential sorting on the CPU. This effect is reduced, in the *PIOF* algorithm as the distance information are calculated instantly, which means that it is located in the local memory, which improves the performance.

In terms of memory, *PIOF* gives the lowest memory usage. The main reason is that we do not store all the distance information. We store only $\tilde{n}$ distances and consider them as a guidance to calculate the rest of references, that allows to reduce the size of the memory that is needed to be allocated which helps to process more objects in parallel, which improves the total running time.

Table 1. Average indexing time (in minutes), memory usage (in MB) and speedup S_p for PDSS, PDPS and PIOF algorithms on GPU

N	n	$\tilde{n}$	Seq_{C_1}	GPU									
				N_l	M	PDSS	S_p	M	PDPS	S_p	M	PIOF	S_p
1M	1K	10	5	1000	5.4MB	3.9	1.3x	5.4MB	3.5	1.4x	**1.6MB**	**2.4**	**2x**
		100	6	1000	5.4MB	4.98	1.2x	5.8MB	6	1x	**2.3MB**	**5**	**1.2x**
	2K	10	10	1000	10MB	6.3	1.5x	10MB	6	1.7x	**2.5MB**	**3.54**	**2.8x**
		100	12	1000	10MB	8	1.5x	11MB	9	1.3x	**3.5MB**	**7.44**	**1.6x**
10M	1K	10	44	10000	46MB	33	1.3x	47MB	32	1.4x	**5MB**	**16**	**2.7x**
		100	50	10000	46MB	47	1.1x	50MB	62	0.8x	**16MB**	**42**	**1.2x**
	2K	10	82	10000	85MB	57	1.4x	86MB	56	1.5x	**10MB**	**24**	**3.4x**
		100	90	10000	85MB	72	1.3x	89MB	88	1x	**17MB**	**56**	**1.6x**
106M	1K	10	445	10600	49MB	345	1.3x	50MB	330	1.34x	**10MB**	**180**	**2.5x**
		100	513	10600	49MB	498	1x	53MB	498	1x	**10.5MB**	**473**	**1x**
	2K	10	888	10600	90MB	601	1.5x	91MB	570	1.6x	**10MB**	**261**	**3.4x**
		100	1206	10600	90MB	771	1.5x	94MB	996	1.2x	**18MB**	**620**	**1.9x**

Effect of N_l. The number of objects N_l that are processed at the same time affects the indexing time. To study the effect of N_l, we have increased the size of $N_l = 106000$ (10 times) and run the experiments on the 106M dataset for $n = 1000$, $n = 2000$ and $\tilde{n} = 100$. Table 2 shows the indexing time of the N_l value for the *PIOF* algorithm. It is clear that the running time is reduced (comparing to results in Table 1) and we gain more speedup when the number of objects that are processed in parallel increases. We also believe that with more powerful GPU (more cores, more global memory), a much better performance can be achieved.

Table 2. Average indexing time in minutes (Seq_{C_1} and T_{PIOF}), average memory usage in megabytes and speedup S_p on GPU for the *PIOF* for $N_l = 106000$

N	n	$\tilde{n}$	Seq_{C_1}	GPU			
				N_l	Memory	T_{PIOF}	S_p
106M	1K	100	513	106000	170MB	466	1.1x
	2K	100	1206	106000	180MB	600	2.1x

5.2 Indexing using Multi-core CPU

Tables 3 and 4 show the sequential indexing time, the parallel indexing time, the speedup (S_P) and the efficiency ($E = \frac{S_P}{P}$) for the three datasets on 4, 8 and 16 multicore CPU for the *single-disk access* and *multi-disk access* algorithms presented in section 4.2 on the second machine C_2.

We note that the *multi-disk access* algorithm is faster than the *single-disk access* algorithm. In *multi-disk access* algorithm the threads do not need to wait for each other in order to access the hard disk compared to the *single-disk access* algorithm as it was discussed in section 4.2. Hence, its indexing time, speedup and efficiency are better compared to the *single-disk access* algorithm.

Table 3. Average indexing time in minutes (sequential Seq_{C_2} and parallel T_4, T_8 and T_{16}), speedup (S_4, S_8 and S_{16}) and efficiency (E_4, E_8 and E_{16}) for *single-disk access* algorithm

N	n	$\tilde{n}$	Seq_{C_2}	Multi-core CPU (single-disk access)								
				T_4	E_4	S_4	T_8	E_8	S_8	T_{16}	E_{16}	S_{16}
1M	1K	10	4	1.96	0.51	2.03x	1.5	0.33	2.64x	1.35	0.19	2.96x
		100	4.5	1.93	0.58	2.3x	2.1	0.3	2.14x	2	0.14	2.3x
	2K	10	9.3	2.9	0.79	3.2x	2.55	0.45	3.6x	2.2	0.3	4.3x
		100	9.4	4.4	0.53	2.12x	4.4	0.3	2.1x	6.3	0.094	1.5x
10M	1K	10	40.7	17.4	0.58	2.3x	14	0.36	2.9x	10.9	0.23	3.7x
		100	46	18.1	0.63	2.5x	15.3	0.4	3x	12.2	0.24	3.8x
	2K	10	73	29.4	0.6	2.5x	24.3	0.4	3x	17.4	0.3	4.2x
		100	82.4	31.7	0.65	2.6x	27.2	0.38	3x	22	0.23	3.8x
106M	1K	10	402	140	0.0.71	2.9x	109	0.5	3.7x	129	0.193	3.1x
		100	450	187	0.6	2.4x	144.9	0.38	3.1x	135	0.21	3.3x
	2K	10	804	279	0.72	2.88x	171.8	0.58	4.7x	153	0.33	5.2x
		100	912	298	0.76	3x	189.2	0.6	4.8x	174.8	0.33	5.2x

For the two algorithms, we note that when the number of cores increases, the efficiency decreases. The reason for that is the increasing time for reading from and writing to the hard-disk. For reading (*multi-disk access* algorithm), the threads access the hard-disk in a pipeline fashion. Hence, when we have more threads, the pipe-lining queue length increases, which increases the running time. For writing (*single-disk access* and *multi-disk access*), when the number of processors increases, more objects are processed at the same time. The available RAM size is then divided by the available number of processes, which increases the writing rate to the hard-disk as the RAM is filled in at a higher rate. For example, we assigned 1.5 GB of memory for our application. With 16 cores, each thread has only around 96 MB of this memory, which increases the writing rate to the hard disk for each thread. There should be a good balance between the available memory and the number of cores that are used to avoid disk contention and to get a good performance.

Increasing Memory Size for Multi-core CPU. To measure the effect of the memory size on the multi-core CPU algorithm, we indexed the 106M dataset again but with different memory size per thread for $n = 1000$, $n = 2000$ and $\tilde{n} = 100$. In these experiments, we assigned 1.8GB of memory for each thread. Hence, if we have 16 running processes, the total allocated memory would be 28.8GB. Table 5 shows the running time in minutes using the *multi-disk access* algorithm. It is clear from the table (comparing to the results in Table 4) that increasing the memory per thread improves the indexing time. The main reason for that improvement is the decrease of the writing time to the disk.

Table 4. Average indexing time in minutes (sequential Seq_{C_2} and parallel T_4, T_8 and T_{16}), speedup (S_4, S_8 and S_{16}) and efficiency (E_4, E_8 and E_{16}) for *multiple-disk access* algorithm

N	n	$\tilde{n}$	Seq_{C_2}	Multi-core CPU (multi-disk access)								
				T_4	E_4	S_4	T_8	E_8	S_8	T_{16}	E_{16}	S_{16}
1M	1K	10	4	1.8	0.55	2.2x	1.2	0.41	3.3x	0.66	0.375	6x
		100	4.5	1.4	0.8	3.2x	1.2	0.47	3.8x	1.3	0.22	3.5x
	2K	10	9.3	3.9	0.6	2.4x	2.4	0.48	3.8x	1.2	0.48	7.7x
		100	9.4	3	0.75	3x	2.46	0.5	4x	2.94	0.2	3.2x
10M	1K	10	40.7	10.5	0.98	3.8x	5.4	0.94	7.5x	5.34	0.48	7.7x
		100	46	13.2	0.86	3.4x	7.8	0.74	6x	7.5	0.38	6x
	2K	10	73	19.8	0.93	3.7x	10.5	0.88	7x	7.8	0.59	9x
		100	82.4	22.8	0.9	3.6x	13.2	0.77	6x	10.68	0.5	8x
106M	1K	10	402	108	0.93	3.7x	85	0.59	4.7x	42.6	0.59	9.4x
		100	450	132	0.85	3.4x	75	0.75	6x	66	0.43	6.8x
	2K	10	804	204	0.98	3.9x	120	0.83	6.6x	84	0.59	9.4x
		100	912	240	0.95	3.8x	126	0.9	7x	120	0.48	7.6x

Table 5. Average sequential Seq_{C_2} and parallel T_{16} indexing time in minutes, speedup S_{16} and efficiency E_{16} on multi-core CPU for *multi-disk access* algorithm with 1.8 GB per thread

N	n	$\tilde{n}$	Seq_{C_2}	Multi-core CPU (multi-disk access)		
				T_{16}	E_{16}	S_{16}
106M	1000	100	450	50	0.6	9x
	2000	100	912	80	0.7	11x

5.3 Comparing GPU to Multi-core CPU

We note that, when we compare the speed up of the *PIOF* algorithm on GPU (Table 1) to the speedup of the *multi-disk access* using 4 cores CPU (Table 4), the CPU provides much better performance at a lower cost. We believe that there are two reasons. The first reason is that the data cannot be loaded on the GPU directly. CUDA does not support system calls. Hence, the reading should be done through the CPU. This results in a two-layer process, the data is read in the RAM, then transferred to the GPU. While for the multi-core CPU, it is a one layer process, the data is read to the RAM and processed directly.

The second reason is the algorithm itself. To get a good performance on the GPU, the GPU kernel should be as compact as possible to make all the data on the registers of the cores, which is not possible in the case of our algorithm. However, with some algorithm optimization, we managed to get a modest speedup, compared to the sequential algorithm as we show in the *PIOF* algorithm.

6   Conclusion

Opening the GPUs to general purpose computation allows to perform parallel computation on a powerful platform. We present different permutation-based indexing algorithms for multi-core GPU and CPU architectures. We show that using the multi-core

CPU a much better performance can be achieved compared to the GPU at a lower cost due to the limitation of the GPU platform and the complexity of permutation-based indexing. The main bottleneck of the permutation-based indexing is the sorting of the ordered lists as it is done many times. Due to the architecture of the GPU, the sorting process on GPU is not efficient. However, with some algorithm optimization, we managed to get a modest speedup. The evaluation is performed using standard and large datasets. We now work at a hybrid version of using multicore CPU and GPU at the same time. We are also working on a distributed version to improve the indexing performance.

Acknowledgment. This work is jointly supported by the Swiss National Science Foundation (SNSF) via the Swiss National Center of Competence in Research (NCCR) on Interactive Multimodal Information Management (IM2) and the European COST Action on Multilingual and Multifaceted Interactive Information Access (MUMIA) via the Swiss State Secretariat for Education and Research (SER).

References

1. Zezula, P., Amato, G., Dohnal, V., Batko, M.: Similarity Search: The Metric Space Approach. Advances in Database Systems, vol. 32. Springer (2006)
2. Gonzalez, E., Figueroa, K., Navarro, G.: Effective proximity retrieval by ordering permutations. IEEE Trans. on Pattern Analysis and Machine Intelligence 30(9), 1647–1658 (2008)
3. Amato, G., Savino, P.: Approximate similarity search in metric spaces using inverted files. In: Proceedings of the 3rd International Conference on Scalable Information Systems, InfoScale 2008, pp. 28:1–28:10. ICST, Brussels (2008)
4. Mohamed, H., Marchand-Maillet, S.: Quantized ranking for permutation-based indexing. In: Brisaboa, N., Pedreira, O., Zezula, P. (eds.) SISAP 2013. LNCS, vol. 8199, pp. 103–114. Springer, Heidelberg (2013)
5. Esuli, A.: Mipai: Using the pp-index to build an efficient and scalable similarity search system. In: Proceedings of the 2009 Second International Workshop on Similarity Search and Applications, pp. 146–148. IEEE Computer Society, Washington, DC (2009)
6. Tellez, E.S., Chávez, E., Navarro, G.: Succinct nearest neighbor search. Inf. Syst. 38(7), 1019–1030 (2013)
7. Amato, G., Gennaro, C., Savino, P.: Mi-file: using inverted files for scalable approximate similarity search. Multimedia Tools and Applications (2012)
8. Lopresti, M., Miranda, N., Piccoli, F., Reyes, N.: Solving multiple queries through a permutation index in GPU. Journal Computacion y Sistemas 17(3), 341–356 (2013)
9. Sanders, J., Kandrot, E.: CUDA by Example: An Introduction to General-Purpose GPU Programming. 1st edn. Addison-Wesley Professional (2010)
10. Dagum, L., Menon, R.: Openmp: An industry-standard api for shared-memory programming. IEEE Comput. Sci. Eng. 5(1), 46–55 (1998)
11. Bolettieri, P., Esuli, A., Falchi, F., Lucchese, C., Perego, R., Piccioli, T., Rabitti, F.: CoPhIR: a test collection for content-based image retrieval. CoRR abs/0905.4627v2 (2009)

An Efficient DTW-Based Approach for Melodic Similarity in Flamenco Singing

J.M. Díaz-Báñez and J.C. Rizo

Universidad de Sevilla, Sevilla, Spain
{dbanez,juarizmas}@us.es

Abstract. We study melodic similarity in flamenco singing by using the Dynamic Time Warping (DTW) distance. Given two melodic contours, the score of the alignment of the two melodies is taken as a similarity measure. Concretely, we consider a particularly representative flamenco repertoire, the *tonás*, a cappella flamenco singings with free rhythm and high degree of complex ornamentation. We show that the DTW-distance discriminates correctly variations between the styles. In order to speedup the quadratic time and space complexity of the standard DTW, our strategy is to perform an efficient segmentation on the pitch contour before applying dynamic programming. We show that our method achieves better results (both in efficiency and accuracy) than other existing DTW-based similarity measures.

Keywords: Melodic Similarity, Alignment, Segmentation, Flamenco Music.

1 Introduction

Alignment algorithms are widely used as a measure to compare similarities of sequences of symbols. For instance, in some domains such as speech recognition there have been many distances proposed all of them providing some alignment between two sequences. One of the most popular technique is the dynamic time warping distance (DTW) [29,27]. Another application field is bioinformatics [5], where the alignment algorithm is used to find patterns in protein sequences. Also, alignment techniques have been adapted to musical similarity [21,26,20]. One of the reasons that make suitable their use in music is that the structural alignment is a prominent model in cognitive science for human perception of similarity [11].

The objective of DTW is to compare two sequences $X := x_1, x_2, ..., x_n$ and $Y := y_1, y_2, ..., y_m$ where X and Y are feature sequences sampled at equidistant points in time. To compare two different features, one needs a local cost measure. Typically, the local cost $c(X, Y)$ is small (low cost) if X and Y are similar to each other, and otherwise $c(X, Y)$ is large (high cost). Evaluating the local cost measure for each pair of elements of the sequences X and Y, one obtains the cost matrix defined by $C(n, m) := c(x_n, y_m)$. Then the goal is to find an alignment between X and Y having minimal overall cost.

A.J. Machado Traina et al. (Eds.): SISAP 2014, LNCS 8821, pp. 289–300, 2014.
DOI: 10.1007/978-3-319-11988-5_27 © Springer International Publishing Switzerland 2014

An (n,m)-warping path $p = (p_1, ..., p_L)$ defines an alignment between two sequences X and Y by assigning the element x_n of X to the element y_m of Y subject to $\max(n,m) \leq L \leq n+m-1$. The warping path is a sequence of elements from the matrix where each element $p_k = (i,j)$ must meet several criteria: (i) boundary conditions to restrict the searching space for warping paths, (ii) monotony: the element pairing is monotonous with respect the time and (iii) continuity to ensure that neighboring elements correspond to adjacent cells in the matrix. The total cost $c_p(X,Y)$ of a warping path p between X and Y with respect to the local cost measure c is defined as $c_p(X,Y) := \sum_{\ell=1}^{L} c(x_{n_\ell}, y_{m_\ell})$. Then, an optimal warping path between X and Y is a warping path p^* having minimal total cost among all possible warping paths. The DTW distance $DTW(X,Y)$ between X and Y is then defined as the total cost of p^*: $DTW(X,Y) := c_{p^*}(X,Y) = \min\{c_p(X,Y)$ s.t. p is an $(n,m)-$warping path$\}$. As most alignment algorithms, the calculation of the DTW distance uses a dynamic programming approach. The element $C(i,j)$ in the matrix contains the score of the optimal alignment up to x_i and y_j and therefore, $C(n,m)$ contains the score of the optimal alignment of the complete sequences. Thus, the total alignment is obtained by tracing back from $C(n,m)$ to $C(0,0)$ in $O(nm)$ time and space. The quadratic time and space complexity of DTW creates the need for methods to speed up dynamic time warping. There have been many efforts spent on speeding the similarity search under DTW, varying in the assumptions on the DTW variant used (e.g., global or local constrains).Two of the most commonly used constraints are the *Sakoe-Chiba band* [29] and the *Itakura Parallelogram* [13], where alignments of cells can be selected only from a specific region. However, the usage of these global constraints can be problematic, since the optimal warping path may traverse cells outside the specified constraint region. This fact may lead to undesirable or even completely useless alignment results. Thus extensive research has been performed on how to accelerate local constrains DTW computations, in particular for one-dimensional (or low-dimensional) real-valued sequences. Some effective strategies for time series are data abstraction and lowerbounding,[16,17,2], multiscale DTW [1,30], among others. We refer to [27] a comprehensive account on DTW and related pattern recognition techniques in the context of speech recognition.

In this paper we introduce a modification of DTW, the Segmentation DTW approximation (SDTW), which is suitable for computing melodic similarity in presence of baroque ornamentation, specifically in flamenco singing. Thus we suggest using the DTW distance as a low-level melodic similarity measure for flamenco music. This measure can be used for automatic classification as well as for musical studies to characterize variations in the flamenco styles. We prove that the SDTW approach is an efficient approximation algorithm for melodic similarity yielding stable alignments in the presence of baroque ornamentation in a capella flamenco singing. The strategy is based on a sweeping algorithm to reduce the input size in the classical DTW approach. Using a set of melodies from the flamenco repertory, we evaluate the strategy and show that it provides better results than other approximation algorithms when trying to automatically distinguish different styles and variants.

The structure of this paper is as follows. Due to the novelty of the application area, next section gives a brief overview of the flamenco music and its computational analysis. Afterwards our strategy is presented. Next, we state the corpus for the evaluation approach and report the empirical evaluations demonstrating the practicality of our algorithm. The paper finishes up with a conclusion section.

2   The Flamenco Melody and Its Computational Analysis

Over the last few years the use of computational tools in music research has grown rapidly, in music technology and especially in music information retrieval (MIR). Currently, we can find a large number of applications of music technology in the literature. Surprisingly enough, most of research and applications were done for Western music, either popular music or classical music from the common practice [4]. The challenges posed by the research in ethnic music are significant because of the particular musical features of a given tradition, which in many cases are markedly different from the Western one. In this respect, flamenco music is a case in point. Flamenco is a music tradition originally from Andalusia in southern Spain. Flamenco music is a non-Western, oral tradition with very particular musical features. We refer to the books [24,28] for a comprehensive study of styles, musical forms and history of flamenco.

Many fundamental problems in flamenco music are open. A simple question such as musical transcription is by no means solved in flamenco music. As an oral tradition, performers never had the need to transcribe. Furthermore, flamenco has started to be studied from academia recently and available scores are scant, mostly limited to guitar. Therefore, recent algorithms have been developed to automatically transcribe flamenco a cappella singing from audio input [8]. Other open problems in flamenco music are melodic and rhythmic similarity, style classification, singer identification, among others.

The most basic element of flamenco is the voice. A lone singer, accompanied only by handclaps, finger snapping or table rapping, or simply the naked voice, can be the quintessence of pure flamenco. In the flamenco jargon, singing is called *cante*, songs are termed *cantes* and styles are *palos*. in this paper we use this terminology. The musical characteristics are major contributors to what makes the art unique. Regarding flamenco melodies, a notable feature is the abundant use of ornamentation, melisma (multiple notes sung on a single syllable) and the apparent lack of a steady rhythm. The last feature is due to the fact that the melodic rhythm of the cante generally does not strictly follow the meter or *compás* played by the percussion and guitar, although the melodic phrase must synchronize with the compás at the end of each cycle. Moreover, although there is an established melodic pattern for each cante (melodic skeleton), the melodies are performed with variations on the base pattern that depend on the abilities and esthetic preferences of the singer or the school of cante to which the performer adheres. Thus two cantes belonging to the same style may sound very different to an unaccustomed ear. As a consequence, from the music technology point of view, the elaborate ornamentation of flamenco makes it difficult to automate the separation of the notes of the melodic pattern from the melismatic ornamentation. In fact, the classification of flamenco cantes is subject to many difficulties, and such a classification is not yet clearly established in the flamenco literature. In this scenario, the finding of suitable and efficient similarity measures is a key challenge.

2.1   Previous Work in Flamenco Similarity

Recently, an interdisciplinary research on flamenco is being conducted by the COFLA project (http://mtg.upf.edu/research/projects/cofla). In the group there are researchers of various disciplines, including mathematics, engineering, and computer science, who strive for understanding a musical tradition as complex and rich as flamenco music. The first computational approach to characterize and study melodic similarity among various flamenco styles was done in [3]. After removing ornamentations, the melodic skeletons were estimated. Then distance measures among songs from styles were calculated based on the underlying melodic figures and a phylogenetic graphs were created and analyzed regarding style organization and historic evolution. A more detailed description is obtained in [22], where several style-specific features defined by experts in the field were stated. Moreover, in the paper [22] a linear combination of the melodic contour and a mid-level distance is proposed. In both papers the melodic similarity is calculated by using the edit distance. Finally, results on detecting flamenco as a genre among other music styles and traditions are obtained when combining global melodic features with instantaneous spectral and vibrato descriptors in [18].

3   Our Approach

The main goal of this paper is to know if an efficient and effective algorithm for flamenco melodic similarity can be derived from the DTW approach. On one hand, regarding melodic similarity methods, the representation of the melody strongly depends on the target set. Many approaches represent notes by encoding only pitch and duration, that is, a two-dimensional representation. Others, only consider pitch to represent the melodic contour, i.e., an one-dimensional sequence. In any case, some melodic transcription is needed for a specific level. Since scores are not available (the manual transcription in flamenco singing has serious difficulties as high degree of ornamentation, very time consuming, subjectivity, etc.) the first step is to extract an automatic transcription from the audio files. We will consider an automatic transcription system (SmsTools) [8] that provides melodic descriptors at different levels: low-level feature extraction, frame-based descriptor extraction (e.g., energy and fundamental frequency), note segmentation (based on location of note onsets), and note labelling. For more technical details, see [8] and the references therein. Concretely, we will consider two levels, a low level representation, the fundamental frequency ($f0$) related to the pitch, and a high level representation in which a segmentation into notes is given (*note*). In both cases, we will use a one-dimensional sequence.

On the other hand, in order to improve the time complexity of the standard DTW algorithm, it is possible to consider two strategies to design a melodic similarity measure under DTW:

1. Apply some DTW-aproximation algorithm to compute an approximated matching path.
2. Compute exact DTW matching on an approximation of the input obtained by some segmentation algorithm.

We adopt the second approach. Thus, we need an efficient algorithm to discretize the fundamental frequency envelope into a short sequence of steps, so that the similarity performance remains unchanged. It would be desirable that the time complexity of the segmentation algorithm be subquadratic. Unfortunately, the automatic segmentation algorithm proposed in [8] is very time consuming (no complexity study is done in [8] but one can verify experimentally that the spent time is about the length of the piece). In other context, the second approach has been used for computing time series similarity [14]. In fact, the so called Piecewise Dynamic time Warping (PDTW) operates on a higher level abstraction of the data by computing a piecewise representation of the input. The time series is divided into a fixed number of frames and the mean of each frame is calculated. However, this method is out-of-use for music similarity since the variation of the melodic contour is not captured when we set a priori the position of the segments and the basic features (peaks) may be lost. Many other segmentation procedures have been considered in the data mining literature. For instance, in [15] the *sliding window algorithm* computes a piecewise linear approximation of the time series. See [10] for a review on segmentation methods on time series data mining.

In this paper, we will show that a simple sweeping algorithm is enough to capture the melodic skeleton of the flamenco singing and then it can be used to measure melodic similarity as a way to classify flamenco styles and variants. Given two sequences of length $O(n)$, our segmentation algorithm spends $O(n)$ time by sampling the data down to $O(N)$ points, $N << n$. Thus, after performing dynamic programming within DTW our strategy leads to an $O(N^2)$-time complexity algorithm that is more efficient than the classical DTW while maintaining a similar accuracy.

3.1   The Fitting Algorithm: A Linear Time Segmentation

Our segmentation algorithm is based on the following approximation problem:

The Step Function Approximation Problem: Given a set S of n points in the plane, and a real number $\alpha \geq 0$, construct a step function R to fit S such that the error of R with respect to S is not larger than α and the number of links of R is minimized.

An optimal $\Theta(n)$ algorithm for this step-function approximation problem was proposed in [6] when the error is defined as

$$e(R,S) = \max_{p \in R, q \in S} d_v(p,q)$$

where $d_v(p,q)$ denotes the vertical distance between p and q. We apply this algorithm on the pitch contour to obtain the shortest segmentation for a fixed range threshold α. Since the melodic pitch range in flamenco is normally limited to one octave, values of threshold equal to one or half semitone seems adequate to capture the melodic skeleton.

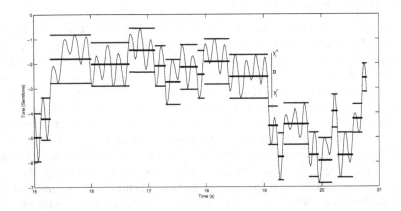

Fig. 1. The fitting algorithm

For completeness, we briefly describe the algorithm, that we call the *fitting-algorithm*: Given a set of points $P = \{p_1, p_2, \cdots, p_n\}$ in the plane and an error tolerance α, plot vertical segments V_i of length $D = 2\alpha$ centered at each point p_i, refer to Figure 1. Our constraint that each point be within α of the step function is equivalent to saying that the step function intersects each of these segments of length $D = 2\alpha$. Sweeping from left to right, the algorithm greedily tries to intersect as many consecutive segments as possible, before starting a new step and repeating this procedure. A vertical segment V_i defines a y interval $[y_i^-, y_i^+]$ where y_i^- and y_i^+ denote the y coordinates of the lower and upper endpoints, respectively. Sweeping from left to right, we maintain the intersection Δ of the y intervals of the vertical segments until we reach a segment V_i whose y interval does not intersect Δ, in which case we terminate the current step, and start a new step at V_i setting $\Delta = [y_i^-, y_i^+]$. Clearly, this algorithm is simple, runs in $O(n)$ time, and it constructs the optimal step function with error tolerance α.

3.2 The SDTW Strategy

As mentioned before, instead of running the classical DTW matching on the melodic contour directly, our strategy (SegmentationDTW, SDTW for short) obtains an approximation on the input by means of a segmentation algorithm before applying the DTW distance between two melodies. The stages of the SWDT strategy are the following:

1. *Frame-based descriptor extraction (fundamental frequency).* We use SmsTools [8] to compute the low-level feature (f0) from the audio file. The audio signal is first cut into frames and by following a frame-by-frame procedure, its fundamental frequency (f0) estimated. The fundamental frequency estimation algorithm is based on the computation of amplitude correlation in the frequency domain.

2. *Pre-treatment of the fundamental frequency f0.* By removing silences (unvoiced regions) and spurious f0 values, only sung regions are considered.

3. *Standardization of the fundamental frequency to a reference frequency.* Since we are analyzing singing voice performances, the reference frequency (with respect to 440 Hz) is unknown. In order to locate the main pitches, an initial estimation of the tuning frequency (i.e. the reference frequency used by the singer to tune the piece) is considered and an equal-tempered scale system was assumed. This tuning frequency is computed by minimizing the estimated instantaneous pitch error weighted average. The weights are computed by combining energy and first and second pitch derivatives.

4. *Segmentation algorithm.* We consider two f0-based note segmentation procedures: the SmsTools system and the fitting algorithm. In the SmsTools system, an iterative approach for note segmentation and labelling can also be used from the extracted f0 and energy. See [8] for more details.

5. *Distance matrix computation and DTW algorithm.* The global warp cost [23] of the optimal path serves as measure for the melodic similarity.

4   The Evaluation

We evaluate our approach on a music collection of 24 sung excerpts representative of two different a cappella singing styles, (*Deblas* and *Martinetes*), selected from the accessible *TONAS dataset* [1]. This monophonic collection was built by the Cofla group in the context of a study on similarity and style classification of flamenco a cappella singing styles. We refer to [22] for a comprehensive description of the considered styles and their musical characteristics. The files were manually transcribed to generate the ground truth.

Four strategies have been considered in this paper to evaluate SDTW:

1. Strategy A: Classical DTW on the non-segmented (f0) set of 24 melodies.
2. Strategy B: Classical DTW on the segmented set by using the SmsTools-based segmentation algorithm.
3. Strategy C: Sakoe-Chiba band algorithm on the non-segmented (f0) set for different values of the band width r.

---

[1] For details on the TONAS dataset see at
http://mtg.upf.edu/download/datasets/tonas

4. Strategy D: Classical DTW on the segmented set by using the fitting segmentation algorithm for different values of the threshold α.

In order to assess how well the different strategies perform, we will classify the cantes in the corpus by using the k-nearest neighbor classifier. We consider each of the 24 excerpts as a query and the algorithm returns the k closest items according to the evaluated distance. The choice of k is critical. Due to the lack of a deep study on this issue for flamenco music we use the empirical rule-of-thumb from [7], that is, k equal to the square root of the number of instances and then $k = \lfloor \sqrt{12} \rfloor = 3$. Precision and recall measures will be computed for that classifier. We complemented these measures with a clustering analysis carried out through phylogenetic techniques. Finally, we explored the relation between the distance matrices by performing a Mantel test [19]. With respect to efficiency, measured with the time it takes the computer to complete the overall similarity algorithm, all tests have been carried on the same computer.

4.1 Precision and Recall for Style Classification

We compute standard information retrieval evaluation measures [25] for each of the queries. Let us denote the true positive, true negative, false positive and false negative values as TP, TN, FP and FN, respectively. We consider precision $P = TP/(TP + FP)$, recall $R = TP/(TP + FN)$, F-measure $F = 2P/(P + R)$ and accuracy $A = (TP + TN)/(TP + TN + FP + FN)$. The computed values of theses measures are shown in Figure 2. The results show high values for our SDTW procedure (Strategies B and D for $\alpha = 0.5$)), specially Strategy B. However, as we will show later, Strategy D is more efficient.

		Precision	Recall	f-Measure	Accuracy
Strategy A		0.94	0.67	0.78	0.35
Strategy B		0.95	0.95	0.95	0.50
Strategy C	r=250	0.95	0.75	0.91	0.44
	r=100	0.95	0.95	0.95	0.50
	r=50	0.92	0.91	0.93	0.47
Strategy D	α=2	0.80	0.33	0.47	0.20
	α=1	0.92	0.54	0.68	0.29
	α=0.5	0.94	0.62	0.75	0.33

Fig. 2. Evaluation measures

4.2 Phylogenetic Analysis

A clustering analysis can be carried out via self-similarity [9] and phylogenetic techniques [12]. Distance matrices can be better visualized by employing phylogenetic graphs, a visualization technique borrowed from Bioinformatics. Given a distance matrix from a set of objects, a phylogenetic graph is a graph whose nodes are the objects in the set and such that the distance between two nodes in the graph corresponds to the distance in the matrix. Obviously, this property cannot be held for arbitrary matrices. The phylogenetic graph algorithm provides with an index, the LSFit, expressed as a percentage. This index indicates how accurate the correspondence between the distances in the graph and the distances in the set of objects is. The higher the index is, the more accurate the correspondence between matrix and graph distances is. To actually compute our phylogenetic graphs we used the tool SplitsTree (www.splitstree.org/). In general, clustering and other properties are easier to visualize. The results validate the SDTW strategy. The LSFit for the obtained graphs is about 0.99. The SplitsTree tool gives one outsider (Debla by Chocolate) for strategies A, B, C and D ($\alpha = 1$) and two outsiders (Deblas by Chocolate and Romero) for Strategy D with $\alpha = 0.5$ (Figure 3). The coincidence of the same outsider in all strategies suggests an interesting subject of musicological study and, this is indeed, another possible application of the music technology tools. In any case, in this study the SplitsTree shows a good discrimination between styles.

Fig. 3. Clustering with the fitting segmentation

4.3 Mantel Test

For the sake of completeness, we carried out a Mantel test on the different strategies. We obtained the correlation coefficients $r_M = 0.90$ for Strategy B, $r_M = 0.9137$ for Strategy C with band width $r = 50$, and $r_M = 0.9509$ for Strategy D with thresold $\alpha = 0.5$. Thus, the correlation is high for all strategies. The lowest correlation should be expected for Strategy B, and that was the case. This fact could be due to the *iterative note consolidation* and *tuning frequency refinement* routines applied in the SmsTools segmentation algorithm [8]. Clearly, the advantage is that this strategy uses a more rigorous method segmentation based on musicological aspects and, therefore, dramatically reduces the number of notes to be considered.

4.4 Efficiency

A comparison of different strategies running times is shown in Figure 4. The results correspond to the time spent to complete the overall similarity algorithm on the musical corpus.

		Efficiency (s)		# Steps
		Segmentation	DTW	
Strategy A		0	2290	25387
Strategy B		720	7.78	1271
Strategy C	r=250	0	699.09	25387
	r=100	0	578.11	25387
	r=50	0	478.56	25387
Strategy D	α=2	0.12	13.65	2052
	α=1	0.14	69.25	4724
	α=0.5	0.23	246.44	8923

Fig. 4. The efficiency with and without segmentation

In the last column the steps (notes in the melodic contour) considered as input by the DTW algorithm are indicated. The increase in performance of the SDTW-based alignment in comparison to the classical DTW-based approach (Strategy A) and the Shakoe-Shiba band approximation algorithm (Strategy C) can be visualized in the table. For example, although the best result based on the precision measures was achieved by Strategy B, that is, the SmsTools segmentation-based approach, Strategy D, this is, the fitting segmentation-based procedure is the more efficient. For $\alpha = 0.5$, Strategy D spends half than Shakoe-Shiba with $r = 50$ and the third part of the Strategy B. Note that the fitting-based segmentation can be computed in less than one second, yielding an overall running time improvement over other strategies.

Finally, putting together all evaluation results, we can conclude that the SDTW-based similarity measure constitute an improvement over other classical strategies and is a reasonable approach for the computational study of flamenco melodies when a compromise between efficiency and precision is required.

Acknowledgments. This research is partially supported by project COFLA: Computational Analysis of Flamenco Music, FEDER-P09-TIC-4840 and FEDER-P12-TIC-1362.

5   Conclusions and Future Research

In this paper, we presented a new efficient DTW-based procedure, the SDTW approach, to evaluate melodic similarity yielding stable alignments even in the presence of high degree of complex ornamentation as is the case of the flamenco singing. To evaluate the alignment quality achieved by our strategy, the method has been tested in a flamenco musical corpus extracted from the metadata TONAS, a dataset of flamenco a cappella sung melodies with corresponding manual transcriptions.

The key idea of SDTW is to apply a segmentation procedure before running the DTW algorithm. To this end, we use a new simple, efficient and robust segmentation algorithm, the fitting segmentation. This algorithm spends less than one second to compute the segmentation of the considered melodies collection. We compared the melodic similarity measure proposed in this paper against other classical DTW-based measures and show that SDTW is competitive in both efficiency and accuracy. Our approach is easily extensible to other flamenco styles. Although the measure only considers low-level features, it seems an useful similarity measure for flamenco music and inter-style classification, even more tacking into account the complexity of the automatic extraction of some mid-level descriptors [22]. As open problem, it would be interesting to improve the accuracy results with the more efficient strategy, that is, the fitting segmentation-based approach. For this problem we plan to implement an "auto-tuning" parameter tolerance, being able to do variable along the melodic contour to be segmented.

References

1. Adams, N., Marquez, D., Wakefield, G.H.: Iterative deepening for melody alignment and retrieval. In: Proc. ISMIR, London (2005)
2. Bartoš, T., Skopal, T.: Revisiting techniques for lowerbounding the dynamic time warping distance. In: Navarro, G., Pestov, V. (eds.) SISAP 2012. LNCS, vol. 7404, pp. 192–208. Springer, Heidelberg (2012)
3. Cabrera, J.J., Díaz-Báñez, J.M., Escobar, F.J., Gómez, E., Gómez, F., Mora, J.: Comparative Melodic Analysis of A Cappella Flamenco Cantes. In: Conference on Interdisciplinary Musicology, Thessaloniki, Greece (2008)
4. Cornelis, O., Lesaffre, M., Moelants, D., Leman, M.: Access to ethnic music: Advances and perspectives in content-based music information retrieval. Signal Processing 90(4), 1008–1031 (2010)
5. Criel, J., Tsiporkova, E.: Gene Time Expression Warper: a tool for alignment, template matching and visualization of gene expression time series. Bioinformatics 22(2), 251–252 (2006)
6. Díaz-Báñez, J.M., Mesa, J.A.: Fitting rectilinear polygonal curves to a set of points in the plane. European Journal of Oper. Research 130(1), 214–222 (2001)
7. Duda, R.O., Hart, P.E., Stork, D.G.: Pattern classification. John Wiley & Sons (2012)
8. Gómez, E., Bonada, J.: Towards Computer-Assisted Flamenco Transcription: An Experimental Comparison of Automatic Transcription Algorithms As Applied to A Cappella Singing. Computer Music Journal 37(2), 73–90 (2013)
9. Foote, J., Cooper, M.: Visualizing musical structure and rhythm via self-similarity. In: Proceedings of the International Conference on Computer Music, pp. 419–422 (2001)

10. Fu, T.C.: A review on time series data mining. Engineering Applications of Artificial Intelligence 24(1), 164–181 (2011)
11. Goldstone, R.L., Son, J.Y.: Similarity. Cambridge University Press (2005)
12. Huson, D., Bryant, D.: Application of phylogenetic networks in evolutionary studies. Molecular Biology and Evolution 23, 254–267 (2006)
13. Itakura, F.: Minimum prediction residual principle applied to speech recognition. IEEE Transactions on Acoustics, Speech and Signal Processing 23(1), 67–72 (1975)
14. Keogh, E., Pazzani, M.J.: Scaling up dynamic time warping for datamining applications. In: Proceedings of the Sixth ACM SIGKDD International Conference on Knowledge Discovery and Data Mining, pp. 285–289 (2000)
15. Keogh, E., Chu, S., Hart, D., Pazzani, M.: An online algorithm for segmenting time series. In: Proceedings IEEE International Conference on Data Mining (ICDM 2001), pp. 289–296 (2001)
16. Keogh, E., Pazzani, M.: Iterative deepening dynamic time warping for time series. In: Proceedings of the Second SIAM Intl. Conf. on Data Mining (2002)
17. Keogh, E., Ratanamahatana, C.A.: Exact indexing of dynamic time warping. Knowledge and information systems 7(3), 358–386 (2005)
18. Kroher, N.: Automatic Characterization of Flamenco Singing by Analyzing Audio Recordings. Master thesis, Master Program in Sound and Music Computing, Universitat Pompeu Fabra (2013)
19. Mantel, N., Valand, R.S.: A technique of nonparametric multivariate analysis. Biometrics, 547–558 (1970)
20. Molina, E., Barbancho, I., Gómez, E., Barbancho, A.M., Lorenzo, J.T.: Fundamental Frequency Alignment vs. Note-based Melodic Similarity for Singing Voice Assessment. In: Proceedings of the 8th Conference on Acoustics, Speech, and Signal Processing (ICASSP), Vancouver, Canada (2013)
21. Mongeau, M., Sankoff, D.: Comparison of musical sequences. Computers and the Humanities 24(3), 161–175 (1990)
22. Mora, J., Gómez, F., Gómez, E., Escobar-Borrego, F., Díaz-Báñez, J.M.: Characterization and melodic similarity of a cappella flamenco cantes. In: Proceedings of ISMIR, Utrecht School of Music, pp. 9–13 (2010)
23. Myers, C., Rabiner, L., Rosenberg, A.: Performance tradeoffs in dynamic time warping algorithms for isolated word recognition. IEEE Transactions on Acoustics, Speech and Signal Processing 28(6), 623–635 (1980)
24. Navarro, J.L., Ropero, M. (eds.): Historia del flamenco. Ed. Tartessos (1995)
25. Olson, D.L., Delen, D.: Advanced data mining techniques. Springer Publishing Company, Incorporated (2008)
26. Pikrakis, A., Theodoridis, S., Kamaroto, D.: Recognition of isolated musical patterns using context dependent dynamic time warping. IEEE Transactions on Speech and Audio Processing, 175–183 (2003)
27. Rabiner, L.R., Juang, B.H.: Fundamentals of speech recognition, vol. 14. PTR Prentice Hall, Englewood Cliffs (1993)
28. Ríos Ruiz, M.: El gran libro del flamenco, editorial Calambur (2002)
29. Sakoe, H., Chiba, S.: Dynamic programming algorithm optimization for spoken word recognition. IEEE Transactions on Acoustics, Speech and Signal Processing 26(1), 43–49 (1978)
30. Salvador, S., Chan, P.: Toward accurate dynamic time warping in linear time and space. Intelligent Data Analysis 11(5), 561–580 (2007)

Author Index